The Neocortex
Ontogeny and Phylogeny

NATO ASI Series
Advanced Science Institutes Series

A series presenting the results of activities sponsored by the NATO Science Committee, which aims at the dissemination of advanced scientific and technological knowledge, with a view to strengthening links between scientific communities.

The series is published by an international board of publishers in conjunction with the NATO Scientific Affairs Division

A	**Life Sciences**	Plenum Publishing Corporation
B	**Physics**	New York and London
C	**Mathematical and Physical Sciences**	Kluwer Academic Publishers Dordrecht, Boston, and London
D	**Behavioral and Social Sciences**	
E	**Applied Sciences**	
F	**Computer and Systems Sciences**	Springer-Verlag
G	**Ecological Sciences**	Berlin, Heidelberg, New York, London,
H	**Cell Biology**	Paris, and Tokyo

Recent Volumes in this Series

Volume 194—Sensory Transduction
  edited by Antonio Borsellino, Luigi Cervetto, and Vincent Torre

Volume 195—Experimental Embryology in Aquatic Plants and Animals
  edited by Hans-Jürg Marthy

Volume 196—Sensory Abilities of Cetaceans: Laboratory and Field Evidence
  edited by Jeanette A. Thomas and Ronald A. Kastelein

Volume 197—Sulfur-Centered Reactive Intermediates in Chemistry and Biology
  edited by Chryssostomos Chatgilialoglu and Klaus-Dieter Asmus

Volume 198—Selective Activation of Drugs by Redox Processes
  edited by G. E. Adams, A. Breccia, E. M. Fielden, and P. Wardman

Volume 199—Targeting of Drugs 2: Optimization Strategies
  edited by Gregory Gregoriadis, Anthony C. Allison, and George Poste

Volume 200—The Neocortex: Ontogeny and Phylogeny
  edited by Barbara L. Finlay, Giorgio Innocenti, and Henning Scheich

Series A: Life Sciences

The Neocortex
Ontogeny and Phylogeny

Edited by
Barbara L. Finlay
Cornell University
Ithaca, New York

Giorgio Innocenti
University of Lausanne
Lausanne, Switzerland

and
Henning Scheich
Technical University Darmstadt
Darmstadt, Germany

ITHACA COLLEGE LIBRARY

Plenum Press
New York and London
Published in cooperation with NATO Scientific Affairs Division

Proceedings of a NATO Advanced Research Workshop
on Neocortex: Ontogeny and Phylogeny,
held August 26-31, 1989,
in Alagna, Italy

Library of Congress Cataloging-in-Publication Data

NATO Advanced Research Workshop on Neocortex: Ontogeny and Phylogeny
(1989 : Alagna, Italy)
  The neocortex : ontogeny and phylogeny / edited by Barbara L.
Finlay, Giorgio Innocenti, and Henning Scheich.
      p.   cm. -- (NATO ASI series. Series A, Life sciences ; v.
200.)
  "Published in cooperation with NATO Scientific Affairs Division."
  "Proceedings of a NATO Advanced Research Workshop on Neocortex:
Ontogeny and Phylogeny, held August 26-31, 1989, in Alagna, Italy"-
-Verso t.p.
  Includes bibliographical references and index.
  ISBN 0-306-43808-9
  1. Neocortex--Congresses.   I. Finlay, Barbara L. (Barbara
Laverne), 1950-   . II. Innocenti, Giorgio M. III. Scheich, H.
IV. North Atlantic Treaty Organization. Scientific Affairs
Division. V. Title. VI. Series.
QM455.N38 1989
599'.048--dc20   90-21500
                                                              CIP

© 1991 Plenum Press, New York
A Division of Plenum Publishing Corporation
233 Spring Street, New York, N.Y. 10013

All rights reserved

No part of this book may be reproduced, stored in a retrieval system, or transmitted
in any form or by any means, electronic, mechanical, photocopying, microfilming,
recording, or otherwise, without written permission from the Publisher

Printed in the United States of America

PREFACE

Of the three organizers of this NATO Advanced Research Workshop on "Neocortex: Ontogeny and Phylogeny", one derived most of his knowledge about neocortex from studies on birds, another had never studied any animal but the cat and could probably recognize not more than ten animal species, and the third had very limited experience with mountaineering. They had in common the belief that evolutionary thinking permeates what biologists do, but that evolution of species and structures cannot be directly experimentally addressed. Although the fossil record can provide some major insights, the inroad to the evolution of the brain is indirect, via comparative anatomy and developmental biology. By identifying similarities and differences between brain structures in the species at hand, comparative anatomy generates hypotheses of evolutionary transformations. By understanding the rules of morphological transformation, developmental biology can, in principle, estimate the likelihood that a given transformation may have actually occurred. The meeting was a way to check if this notion is viable, by gathering together scientists from these two fields.

Standing, left to right: F.Ebner, V.Caviness, M. Weisskopf, B. Fritszch, N. Swindale, J. Walter, H. Karten, J. Pettigrew, E.Welker, M. Cynader, D.Frost, L. Lopez-Mascaraque, P.Katz, H. Jerison, E. Soriano, Mayor of Alagna, Dr. G. Guglielmina, and associate, H. Van der Loos, B. Finlay, H. Scheich, C. Ruela.

Seated: S. Pallas, T. Lohmann, J. De Carlos, F. Valverde, G. Innocenti, M. Diamond

"Gathering" does not accurately describe what really happened. NATO guidelines for the choice of a site for the meeting are strict: it should favor close interaction of participants in and out of the lecture hall and discourage their dispersion. This charge was taken very seriously by the organizers. Therefore we chose the Institute Angelo Mosso at the Col d'Olen. The Institute is at nearly 3000 m above sea level, on the Italian side of the Monte Rosa, the second highest group of mountains in the Alps, and about 1000 m above tree level, about 2000 m above the nearest village. Direct access involves walking up from the villages, the use of helicopter or parachuting. Indirect access involves taking three trams from Alagna to Punta Indren (3300 m) and walking down on a sometimes exposed and occasionally strenuous mountain path. And this is how participants were herded to the Institute in a windy, foggy and snowy day in August 1989. Once at the Institute none of the participants ever expressed the slightest intention of not coming to the lecture hall, which for the first few days was the only warm (moderately so) place on site. None of them expressed the desire for more mundane activities which the villages, 2000 m below, might have offered. Variable electrical supplies allowed the participants to develop camaraderie and improve their aerobic capacity as they packed blackboards, projectors, and ultimately car batteries to and from the nearby Rifugio Vigevano (where participants, according to NATO guidelines, had meals together) and the Angelo Mosso Institute.

The location did prove to be exceptionally appropriate for intense, intellectual interaction in an unusual and spectacular site. The Institute was built at the beginning of the century with the financial help of several European countries and for a number of years enjoyed great reputation as a laboratory for the study of high altitude physiology. It is foreseen that a Foundation, created by the efforts of the Mayor of Alagna (Dr. B. Guglielmina) under the patronage of the Department of Anatomy and Physiology and the University of Turin (Director, Professor G. Losano) will in the future revitalize the Institute by helping those who may want to use it for scientific encounters.

The desired merging of comparative neuroanatomy and developmental neurobiology is directly reflected by a number of papers which dealt specifically with theoretical issues, which are the first section of this volume. Other papers more prudently remained on one or the other side of the border, and connections between them were identified in an editorial meeting at Cornell University in Ithaca in March, 1990. Although these connections may appear artificial to everybody but ourselves, they are offered in the spirit of the meeting.

The Editors

ACKNOWLEDGEMENTS

The production of this conference in its unusual location naturally incurred many debts of gratitude. We thank Dr. Mayor G. Guglielmina of Alagna for his warm welcome, and Prof. Losano, Director of the Angelo Mosso Institute, for making such spectacular facilities available. Mr. and Mrs. Carestia, caretakers of the Institute, worked heroically to keep the group warm, and provisioned, as did the staff of the Vigevano hut. We thank the various members of the conference, particularly Doug Frost, for their essential help with the porterage.

The conference mailings and the production of this book were entirely dependent on Kim Stockton, whose extreme competence and patience cannot be too highly praised. Lael Hinds, Brad Miller, Kim Jordan, and Larry Chou were helpful with editing and proofing. Edie Clark and Fred Horan were critical for financial and computer mechanics.

Finally, we thank NATO for the support of this conference, and for its support of this kind of international interchange.

CONTENTS

EVOLUTIONARY AND DEVELOPMENTAL SYNTHESES

Introduction 3

Fossil Brains and the Evolution of the Neocortex 5
   H.J. Jerison

Critical Cellular Events During Cortical Evolution: Radial Unit Hypothesis 21
   P. Rakic

Control of Cell Number and Type in the Developing and Evolving Neocortex 33
   B.L. Finlay

Pathways Between Development and Evolution 43
   G.M. Innocenti

Introduction to Sections 2 and 3 54

COMPARATIVE ASPECTS OF FOREBRAIN ORGANIZATION

The Dorsal Ventricular Ridge and Cortex of Reptiles in Historical and
Phylogenetic Perspective 59
   A.H.M. Lohman and J.A.J. Smeets

Multiple Origins of Neocortex: Contributions of the Dorsal Ventricular Ridge 75
   T. Shimizu and H.J. Karten

Aspects of Phylogenetic Variability of Neocortical Intrinsic Organization 87
   F. Valverde

On the Coincidence of Loss of Electroreception and Reorganization of Brain Stem
Nuclei in Vertebrates 103
   B. Fritzsch

The Design of Striate Cortex 111
   N.V. Swindale

Representation Geometries of Telencephalic Auditory Maps in Birds and Mammals . . . 119
   H. Scheich

Flying Cats and Flying Primates: Evolutionary Surprises From Neurobiology 137
    J.D. Pettigrew

Emergence of Radial and Modular Units in Neocortex 159
    M.W. Diamond and F.F. Ebner

NEUROEMBRYOLOGY OF THE NEOCORTEX

Ontogeny and Structure of the Radial Glial Fiber System of the Developing
Murine Cerebrum 175
    V.S. Caviness, Jr., J.-P. Mission, T. Takahashi and J.-F. Gadisseux

Nonpyramidal Neurons in the Mammalian Hippocampus: Principles of
Organization and Development 185
    E. Soriano, J.A. Del Rio and I. Ferrer

Morphological Characterization of ALZ-50 Immunoreactive Cells in
the Developing Neocortex of Kittens 193
    J.A. De Carlos, L. López-Mascaraque and F. Valverde

Guidance of Chick Retinal Axons in Vitro 199
    J. Walter and F. Bonhoeffer

Cross-Modal Plasticity in Sensory Cortex 205
    S.L. Pallas

Visual Responses on Neurons in Somatosensory Cortex of Hamsters With
Experimentally Induced Retinal Projections to Somatosensory Thalamus 219
    C. Métin and D.O. Frost

Brain Maps: Development, Plasticity and Distribution of Signals Beyond 229
    H. Van der Loos, E. Welker, J. Dörfl and P. van Hoogland

The Possibility of Gaba-Ergic Innervation in Plasticity of Adult Cerebral
Cortex 237
    E. Welker, H. Van der Loos, J. Dörfl and E. Soriano

Transient Receptor Expression in Visual Cortex Development and the
Mechanisms of Cortical Plasticity 245
    M. Cynader, C. Shaw, F. van Huizen and G. Prusky

Contributors 255

Index 257

1. EVOLUTIONARY AND DEVELOPMENTAL SYNTHESES

EVOLUTIONARY AND DEVELOPMENTAL SYNTHESES: INTRODUCTION

The border between the three fields represented at the meeting, evolution, comparative anatomy-physiology and development, was crossed explicitly in chapters in this section. The main questions addressed in these studies are listed below. The answers suggested should be viewed as appetizers, and reflect our not necessarily unprejudiced understanding of the authors' views.

To what extent did corticalization really occur in evolution ?

Jerison was very convincing in drawing both borders and relationships between comparative anatomy and evolutionary anatomy. Some aspects of the latter are accessible directly, even in the case of the nervous system, by the analysis of endocasts of fossil skulls. The results of these studies are reassuringly consistent with the belief, based on comparative anatomical data, that in the course of evolution, the tangential expansion of neocortex was disproportionately greater than the volumetric increase of other brain structures. This provided a solid ground on which the explanatory power of various developmental mechanisms can be tested.

How does neocortex increase tangentially in evolution?

Two mechanisms for the increase in the tangential extent of neocortex are conceivable on the basis of what we know of cortical development. One involves tangential increase of the germinative layer as suggested by Rakic, Finlay, and Innocenti, perhaps by prolonging the period of symmetrical divisions. Alternatively, an increase in the number of radial glial channels could lead to more widespread tangential distribution of the migrating neurons into the cortical plate (Innocenti). It seems probable that the tangential expansion should increment the cortex by neuronal groups. The question is what kind of multicellular "unit" is involved: the column, the area, the proliferative unit Rakic identified or still something else?

How did neocortex increase radially in evolution?

Increase in the radial extent of neocortex and thickening of the supragranular layers is seen in larger cortices and probably occurred in evolution (Hofman, 1989). The developmental mechanism for this could simply be an increase in the time of asymmetrical neuronal division and of total neuronal migration (Innocenti) , without the need for any other changes in speed of neuronal migration or cycles of neuronal generation. Allocortex has shorter time of generation than isocortex (Rakic, 1988) and cytoarchitectonic differences between these cortical subdivision to some extent mimic those which occurred in evolution .

Does a bigger cortex violate neurotrophic constraints?

Tangential expansion of the cortex raises a problem for the relationship of the cortex to the rest of the brain. How do cortical neurons survive if i) like other neurons, they have trophic dependence on targets for their survival and ii) the target of cortical neurons (i.e. the rest of the brain) increases less than cortical volume as the corticalization in evolution requires ? Finlay argues that the cortex itself is the supportive target for cortical neurons.

How do new cortical areas appear?

The ground for this question may not be as firm as in the case of tangential expansion (above) partially because of difficulties intrinsic to the definition of an area, and because probably not all areas have been identified in any species, and because we cannot guess the number of areas in the ancestors of the extant mammals. However, nobody would seriously claim that, for example, insectivores and primates have a similiar number of areas (no matter how defined) in their frontal cortex. Therefore the question appears to be legitimate, to the extent that "evolutionary trajectories" linking the extant mammals in the direction of increasing cortical complexity can be useful to retrace cortical evolution (Innocenti).

Much discussion was on the role of thalamus (and thalamopetal afferents) in the specification of cortical areas (Rakic, Finlay, Van der Loos). From the evolutionary point of view the critical issue here seems to be whether intrinsic differences exist between areas before they are contacted by thalamic afferents, i.e., whether the evolution of the cortex precedes that of the thalamus. If in contrast the thalamus specifies cortex, then would an increase in the number of thalamic nuclei precede and determine that in the number of cortical areas?

An alternative to this possibility, occurred to us after the meeting. The thalamic specification of cortex could use a combinatorial code. Each area could be specified by the combination of afferents from two or more thalamic nuclei and differently depending on the nuclei. Which combination of thalamic nuclei would a newly emerged part of cortex receive may be decided on the basis of some general "algorithm" regulating the mapping of the thalamus on the cortex. This would account for the otherwise unexplicable fact that one thalamic neuron may project with bifurcating axons, and therefore possibly differentially specify two or more cortical area. This model is in agreement with aspects of thalamocortical organization which have been well analyzed in adults (Caviness and Frost, J. Comp. Neurol., 1980; Frost and Caviness, 1980). Clearly more information is needed than currently available, on the development of thalamocortical connectivity.

How does a new area get wired up?

A newly formed area or part of an area could connect to preexisting other cortical areas and to subcortical structures through the maintenance of juvenile, transient or exuberant projections. However these projections are patterned, probably according to rules of axonal growth. This pattern could impose constraints on the connectivity of a newly emerged area. Innocenti argued that some of these constraints may be relaxed by increasing heterochronically the time of axonal growth which could have the consequence of increasing the amount of transient projections . This would have the consequences that the amount of transient projections would not be progressively depleted in larger brains, and over evolutionary time.

REFERENCES

Caviness, V.S. and Frost, D.O. (1980) Thalamic projections to the neocortex in the mouse I. Organization in the tangential dimension of the cortex. *J. Comp. Neurol.,* 194: 335-368.
Frost, D.O. and Caviness, V.S. (1980) Thalamic projections to the neocortex in the mouse II. Organization of the radial dimension of the cortex. *J. Comp. Neurol.,* 194: 369-390.
Hofman, M.A. (1989) On the evolution and geometry of the brain in mammals. *Progress in Neurobiology,* 32: 137-158.
Rakic, P. (1988) Specification of cerebral cortical areas. *Science,* 241: 170-176.

FOSSIL BRAINS AND THE EVOLUTION OF THE NEOCORTEX

Harry J. Jerison

Department of Psychiatry and Biobehavioral Sciences
UCLA School of Medicine
760 Westwood Plaza
Los Angeles, California 90024

The fossil record of neocortex is based on the impression of the rhinal fissure and the olfactory bulbs on the cranial cavity of fossil skulls. A cast molded by this cavity is called an "endocast," and in fossil mammals it looks enough like a brain to be called a "fossil brain." Neocortex is identified (and in a sense defined) as the forebrain region of the endocast that is dorsal to the rhinal fissure and posterior to the olfactory tubercle. Errors in identifying neocortex in fossil endocasts are likely to be about the same as in living brains when these superficial markings rather than histological evidence of lamination are the basis of the identification.

Fossil endocasts provide the only direct evidence on the evolution of the brain. Figure 1 is a picture of a natural endocast of this kind. Endocasts can be made in the laboratory by preparing a skull and using it to mold a latex or other rubber-like cast, and eventually preparing a plaster cast. The endocast in Figure 1 is natural in the sense that it was made when the skull had been cleaned and the cavity packed by natural agents, and the packing hardened into rock during a period of fossilization. It is from a small hooved animal called *Bathygenys reevesi* (Wilson, 1971), which lived in the Big Bend region of Texas about 35 million years ago (mya). Although no more than a piece of rock, it is unmistakably a picture of the brain of *Bathygenys* as it was in life.

Comparisons between brains and endocasts in living mammals indicate that only minor errors occur in treating endocasts as brains that cannot be dissected further. Hundreds of "fossil brains" have been collected throughout the world, either as natural endocasts or as latex or plaster casts made from fossil crania (Edinger, 1975; Jerison, 1973; Radinsky, 1979).

We can, of course, do much more with real brains; most importantly for our purpose, we can relate microscopic and molecular features of the brain to measurable superficial features of the sort that are visible on endocasts. If we accept the "uniformitarian" hypothesis, we can then analyze the fossil history of the brain from data on endocasts, because of the uniformitarian assumption that brain-endocast relationships in living species were also true in fossils. The measurable features of fossil endocasts would then provide evidence on the probable features of the brains that they represent.

The quality of an endocast as a model of the brain differs in different taxa. In fish, amphibians, and reptiles the model is usually poor and may be useful only for gross estimates of total brain size. In mammals and birds, on the other hand, endocasts normally provide detailed pictures of the external surface of the brain. Olfactory bulbs, forebrain, and hindbrain are readily identifiable, as are most of the cortical gyri and sulci that are seen when a brain is first removed from the skull.

In large brained living mammals, such as cetaceans, elephants, great apes, and man the brain's convolutions are usually not well represented in endocasts, and a rhinal fissure cannot be identified. But the fissure is unmistakable in carnivores and ungulates (hooved mammals). The quantitative analysis of neocorticalization presented later is based on fossil and living species from the latter orders.

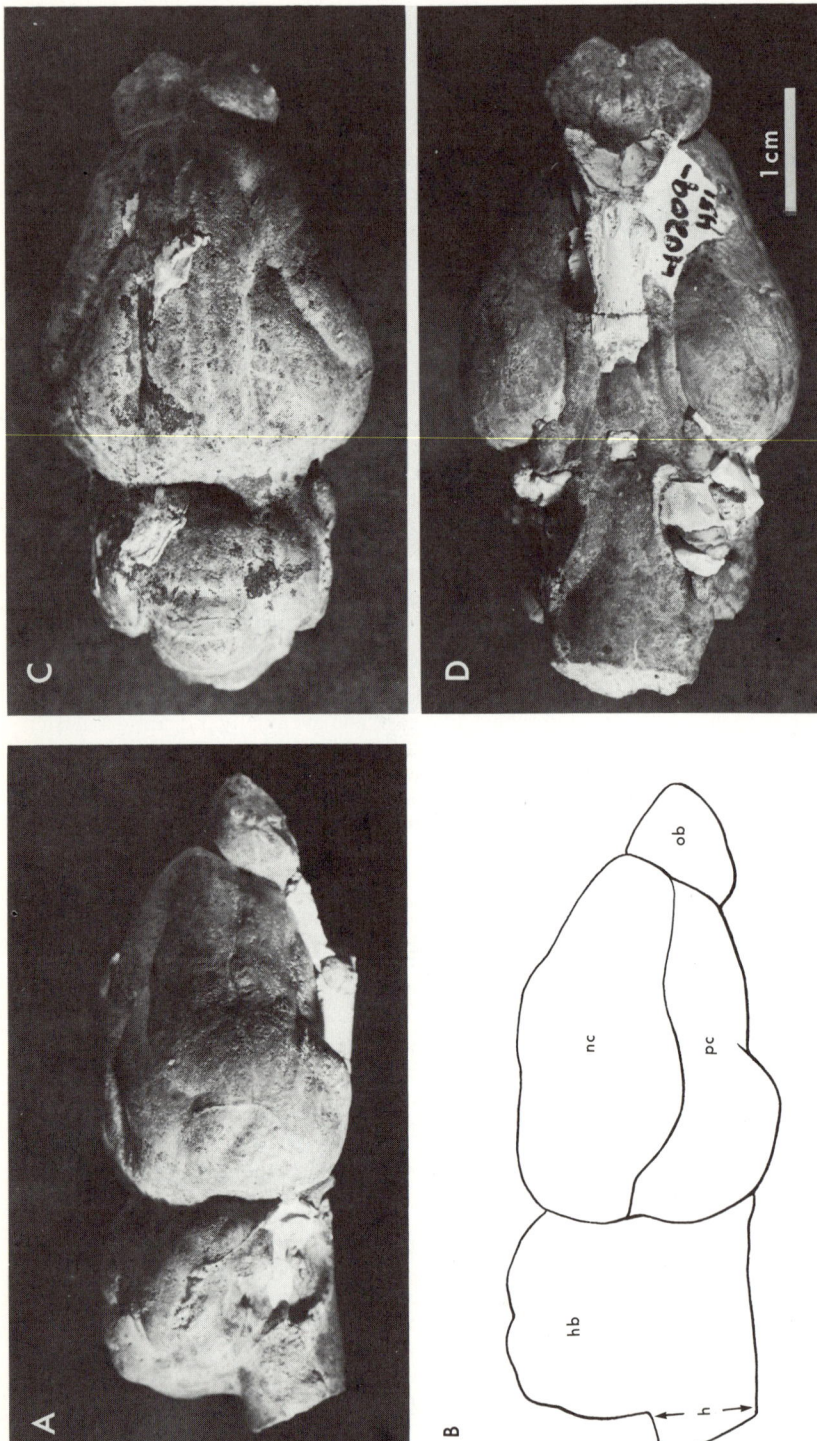

Figure 1. Natural endocast of *Bathygenys reevesi*, a Lower Oligocene oreodon from the Big Bend region of Texas (Wilson, 1971). Fossil bone appears as white; endocast is extremely hard rock, orginally compressed and petrified sand that had been packed into the skull by natural forces before fossilization. A. Lateral view. B. Lateral view divided by areas to show portions allotted to neocortex (nc), paleocortex (pc), hindbrain (hb), olfactory bulbs (ob), and the height of the foramen magnum (h). C. Dorsal view. D. Ventral view. (Specimen UT 40209-431; Courtesy of J.A. Wilson and the Department of Paleontology, University of Texas.)

It is easy to see qualitative differences among species in the amount of neocortex, and it is usually assumed (correctly, as we shall see) that there have been progressive increases in neocortex during the evolution of the mammals. This evolutionary conclusion when based on fossil evidence (Radinsky, 1979) is completely acceptable, but it has more often been proposed and justified by a comparative analysis of living brains. The latter usage is unfortunate, requiring the discredited idea of a *scala naturae* in which, e.g., insectivores are the primitives and primates represent a newer and more advanced grade. (For critical discussion, see Hodos & Campbell, 1969.)

Neocorticalization is a concept in comparative neuroanatomy, which describes, e.g., the fact that primates have relatively and absolutely more neocortex than insectivores. The statement is: Primates are more neocorticalized than insectivores. There was, of course, no transformation of living insectivores into living primates under natural selection. But there is an evolutionary translation of the statement: The ancestors of living insectivores were members of an order (or other taxon) of mammals that probably included at least one species from which primates evolved. This species has not been identified, but as evolutionists positing a relationship between insectivores and primates as "sister" groups, we have to assume that it existed. Neocorticalization is then understood as part of the differentiation of daughter species from parent species: One daughter species of a fossil insectivore, which had relatively more neocortex than the parent species, was the ancestral primate species.

These statements are almost parodies of evolutionary analysis, but they approximate a correct analysis. The ancestral "insectivores" are recognized by systematic biologists as a waste-basket order, difficult to analyze by modern methods of classification. Similarly, the ancestral primates are hard to identify with certainty, and there is fairly complete consensus only about species that appear later in the fossil record — at the beginning of the Eocene epoch (see below). A sense of these difficulties is available from the various contributions in Benton (1988). Despite the difficulties, a "cladistic" analysis of relationships among living species based on nervous system characteristics is quite consistent with analyses based on molecular and gross data, as well as data from the fossil record (Johnson et. al., 1982; Northcutt, 1985).

The concept of neocorticalization can also be used in another sense, as describing the history of a trait in successive populations of species. If we sample a broad range of species across geological time and determine that later species had relatively more neocortex than earlier species, we could state that neocorticalization had occurred, even though we would not be able to determine its phylogenetic history. Such a discovery would be enough to suggest that there was a selective advantage in an increase in neocortex. The analytic model to be applied might be an elaboration of simple models of phenotypic evolution *within* lineages (e.g., Lande, 1976). It would require theorizing about selective advantages "above the species level" (Stanley, 1979), about broad evolutionary "landscapes" which are contexts for interaction among species.

Neocorticalization in this sense can be quantified as a feature of the history of the mammals. It is more or less evident from a simple inspection of endocasts arranged in temporal order (Figure 2), but the quantitative effect is seen more clearly with the help of some statistical analysis.

MAMMALIAN NEOCORTICALIZATION

In their early history the mammals were small, probably insectivorous, creatures. There are one or two fairly complete skeletons for a few species from as long ago as the Upper Triassic and lower Jurassic, about 220 mya, and although they are not related to any living species, their appearance is suggestive of shrews and mice (see Carroll, 1988; Savage & Long, 1986). These animals, contemporaries of early dinosaurs, were probably adapted for life in nocturnal niches (Crompton et. al., 1978; Jerison, 1973; Kemp, 1982), and if nocturnicity was critical for the transition from reptiles to mammals, the correlated selection pressures on the evolving brain could explain the "advance" to a mammalian grade of brain morphology.

The earliest mammalian brain is known from an Upper Jurassic endocast of *Triconodon mordax*, and is about 150 million years old. The lateral surface of this endocast is not well enough preserved to indicate whether or not there was a rhinal fissure, and positive evidence on the presence of neocortex is, therefore, not available. In encephalization, however, its brain was comparable to that of small-brained living species such as opossums and hedgehogs, in which neocortex is present. It is, therefore, likely that neocortex had appeared at least 150 mya. The best assumption from presently available information is that neocortex is, in fact, part of the suite of traits that characterized the mammals from the beginning of their evolution, at least 50 million years earlier (Kielan-Jaworowska, 1986).

Mammals in which the endocasts are sufficiently complete to show a rhinal fissure, if present, are about 75 million years old. They are from a unique assemblage of late Cretaceous mammals (Kielan-Jaworowska, 1983, 1984, 1986), which includes early placentals. The most common mammals of the time, the multituberculates, were unrelated to any living species. Superficially, multituberculates probably looked like living rodents, or insectivores. Their life span as an order was about 120 million years, between about 150 to 30 mya (Kielan-Jaworowska, personal communication; Savage and Long, 1986), a very long span for a mammalian group. The specimen in which the endocast is best known, *Chulsanbaatar*, weighed no more than about 15 grams, a small mammal even for the Mesozoic, and smaller than most living species of mice. There is a suggestion of a rhinal fissure in its endocast, although there is some disagreement about where it is (cf. Jerison, in press; Kielan-Jaworowska et al., 1986). Whether or not one can see a rhinal fissure, from its grade of encephalization, it is very likely that neocortex was present in its brain.

A Digression On Vertebrate History

Table 1 sketches the present consensus on dating the history of the vertebrates. Since neocortex is a uniquely mammalian trait, we can focus on mammals and their immediate ancestors, the therapsids, or mammal-like reptiles, which were the dominant land animals at the end of the Paleozoic and beginning of the Mesozoic Era (250 mya). This digression is an elaboration of the top half of Table 1.

The mammal-like reptiles became extinct during the Jurassic period (215-145 mya), perhaps having lost out in their Darwinian encounters with archosaurians ("ruling reptiles" such as the dinosaurs), which replace them in the history of life. But the replacement was incomplete. At least one therapsid species evolved into an ancestral mammal, and the mammals thrived during the Mesozoic "age of reptiles."

Mammals should be thought of as therapsids that escaped extinction by evolving certain specialized adaptations (probably as nocturnal species), which enabled them to survive in niches that were inaccessible to other "reptiles" of the period (Kemp, 1982; Carroll, 1988). Perhaps from familiarity and long exposure, most of us find the 1773 classification by Linnaeus, when the "Mammalia" were named, intuitively natural, but it would be more consistent with present taxonomic practice to call mammals specialized therapsids, and some systematic biologists do just that (Kemp, 1988).

Mammals and dinosaurs were contemporaries throughout most of the Mesozoic Era. The earliest mammalian fossils are known from the late Triassic period, about 220 mya. During the remaining 155 million years of the Mesozoic, mammals became diversified, evolving some of the major lineages that are present today including the earliest marsupials and placentals. All of the species were small; their giants were the size of domestic cats, and most were shrewlike or mouselike in size and habitus. Savage & Long (1986) have published a well-illustrated popular account of the entire history of the mammals, including a review of their therapsid ancestors, and Carroll (1988) provides an encyclopedic text that covers all of vertebrate history.

Massive extinctions at the Cretaceous-Tertiary (K-T) time boundary, 65 mya, included most large land vertebrates, such as dinosaurs. Adaptive zones that were emptied were evidently invaded and filled by rapidly evolving early Tertiary mammalian groups, which were much more diverse than their Cretaceous predecessors in size, habitus, and presumably in their ecological niches.

We humans are understandably most interested in the history or our own lineage. We are a species *(Homo sapiens)* of the order Primates, one of the oldest of living orders of mammals. Primate fossils have been identified in the late Cretaceous, about 70 mya. Among living primates the suborder of prosimians (lemur-like species) is known from the lower Eocene, about 55 mya. Large numbers of fossils from the suborder of "higher" simian primates, the anthropoids (monkeys, apes, and humans), are known from the Oligocene, more than 30 mya. From their skeletal features, early simian fossils could be described as either monkeys or apes, or, even better, as the ancestors of both lineages before they became separated. Species that were clearly

Table 1

Synopsis of Vertebrate Evolution

Era	Period and Epoch	Age (Years x 10^6)	Fauna (first appearance)
Cenozoic	Quaternary		
	Recent	0.01 - present	
	Pleistocene	1.9 - 0.01	*Homo erectus, H. sapiens*
	Tertiary		
	Pliocene	5.1 - 1.9	Hominids: *Australopithecus, H. habilis*
	Miocene	24 - 5.1	Hominoids
	Oligocene	35 - 24	Anthropoids
	Eocene	55 - 35	Progressive ungulates
	Paleocene	65 - 55	Primates[a] and carnivores
Mesozoic	Cretaceous	145 - 65	First definite rhinal fissure
	Jurassic	215 - 145	Earliest mammalian endocast
	Triassic	250 - 215	Earliest mammals
Paleozoic	Permian	285 - 250	
	Carboniferous	360 - 285	Reptiles
	Devonian	410 - 360	Bony fish & amphibians
	Silurian	440 - 410	Jawless fish
	Ordovician	505 - 440	First vertebrates
	Cambrian	590	Large shelled invertebrates

[a]Primate teeth have been reported in late Cretaceous deposits.

apes (dryopithecines) are known from the Miocene, about 20 mya, and hominids, which appeared during the Pliocene, are well-known from about 3.5 mya.

The present consensus is that an ape lineage, not known as fossils but surviving as chimpanzees, became separated from the hominid lineage relatively recently, about 5 mya (Andrews, 1988; Pickford, 1988). The earliest hominids are known as fragmentary fossils as long ago as that, although the good fossil record, including the record of the brain, begins about 3.5 mya, with the genus *Australopithecus*. The genus *Homo* is about 2 million years old, first appearing as habilines (*H. habilis*), which were evidently replaced by the pithecanthropines (*H. erectus*), known from about 1.6 mya. "Modern" humans (*H. sapiens*) appeared about 250,000 years ago, and at least with respect to brain size there have been no important changes since then. The Neandertal and the more conventionally "modern" human fossils are not easily distinguished from one another by brain criteria (Holloway, 1981), although there is a lively debate about their possible differences in the capacity for language (Lieberman, 1984; Wind, 1976).

Despite our special interest in primate evolution, it is easier to analyze the evolution of the neocortex in other orders of mammals, in which the convolutional pattern of the brain is reflected more clearly in the endocast. The groups that we will concentrate on are ungulates and carni-

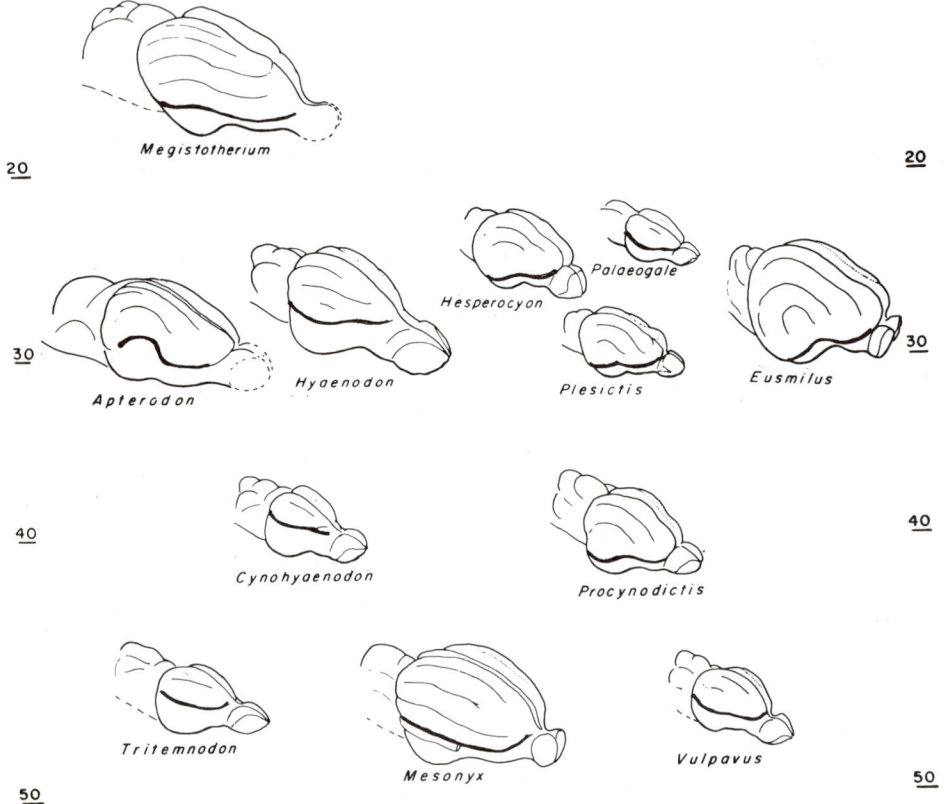

Figure 2. Carnivore brain evolution as revealed in fossil endocasts. This illustration was prepared under the direction of L. Radinsky; emphasis added by the author (darker line on each endocast) to indicate rhinal fissure. Vertical scales are millions of years ago; earliest fossils at bottom and latest at top. (From L. Radinsky, Brains of early carnivores. *Paleobiology*, 1977, Vol. 3, p. 345. Reprinted by permission.)

vores. As mentioned earlier, Cretaceous mammals were small. One of the major evolutionary changes at the K-T boundary was the appearance of much larger species of mammals, and even at the beginning of the Tertiary there are mammalian fossils that must have weighed several hundred kilograms (kg) in life. These were large herbivorous ungulate species that were "archaic" in that they were members of taxonomic orders that are now extinct, perhaps replaced by presently living ("progressive") orders of ungulates. The earliest carnivores were smaller but also included larger bodied animals than those of the Cretaceous. They were not necessarily from the living order Carnivora; in fact, the oldest specimen in our sample, the Paleocene *Arctocyon primaevus* (about 60 mya and about 85 kg body weight), was an ungulate of the extinct order Condylarthra (see Prothero et al., 1988, on classification of fossil ungulates). From its dental anatomy *A. primaevus* appears to have had at least partly carnivorous habits, perhaps as a scavenger.

A completely carnivorous archaic order, the Creodonta, appeared during the late Cretaceous, and its last known species became extinct during the Miocene or Pliocene. Although the living carnivores, order Carnivora, appeared somewhat later than the creodonts, during the Upper Paleocene, species from the two orders coexisted during much of the Tertiary.

There were also several archaic ungulate orders that evolved in isolation in South America. Successful until the end of the Tertiary, about 2 mya, these ungulate orders became extinct during the Pleistocene epoch. Their extinction is usually attributed to the re-establishment of the land bridge to North America (3 mya) and the successful invasion of their "neotropical" adaptive zone by better adapted North American "holarctic" species.

Most of the living orders of mammals had appeared by the end of the Paleocene or in the Lower Eocene, about 55 mya. These included the odd-toed and even-toed ungulates (Perissodactyla and Artiodactyla, e.g., horse and cow) and the living carnivores.

In analyzing our data on neocortical evolution, I will distinguish between "archaic" and "progressive" groups on the basis of their survival. All living orders of mammals (including "primitive" insectivores and didelphids) are defined as "progressive," and all extinct orders are "archaic." As we shall see, this simple and objective dichotomy is also useful for the analysis of neocorticalization.

NEOCORTICALIZATION: A QUANTITATIVE EXERCISE

Neocorticalization refers to an increase in the amount of neocortex relative to other brain structures. The most elaborate analysis has been based on comparisons of insectivore brains with primate brains by Stephan and his colleagues (e.g., Stephan et al., 1981), and I will review some of the results of this analysis later. They use neocorticalization as a comparative rather than evolutionary concept, and they show that there is much more neocortex, both absolutely and relatively, in any primate brain than in any insectivore brain. Primates are more neocorticalized than insectivores. Within the insectivores and within the primates there appear to be no significant differences in neocorticalization. The amount of neocortex in the human brain, for example, is about the amount to be expected in a 1350 gram (g) primate brain.

The evolutionary concept of neocorticalization is that an increase in neocortex can be recognized in comparisons between earlier and later species, i.e., as a function of geological time. My evolutionary analysis is of this sort, "above the species level," that is, I analyze change that can be identified across species by comparing neocorticalization in different slices of time, independently of the species that contribute to the slices. I build, primarily, on the work of Leonard Radinsky (1979), who published qualitative analyses of most of these data (Figure 2). Mine is a quantitative analysis.

Using a fairly broad sample of species of carnivorous and hooved mammals, fossil and living, I consider how the relative size of the neocortex as visible on an endocast has changed with the passage of time. The data are the sizes (area) of neocortex and other parts of the endocasts, and the analysis is based on the relationships among these measures.

There are some technical difficulties. First, many important structures that are connected with or are part of neocortical systems in the brain are either invisible or partly obscured in endocasts. These include "old cortex" such as hippocampus and piriform lobes, and the large fraction of the neocortex that is folded into and hidden in the depths of the convolutions. Second, most visible structures on endocasts are surfaces on an irregular solid, which are difficult to measure directly. Computerized measurements of these surfaces are now possible (Toga et. al, 1989), and I will perform them in the future. The two-dimensional analysis that I present now is, therefore, preliminary to a 3-D analysis, and I present it to indicate the kinds of questions that can be put quantitatively. The answers that I report are, therefore, provisional.

Methods, Material, and Problems

For this exercise I used lateral projections of endocasts (see Figure 1) to measure the height of the foramen magnum ("medulla" of endocasts) and four "brain" areas: neocortex, paleocortex, hindbrain, and olfactory bulbs. These were to be related to geological age in millions of years. The complete data set is in the Appendix.

Except for one measurement of the foramen magnum on a deer skull (*Odocoileus*), all of the analysis was on specimens represented by drawings or photographs that have been published; 55 of these are from Radinsky (1973, 1975, 1976, 1977, 1981). Radinsky's sample included 38 Carnivora, 7 Creodonta, 4 Condylarthra, 5 species from extinct South American (neotropical) ungulate orders, and one Eocene perissodactyl (*Hyrachyus*, an ancestral rhinoceros). To these I added four other progressive ungulates (artiodactyls): *Bathygenys* of Figure 1, and two fossil and one living species from Jerison (1973). The complete sample consisted of 35 fossil species and 24 living species, a total of 59 species.

The variables were geological age, the four brain areas, and the height of the foramen magnum. The most straightforward variable is geological age, about which there is broad agreement

(Radinsky, 1978; Savage and Long, 1986). The area of neocortex is an area as shown in Figure 1. It is not the true area because it is a two-dimensional projection of a solid, and because in convoluted brains a significant amount of neocortex is buried in the sulci. The amount of hidden neocortex is strongly correlated with the superficial visible area in living species (Jerison, 1982), however, and our measure of neocortex almost certainly can provide appropriate estimations of neocorticalization. The area of the olfactory bulbs on the endocast is, similarly, a reasonably good estimator of the volume of the bulbs on the brain, and the height of the foramen magnum estimates the height of the medulla on the brain.

Although I present data on paleocortex and hindbrain, these cannot be interpreted as brain data. In neocorticalized species paleocortex always appears reduced on lateral projections, because as neocortex expands paleocortical tissue becomes hidden in the rhinal fissure, and the piriform lobe and other "old cortex" structures are displaced ventrally. A lateral view of the brain loses data about the ventral surface, and paleocortex will always be underestimated. The visible area of hindbrain, which consists of "medulla" (foramen magnum), pons, and cerebellum, will also be underestimated as species become neocorticalized, because neocortex expands posteriorly to cover some of the cerebellum. These problems are not solvable with present methods.

Another set of problems is related to the evolution of diversity in body size. Species differ in body size for many reasons, most of which are only marginally related to brain functions. Nevertheless, body size must be taken into account in the analysis of the evolution of the size of brain structures, because of the allometric relationships among the sizes of organs. Other things being equal, larger species have larger brains and more neocortex than smaller species, but it would be silly to characterize this difference as neocorticalization. The crucial feature is obviously the amount of neocortex *relative* to the amount expected in species of a given body size. For this exercise I used the foramen magnum to estimate body size (Radinsky, 1967). In the data of Stephan, et al. (1981), the volume of the medulla is the brain-measure most highly correlated with body size (log-units: $r = .983$), and insectivores and primates are not significantly different from one another in their medulla-body size relationships. A single equation, independent of taxonomic order, estimates body size from the size of the medulla in both insectivores and primates, despite the substantial difference between these groups in encephalization. Harman (1957) reported a similar result for a mixed sample of primates, ungulates, and carnivores (see Jerison, 1973). This is the justification for using the endocast foramen magnum, which is the "medulla" on an endocast, as the statistical control for body size in the carnivores and ungulates in our sample.

The statistical control is accomplished by using residuals from linear regressions of the log-transforms of the brain measures against foramen magnum height as the independent variable. Because the scaling is logarithmic the residuals are actually ratios of the areas of the brain relative to the "expected," or "average," areas as estimated by the regression analysis[1]. In presenting my results graphically I call these residuals, "quotients." A neocortical quotient of 1.0 means that the amount of neocortex is exactly as large as expected; 0.5 means that it is half as large: 2.0 means twice as large, and so on. If there was an evolutionary increase or decrease in the relative amount of a particular part of the brain in mammals, it would appear as a positive or negative change in the quotient with the passage of time.

RESULTS AND DISCUSSION

A summary of the relationships among the measurements (raw data in the Appendix) in our sample of 59 species is presented in Table 2. This is a matrix of correlation coefficients from which one can read the extent to which measurements are related to one another. The only brain measurement that became greater (had a positive correlation coefficient with age) during the

---

[1]Logarithmic scales are the natural scales for biological size. This is a consequence of the role of cell division in determining organismic size. A human brain size could be generated by having one or two more cell divisions for brain cells beyond a chimpanzee grade of brain size. The problem of scaling to reflect natural phenomena was discussed many years ago by Stevens (e.g., Stevens, 1946), and logarithmic scales for biological size can be understood in the light of that discussion (cf. Schmidt-Nielsen, 1984).

Table 2

Correlation Coefficients among Endocast Measures and Geological Age

	Age	Neocortex	Paleocortex	Hindbrain	Olf. Bulbs	Foramen
Age	1.000					
nc	0.474	1.000				
pc	-0.365	0.332	1.000			
hb	-0.255	0.582	0.822	1.000		
ob	-0.092	0.599	0.747	0.820	1.000	
h	-0.175	0.627	0.794	0.908	0.837	1.000

Note. — Data on brain parts in log-units. At N = 59, r > |0.26| is significantly different from 0 (p < 0.05). Abbreviations as in Figure 1. Raw data in Appendix A.

Tertiary was the neocortex, r = 0.47, p < .05. As mentioned earlier, the only other interpretable measure is of the olfactory bulbs, and their size was evidently uncorrelated with geological age, r = -0.09, p > 0.05.

About the other measures, the "significant" negative correlation between paleocortex and time is very likely a consequence of the increased amount of neocortex, which would have covered, and forced a ventral rotation of, the paleocortical surface. The same kind of effect would have affected the hindbrain projection, which was also negatively, but not significantly, correlated with geological age. Neither relationship is meaningful for an analysis of the evolution of the brain. The matrix of correlation coefficients is, nevertheless, interesting in that all of the parts of the brain are positively correlated with one another. This almost certainly is due to the correlation of both neocortex and paleocortex areas with total brain size. Even if neocortex masks paleocortex in more recent species, the absolute size of measured paleocortex is larger in large brains than in small brains as is the absolute size of neocortex, and the correlation reflects the fact that both are correlated with the hidden variable of gross brain size.

Body size, at least as reflected in the height of the foramen magnum was not correlated with geological age (r = -0.175, p > .05). This may be partly an artifact of the living species in the sample, however, which were mainly small carnivores such as civets and mongooses. There is an old paleontological rule called "Cope's Law" to the effect that body size tended to increase in evolution. It is appropriate, therefore, to note that the 35 fossil species, when analyzed as a separate group, did show a slight positive correlation between the height of the foramen magnum and geological age (r = 0.32, p < .05 one-tailed — appropriate for testing "Cope's Law").

Regardless of evolutionary trends in body size, the well known allometric relationship between brain size and body size requires that we control for body size when we analyze other gross quantitative features of the brain. To examine the brain's evolution independently of its known correlation with body size, we can analyze the correlations of the residuals for each brain measure when the foramen magnum height is "partialled out," that is, controlled statistically. I did this with the analyzable data, namely, on neocortex and olfactory bulbs. The results were striking. There was a significant positive correlation between the neocortical residuals and geological age (r = 0.75) and no significant correlation between the olfactory bulb residuals and geological age (r = 0.10). The functional relations and the data points are presented in Figures 3 and 4 as neocortical and olfactory bulb "quotients," as described above.

The graphic analysis in Figure 3 displays three results. There is, first, the fact of neocorticalization. Later species tended to have significantly more neocortex than earlier species. This is the meaning of the positive slope of the regression line. Second, species from archaic orders tended to have less neocortex than species from progressive orders. Thus, three of the four archaic ungulate species, five of the seven archaic carnivore (creodont) species, and four of the five Neotropical ungulate species (also archaic in that their orders are extinct) are below the regression line. Twelve of our 16 "archaic" species thus had less neocortex than would be predicted for their geological age by an unbiased regression analysis. For those who enjoy playing with statistics, a chi-square analysis contrasts this with an expected even split; chi-square = 4, df=1, p < 0.05.

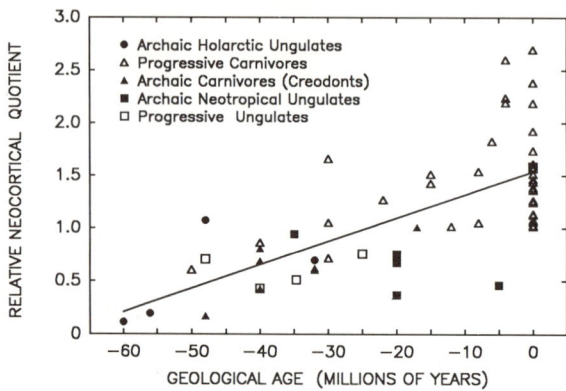

Figure 3. Change in relative neocortical surface area on a lateral projection of the endocast (cf. Figure 1) as a function of geological age. Neocortical quotient is residual from regression of neocortex area on foramen magnum height and is interpreted as a ratio of measured neocortical area relative to area expected on the basis of body size in this sample. "Progressive" change noted here (positive slope of regression line) indicates increased neocorticalization over time. Each point is a species. For raw data, see Appendix.

The third result is in a comparison between progressive and archaic species limited to fossil Carnivora *versus* Creodonta and is in two parts. First, the Carnivora points appear generally to be higher than the Creodonta. Second, the Carnivora points seem to show more "progress" over time than do the Creodonta. It is not really possible to test the first part properly, because the species are from different geological times, and there is no obvious way to control the time-variable. The second part, however, can be tested by simple regression analysis. The correlation between age and neocortical quotient for 15 Carnivora species was $r = 0.72$, $p < .01$. For seven Creodonta it was $r = 0.42$, $p > .05$.

The result is important for evolutionary analysis of the relations between true carnivores and creodonts and is relevant for a disagreement that I had with Radinsky on how to interpret the data. The argument is summed up by Carroll (1988) as follows:

> Romer (1966) and Jerison (1973) stigmatized the creodonts as archaic and small brained, but Radinsky (1977) demonstrated that relative brain size increased as rapidly among creodonts as it did in the early members of the Carnivora, together with an increase in the extent of the neocortex (Carroll, 1988, pp. 478-479).

Viewing Figure 2, above, from Radinsky, one can see why he reached his conclusion. But the quantitative analysis supports Romer's view as mentioned by Carroll. (My contribution in 1973 was mainly to quote Romer and to provide very limited quantitative data. The present confirmation of our older view is possible because of the data collected by Radinsky, which permitted a statistical test.)

Having presented these results and their interpretation it is appropriate to repeat the earlier caveat. The analysis is preliminary, based on 2-D data. It should be possible to check it with a larger sample of fossil material and with a 3-D analysis of that sample.

Unlike neocortex, the olfactory bulbs did not change in relative size with the passage of time (Figure 4). The correlation between the olfactory bulb quotient and geological time was $r = 0.1$, which is not significantly different from 0. This is an important point for several reasons. First, it shows that our approach is fine enough to discriminate between the presence and absence of change. Second, this result is instructive about the meaning of "primitive" and "progressive" in the analysis of brain evolution.

Although careful students (Baron et al., 1983; Sarnat and Netsky, 1974) do not make the error, neurobiologists often assume that having large olfactory bulbs is a primitive mammalian

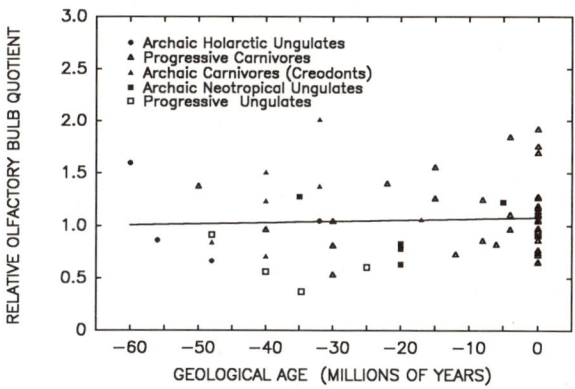

Figure 4. Absence of change in relative area surface quotient of olfactory bulbs on a lateral projection of the endocast (cf. Figure 1) as a function of geological age (Cf. Figure 3). For raw data, see Appendix.

trait and that the olfactory bulbs became relatively smaller as the mammals evolved. Figure 4 corrects this error by showing that olfactory bulbs have been a stable feature of the brain in Tertiary carnivores and ungulates. The misconception is a bit of primate chauvinism. Primates (at least the anthropoids) are neocortical specialists, but they are deficient mammals in olfactory development. A reduced role for olfaction is part of the adaptive mosaic of "higher" primates and is not a broad feature of mammalian evolution. Neocorticalization, on the other hand, appeared as a general trend in many mammalian groups and its relative absence in, e.g., the insectivores and many marsupials is correctly recognized as a primitive feature in these groups.

CONCLUSIONS

The fossil evidence indicates that there was neocorticalization in mammals and that it can be detected even in samples as small as 15 species of Carnivora. Evolution of the carnivores involved neocorticalization in another sense, in that the two great orders of Tertiary carnivores, namely, Creodonta and Carnivora, differed in the extent of neocorticalization. This could have been a factor in the survival of true carnivores. In any event the history of neocorticalization indicates that there was almost certainly some benefit derived from the expansion of neocortex. The fossil evidence, therefore, confirms conventional wisdom in neurobiology that it was a progressive thing for mammals to evolve neocortex and (perhaps within limits) more is better.

If our conclusions on neocortex are expected, those on the olfactory bulbs are not. These structures are surprisingly constant features in mammalian brains. There is no reason to have predicted that they would not evolve to large or small size relative to other parts of the brain, depending on the extent of olfactory specializations in particular species. But according to present evidence from fossil brains, the olfactory bulbs have been constant and relatively unchanging features that make a brain a mammal's brain. They are not unusually enlarged in any species; most mammals are olfactory specialists. Evolutionary changes in the olfactory bulbs occurred mainly in a negative way, by reduction. The reduced state of olfactory bulbs in humans and other primates (and their complete absence in some cetaceans) merely reflects the extremes of diversity that are possible as the brain evolved to control the activities of mammals in the variety of niches in which they function.

It would be consistent with the fossil evidence to emphasize neocortical functions and assume that the evolutionary advantage conferred by these functions was the engine driving *progressive* brain evolution in mammals. But it would be a mistake to make much of such an idea of progress. It is true that neocorticalized species are more prevalent now than in the distant past. It must also be true that some fitness is associated with this aspect of the brain's evolution. But there are many successful living species that are at a very ancient grade of mammalian neocorticalization. Hedgehogs in Europe and opossums in America are outstanding examples, because they are so fit in the evolutionary sense. They may litter our highways because of their "stupidity" in refusing to yield the right of way to cars and trucks, but the litter is part of the evidence of their reproductive success. And they manage this at a grade of neocorticalization and encephalization that some mammalian species reached 150 million years ago.

ACKNOWLEDGMENTS

It would not have been possible to write this chapter without the major body of data published by the late L. B. Radinsky, premier scholar of endocast morphology. I thank Dr. Alan Charig, Dr. Alan Gentry, and Dr. Jerry Hooker of the British Museum (Natural History) for access to their fossils and their library, and Dr. Donald R. Prothero for advice on some paleontological and geological questions. I wish to thank, especially, Professor Zofia Kielan-Jaworowska, now at the Paleontological Museum, University of Oslo, for her kindness and graciousness when I visited her in Warsaw, when I had the unique opportunity to examine her extensive collection of Mesozoic materials at her laboratory at the Paleobiological Institute of the Polish Academy of Sciences.

This paper is a shorter, up-dated version of Jerison (in press), which was prepared at the request of Dr. E.G. Jones.

REFERENCES

Andrews, P. (1988) A phylogenetic analysis of the Primates. In Benton, M.J. (ed.) *The phylogeny and classification of the tetrapods*, Vol. 2: Mammals, Systematics Association Special Volume No. 35B, pp. 143-75. Clarendon Press, Oxford, U.K.

Baron, G., Frahm, H.D., Bhatnagar, K.P. and Stephan. H. (1983) Comparison of brain structure volumes in Insectivora and Primates. III. Main olfactory bulb (MOB). *Journal für Hirnforschung*, 24: 551-558.

Benton, M.J. (ed.) (1988) *The phylogeny and classification of the tetrapods*, 2 Vols. Systematics Association Special Volume No. 35A,B, Clarendon Press, Oxford, U.K.

Carroll, R.L. (1988) *Vertebrate paleontology and evolution*. New York, Freeman.

Crompton, A.W., Taylor, C.R., and Jagger, J.A. (1978) Evolution of homoeothermy in mammals. *Nature* (London) 272:333-6.

Edinger, T. (1975) Paleoneurology, 1804-1966: An annotated bibliography. *Advances in Anatomy, Embryology and Cell Biology*, 49: 12-258.

Harman, P.J. (1957) *Paleoneurologic, neoneurologic, and ontogenetic aspects of brain phylogeny*. James Arthur Lecture on the Evolution of the Human Brain. New York: American Museum of Natural History.

Hodos, W. and Campbell, C.B.G. (1969) *Scala naturae*: Why there is no theory in comparative psychology. *Psychological Review*, 76: 337-350.

Holloway, R.L. (1981) Volumetric and asymmetry determinations on recent hominid endocasts: Spy I and II, Djebel IHROUD I, and the Sale' *Homo erectus* specimens, with some notes on Neanderthal brain size. *American Journal of Physical Anthropology*, 55: 385-393.

Jerison, H.J. (1973) *Evolution of the Brain and Intelligence*. New York, Academic Press, xiv+482 pp.

Jerison, H.J. (1982) Allometry, brain size, cortical surface, and convolutedness. In Armstrong, E. & Falk, D. (eds.). *Primate Brain Evolution: Methods and Concepts*. pp. 77-84. New York, Plenum.

Jerison, H.J. (in press). Fossil evidence on the evolution of the neocortex. In Jones, EG and Peters, A (eds) *Cerebral Cortex*, Vol. 8. New York, Plenum.

Johnson, J.I. Kirsch, J.A.W., and Switzer, R.C. (1982) Fifteen characters which adumbrate mammalian geneology. *Brain, Behavior and Evolution*, 20: 72-83.

Kemp, T.S. (1982) *Mammal-like reptiles and the origin of mammals*. London and New York, Academic Press.

Kemp, T.S. (1988) Interrelationships of the Synapsida. In Benton, M.J. (ed.) *The phylogeny and classification of the tetrapods*, Vol. 2: Mammals, Systematics Association Special Volume No. 35B, pp. 1-22. Clarendon Press, Oxford, U.K.

Kielan-Jaworowska, Z. (1983) Multituberculate endocranial casts. *Paleovertebrata*, Montpellier, 13 (1-2):1-12.

Kielan-Jaworowska, Z. (1984) Evolution of the therian mammals in the Late Cretaceous of Asia. Part VI. Endocranial casts of eutherian mammals. *Paleonotologica Polonica*, No. 46-1984: 151-171, Pls. 29-31.

Kielan-Jaworowska, Z. (1986) Brain evolution in Mesozoic mammals. In Lillegraven, J. A. (ed.) G.G. Simpson Memorial Volume. *Contributions to Geology*, University of Wyoming, Special Paper 3:21-34.

Kielan-Jaworowska, Z., Presley, R. and Poplin, C. (1986) The cranial vascular system intaenioloabidoid multituberculate mammals. *Phil Trans. Roy. Soc.* (London), B313: 525-602.

Lande, R. (1979) Quantitative genetic analysis of multivariate evolution, applied to brain: body size allometry. *Evolution*, 33: 402-416.

Lieberman, P. (1984) *The biology and evolution of language*. Harvard Univ. Press, Cambridge, Mass.

Northcutt, R.G. (1985) The brain and sense organs of the earliest vertebrates: Reconstruction of a morphotype. In Foreman, R.E., Gorbman, A., Dodd, J.M., & Olsson, R. (eds.) *Evolutionary biology of primitive fishes*. 81-112. New York, Plenum.

Pickford, M. (1988) The evolution of intelligence: A palaeontological perspective. In Jerison, H.J. and Jerison, I.L. (eds.) *Intelligence and evolutionary biology*. pp. 175-198. Heidelberg, Berlin, New York, Springer-Verlag.

Prothero, D.R., Manning, E.M., and Fischer, M. (1988) The phylogeny of the ungulates. In Benton, M.J. (ed.) *The phylogeny and classification of the tetrapods*, Vol. 2: Mammals, Systematics Association Special Volume No. 35B, pp. 201-34. Clarendon Press, Oxford, U.K.

Radinsky, L.B. (1967) Relative brain size: A new measure. *Science,* 155: 836-838.

Radinsky, L. (1973) Evolution of the canid brain. *Brain, Behavior and Evolution*, 7: 169-202.

Radinsky, L. (1975) Viverrid neuroanatomy. *Journal of Mammalogy*, 56: 130-150.

Radinsky, L. (1976) The brain of *Mesonyx*, a Middle Eocene mesonychid condylarth. *Fieldiana Geology*, 33: 323-337.

Radinsky, L. (1977) Brains of early carnivores. *Paleobiology*, 3: 333- 349.

Radinsky, L. (1978) Evolution of brain size in carnivores and ungulates. *American Naturalist*, 112: 815-831.

Radinsky, L. (1979) *The Fossil Record of Primate Brain Evolution*. The James Arthur Lecture. New York, American Museum of Natural History. 27 pp.

Radinsky, L. (1981) Brain evolution in extinct South American ungulates. *Brain, Behavior and Evolution*, 18: 169-187.

Romer, A.S. (1966) *Vertebrate Paleontology*, 3rd ed., Univ. of Chicago Press, Chicago, Illinois.

Sarnat, H.B. and Netsky. M.G. (1974) *Evolution of the Nervous System*. New York, London, and Toronto, Oxford University Press.

Savage, R.J.G. and Long, M.R. (1986) *Mammal Evolution: An Illustrated Guide*. London: British Museum (Natural History).

Schmidt-Nielsen, K. (1984) *Scaling: Why is animal size so important*. Cambridge, England, Cambridge Univ. Press.

Stanley, S.M. (1979) Macroevolution: Pattern and Process. San Francisco, W.H. Freeman.

Stephan, H., Frahm, H., and Baron, G. (1981) New and revised data on volumes of brainstructures in insectivores and primates. *Folia Primatologica*, 35:1-29.

Stevens, S.S. (1946) On the theory of scales and measurement. *Science*, 103:677-680.

Toga, A.W., Samaie, M. and Payne, B.A. (1989) Digital rat brain: A computerized atlas. *Brain Research Bulletin,* 22: 323-333.

Wilson, J. A. (1971) Early tertiary vertebrate faunas, Vieja Group. Trans-Pecos Texas: Agriochoeridae and Merycoidodontidae. *Texas Memorial Museum Bulletin* 18: 1-83.

Wind, J. (1976) Phylogeny of the human vocal tract. Annals of the *New York Academy of Sciences*, 280: 612-630.

APPENDIX

Endocast Data (mm for Foramen Magnum; mm^2 for other Structures)

Species	Group	Age	Neo-cortex	Paleo-cortex	Hind-brain	Olf. bulbs	Foramen magnum
Arctocyon	A	60	89	499	714	221	12
Mesonyx	A	48	1105	436	967	124	14
Phenacodus	A	56	179	357	630	142	13

Apterodon	A	32	752	374	925	204	15
Vulpavus	C	50	255	141	209	103	8
Procynodictis	C	40	505	182	478	101	10
Hesperocyon	C	30	397	105	234	81	9
Hoplophoneus	C	30	1268	208	832	117	16
Nimraevus	C	30	2004	106	1152	228	16
Mesocyon	C	22	922	271	625	183	11
Enhydrocyon	C	15	1296	208	888	240	13
Phlaocyon	C	15	1198	247	690	190	13
Cynodesmus	C	12	740	134	528	95	11
Newgenus1	C	8	1448	271	751	215	17
Tomarctus	C	8	1269	277	659	185	12
Pseudaelurus	C	6	1628	299	1096	132	13
Aelurodon A	C	4	2128	351	764	169	14
Aelurodon B	C	4	2819	457	1278	419	16
Osteoborus	C	4	3011	482	1020	232	16
Atilax	C	0	939	87	187	87	9
Bdeogale crassicauda	C	0	517	65	270	78	9
Bdeogale nigripes	C	0	787	83	270	126	11
Canis adjustus	C	0	1050	198	396	178	12
Canis lupus	C	0	2455	578	629	355	15
Canis mesomalis	C	0	1235	163	328	177	9
Cerdocyon	C	0	953	213	325	142	11
Cryptoprocta	C	0	957	122	326	78	10
Cynogale	C	0	918	152	270	78	11
Dusicyon	C	0	786	157	284	117	9
Felis concolor	C	0	2319	330	1091	188	18
Fennecus	C	0	653	71	220	66	7
Fossa fossa	C	0	526	83	196	87	8
Genetta	C	0	531	67	259	62	8
Hemigalus	C	0	539	104	274	100	9
Lycaon	C	0	2847	436	517	213	15
Osbornicus	C	0	692	67	196	98	8

Paradoxurus	C	0	543	135	304	78	7
Prionodon	C	0	375	67	228	45	6
Suricata	C	0	405	52	104	44	7
Viverra civetta	C	0	940	303	616	205	11
Viverra tangalunga	C	0	607	156	393	67	9
Viverricula	C	0	567	134	313	103	9
Thinocyon	D	48	44	108	95	38	6
Cynohyaenodon	D	40	188	264	291	118	8
Humbertia	D	40	499	168	473	79	10
Pterodon	D	40	754	494	932	245	15
Hyaenodon crucians	D	32	450	575	665	270	12
Hyaenodon horridus	D	32	901	880	1078	364	18
Megistotherium	D	17	2599	948	2631	502	27
Rynchippus	N	35	1865	259	1710	461	22
Diadiaphorus	N	20	902	654	1195	170	16
Hegetotherium	N	20	272	239	434	83	11
Proterotherium	N	20	748	620	1068	167	15
Pseudotypotherium	N	5	966	668	1284	469	23
Hyrachyus	P	48	872	588	1124	205	16
Anoplotherium	P	40	1017	1131	1375	244	25
Bathygenys	P	35	256	200	293	33	9
Protoceras	P	25	750	506	656	108	14
Odocoileus	P	0	2103	417	722	222	20

Note. - Group codes: A: Order Condylarthra (Archaic Ungulates); C: Order Carnivora (Progressive Carnivores); D: Order Creodonta (Archaic Carnivores); N: Orders Notoungulata and Litopterna (Neotropical Archaic Ungulates); P: Orders Perissodactyla and Artiodactyla (Progressive Ungulates). Age is millions of years ago, based on Radinsky (1978) and Savage and Long (1986). Original drawings for the first 55 species are in Radinsky (1973, 1975, 1977, 1981); those for the last four are from this chapter and Jerison (1973).

CRITICAL CELLULAR EVENTS DURING CORTICAL EVOLUTION: RADIAL UNIT HYPOTHESIS*

Pasko Rakic

Yale University School of Medicine
New Haven, Connecticut 06510 USA

INTRODUCTION

The question of cerebral evolution is among the oldest and most difficult, perhaps because one cannot expect that a single and simple answer can explain complex cellular changes. Furthermore, the genetic, molecular and cell biological mechanisms by which the brain evolves to match an ever changing environment are still largely unexplored. There is, however, little disagreement among contemporary neuroscientists that emergence of distinctly human mental capacities reflect the expansion of the cerebral neocortex (Rakic and Singer, 1988). The present article is concerned with the cellular mechanisms that may underlay the evolution of this remarkable structure and formation of cytoarchitectonic maps in primates.

Examination of the cerebral hemispheres in various living primates shows a large range in the size and pattern of cytoarchitectonic maps that reflect functional specializations. For example, since the time that humans and old world monkeys departed from their common ancestor at the beginning of the Miocene period some 23 million years ago (Fleagle, 1988), the cortex of both species has undergone considerable modification. The total surface area of cortex in humans is not only about ten times larger than in monkeys, the relative proportion of various cytoarchitectonic areas is also quite different, with humans having a larger proportion of association cortex (Blinkov and Glaser, 1968). In addition, humans have some new cytoarchitectonic subdivisions, notably the speech area, that monkeys do not possess. How these differences are introduced at the genetic, molecular, and cellular level is a major challenge of neuroscience.

It seems reasonable to start with the assumption that evolutionary novelties in the neocortical cytoarchitectonic map (e.g., overall size, relative proportions of existing areas and introduction of new areas) occur as structural modifications that are introduced as mutations in an individual common ancestor and passed on to descendents. Most evolutionary biologists believe that we have to rely on the study of ontogeny to uncover the cellular and molecular mechanisms that generate major morphological changes on an initial blueprint. There are many examples from evolutionary biology that support the validity of this approach (e.g., Gould, 1977). In the present article I will try to suggest and interpret possible cellular events that could occur during cortical evolution in the framework of the *radial unit hypothesis* (Rakic, 1988). This hypothesis was derived from a series of neuroembryological studies starting about two decades ago (Rakic, 1971, 1972). The experimental evidence and documentation for most of the factual statements can be found in the primary references listed in the bibliography and in a recent review (Rakic, 1988).

An examination of cellular events during cortical ontogeny indicates that the changes in cytoarchitectonic maps could be explained by a *heterochronic process* or modification of timing and proliferation kinetics in the ventricular zone. The DNA labeling data on the kinetics of cell proliferation (Rakic, 1988) indicate that a relatively small alteration of onset, offset and rate of cell proliferation in a particular embryonic zone can explain the difference in size and pattern of

---

*This is a slightly modified and updated version of the presentation made at the Pontifical Academy of Science, The Vatican, October, 1988.

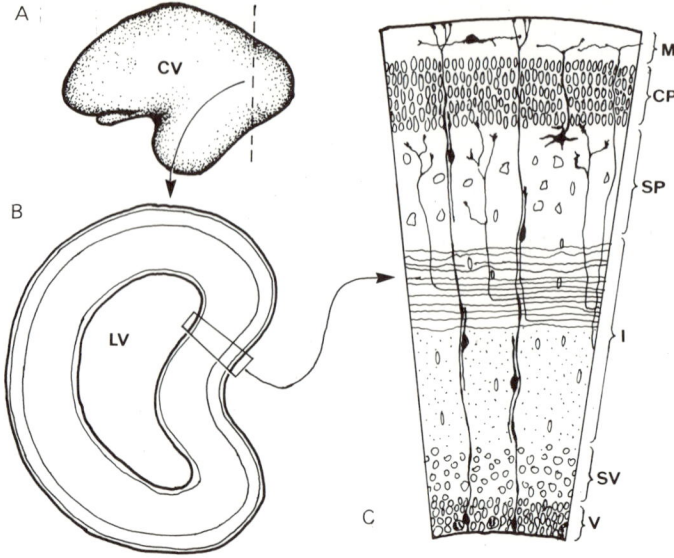

Figure 1. Cytological organization of the primate cerebral wall during the first half of gestation. A. Cerebral vesicles of 60-65 day old monkey fetuses are still smooth and lacks characteristic convolutions that will emerge in the second half of gestation. B. Coronal section across the occipital lobe at the level indicated by a vertical dashed line in A. The lateral ventricle at this age is still relatively large and only the identification of incipient calcarine fissure marks the position of the prospective visual cortex. C. A block of the tissue dissected from the upper bank of calcarine fissure. At this early stage one can recognize all embryonic layers: ventricular zone (V); subventricular zone (SV); intermediate zone (I); subplate zone (SP); cortical plate (CP); and marginal zone (M). Note the presence of migrating neurons (dark bipolar profiles) moving along radial glial fibers which span the full thickness of the cortex. The early afferents from the brain stem and thalamus invade the cerebral wall and accumulate in the subplate zone, where they make transient synapses before entering the cortical plate. Further explanation in text.

cytoarchitectonic maps between *macacus rhesus* and *homo sapiens*. The formulation which describes the introduction of novelty in evolution by the process of heterochrony is described for various morphological features of the body by Alberch et al., 1979. However, understanding phylogenetic or ontogenetic development of the central nervous system presents its own special problems because of the complex interplay of nerve cells, hormones, and various growth factors which control gene expression (Easter et al., 1985; Purves, 1988; Edelman, 1988). In the cerebral cortex this type of cellular interaction probably plays a more significant role than in any other organ, and has to be taken into consideration.

To make this problem manageable I will focus on the relative size and selected anatomical and biochemical differences between two visual areas (V1 and V2): distinct lamination, separate thalamic input, and characteristic distribution of the mitochondrial enzyme cytochrome oxidase (CO). These anatomical and biochemical features of the two areas are remarkably similar in all old world primates, including macaque and human (Allman, 1989; Burkhalter and Bernardo, 1989; Shkol'nik-Yarros, 1971; VanEssen, 1985). How is a cortex with such diverse anatomical and biochemical characteristics in adjacent areas created in phylogeny? What controls their relative size and appearance?

Early cellular events and formation of embryonic zones

To understand the relevance of the radial unit hypothesis for cortical phylogeny it is essential to review early cellular events that occur in the telencephalon during cortical ontogeny. The telencephalic wall in the embryo contains several cellular zones that do not have counterparts in the adult cerebrum (reviewed and updated in Rakic, 1982). The lining of the cerebral ventricle during the entire first half of gestation of both the macaque and human consists of proliferative cells that eventually produce all neurons of the neocortex. These precursor cells form the germinal, or *ventricular zone* (Figure 1). The proliferative cells in this zone are in different phases of

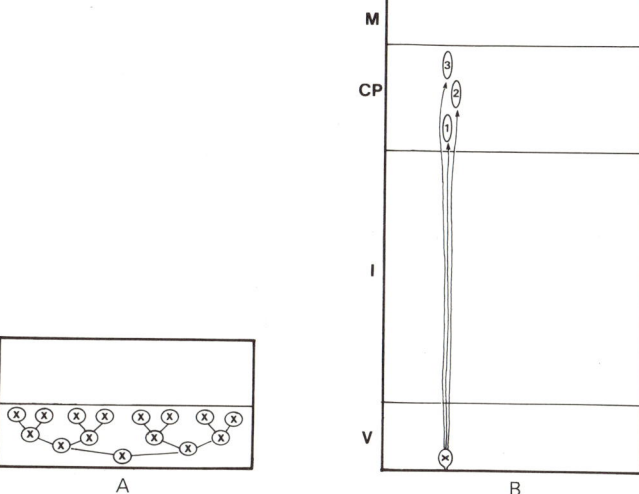

Figure 2. A. Schematic model of symmetrical cell divisions which predominate before the 40th embryonic day (E40). At this early age the cerebral wall consists of only the ventricular zone (V), where all cells proliferate and the marginal zone (M), where they extend their radial processes. Symmetric division produces two progenitors during each cycle and causes lateral spread. B. Model of asymmetrical division which becomes predominant in the monkey after E40. During each asymmetrical division progenation produces one postmitotic neuron which leaves the ventricular zone and another progenitor which remains within the proliferative zone. Postmitotic neurons migrate across the intermediate zone (I) and become arranged in the cortical plate (CP) in reverse order of their arrival (1,2,3). Further explanation in text.

mitotic cycle and have radial processes that protrude towards the pial surface and from the outer cell-free *marginal zone* (Figure 1C). These neuronal precursors, intermixed with dividing radial glial cells, are arranged in a pseudostratified manner and are attached to the ventricular surface by their endfeet. I termed a group of precursor cells that form a pseudostratified column separated by a glial cells *proliferative unit* (Rakic, 1978). Cohorts of cells produced in succession from the same proliferative unit migrate along radial glial fascicles and pass through the *intermediate* and *subplate zones* before entering the developing cortical plate (Figure 1C).

Neurons that arrive at the cortical plate pass by each other and become arranged radially in the form of *ontogenetic columns* (Rakic, 1982). Thus, an ontogenetic column is defined as a cohort of neurons that originate from several progenitors of the same proliferative unit. These columns can be easily recognized in the cortical plate during midgestation (e.g., Figure 1 in Rakic, 1988). The radial unit, migrating pathway, and ontogenetic column together form the *radial unit* that extends from the ventricular to pial surface. Therefore, the developing neocortex can be considered as a mosaic composed of a large number of such radial units. A full description of cytological details, including ultrastructural and immunocytochemical documentation for the underlying developmental events can be found in previously published papers (Rakic, 1971, 1972, 1974, 1976, 1978, 1982, 1988; Rakic et al., 1974; Levitt and Rakic, 1980; Levitt, et al., 1981, 1983; Schmechel and Rakic, 1979a, b).

Formation of radial units

Before E40 all cells in the ventricular zone of the monkey telencephalon are still dividing. Most divisions during this phase are symmetrical indicating that each progenitor produces two additional progenitor cells during each mitotic cycle (Rakic, 1988). As a consequence, this mode of proliferation leads to an exponential increase in the number of cells in the ventricular zone: e.g., each extra round of mitosis before E40 doubles the number of progenitor cells (Figure 2A). This increase in turn may be responsible for the surface enlargement of the cerebral cortex or, in the case of regional differences in mitotic activity, for enlargement of certain areas.

Around E40 some progenitors begin to produce postmitotic neurons that will never divide again (Rakic, 1974, 1988). The inhibition of DNA synthesis in neurons after this point is so

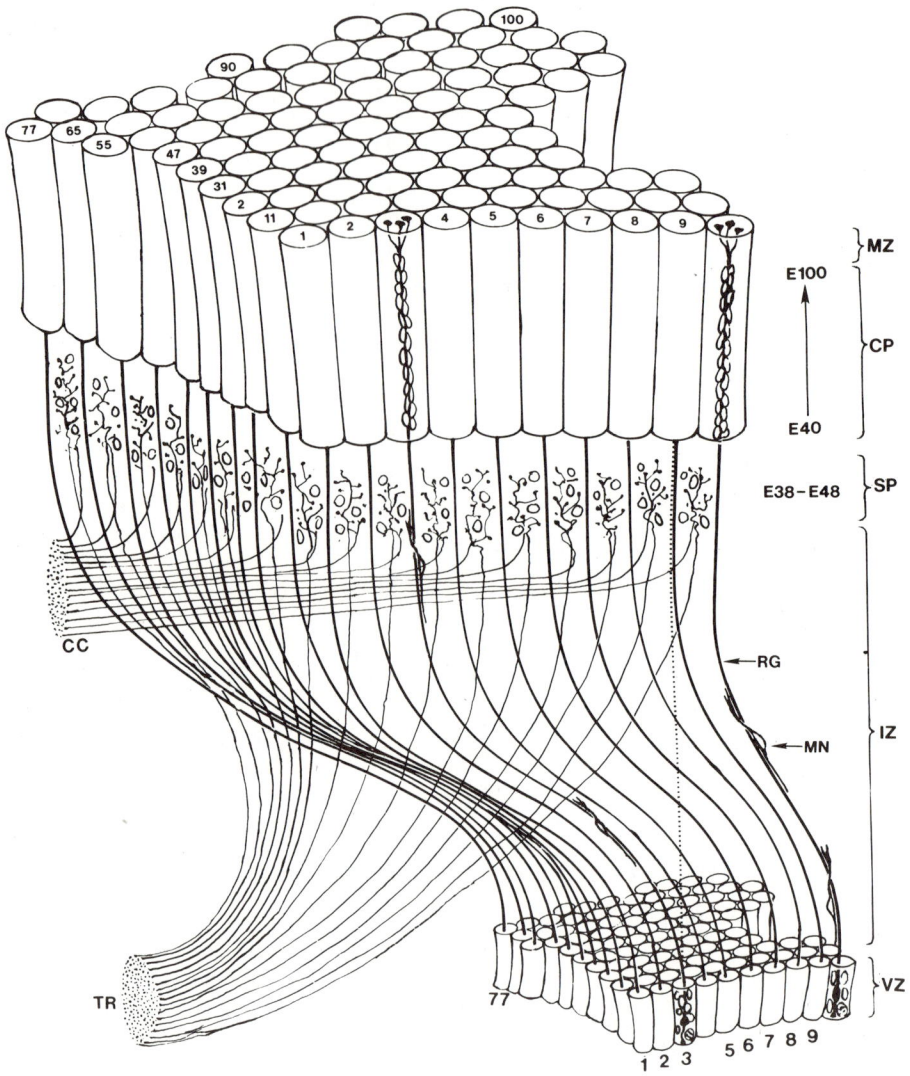

Figure 3. The relationship between a small patch of the proliferative, ventricular zone (VZ) and its corresponding area within the cortical plate (CP) in the developing cerebrum. Although the cerebral surface in primates expands during prenatal development, resulting in a shift between the VZ and CP, ontogenetic columns (outlined by cylinders) remain attached to the corresponding proliferative units by the grid of radial glial fibers. Cortical neurons produced between embryonic age 40 and 100 by a given proliferative unit migrate in succession along the same radial glial guides (RG) and form a single ontogenic column. Each migrating neuron traverses the intermediate (IZ) and subplate (SP) zones, which contain "waiting" afferents from the thalamic radiation (TR) and callosal and intrahemispheric fasciculi (CC). After entering the cortical plate, each wave of migrating neurons bypasses previously generated neurons and assumes a position at the interphase between the CP and marginal zone (MZ). As a result, proliferative units 1-100 produce ontogenetic columns 1-100 in the same relative position to each other. The glial scaffolding prevents a mismatch between proliferative unit 3 and ontogenic column 9 (dashed line). Thus, the specifications of topography and/or modality depend on the spatial distribution of proliferative units, while the phenotype of neurons within each unit depends on its time of origin. (From Rakic, 1988).

powerful that they can not be induced to reenter the cell cycle even in pathological conditions or malignancy (Rakic, 1985). Many precursor cells, after E40, divide asymmetrically and therefore only one of the two daughter cells becomes permanently postmitotic. It detaches its endfoot from the ventricular surface and migrates towards the pial membrane to become a cortical neuron (Figure 2B). The other daughter remains attached to the ventricular zone by the endfoot and continues to divide giving additional pairs of unequal cells: one postmitotic progenitor and one neuron (Rakic, 1988). This pattern of cell division proceeds during the next 30 to 60 days depending on the cortical area (Rakic, 1974, 1976, 1982).

Initial evidence for the radial organization of the developing cortex comes from the light and electron microscopic analysis of the developing cerebral vesicle. The ventricular zone from early stages shows a pseudostratified organization of cells that are deployed radially. Electron microscopic analysis reveals columns of radially oriented bipolar cells that are directly opposed to the elongated radial glial fibers (Rakic, 1971, 1972). This model is presented schematically in Figure 3, and cellular details can be found in primary references (e.g., Rakic, 1971, 1972, 1978; Rakic et al., 1974). The affinity between migrating neurons and the surface of radial glial cells initially implied from electron microscopic observations was also demonstrated in tissue culture analysis (Hatten and Mason, 1986) and several candidate adhesion molecules are currently under study (e.g., Schachner et al., 1985; Edmondson et al., 1988; Edelman, 1983). A series of experiments in which prelabeled ventricular cells from an embryo are transplanted into the cerebral wall of a neonatal animal show that donor neurons move along radial pathways and become arranged predominantly in a radial fashion - settling in the cortex according to their origin and laminar position in the donor (McConnell, 1988).

The radial unit organization of the cerebral cortex has been also confirmed by recent experiments using recombinant retroviral vectors to trace cell lineages from the place of their origin to their ultimate destiny. The viral marker gets introduced into a dividing cell's genome and is passed without dilution to all progeny of an infected cell (Sanes, 1988). Retrovirus injected directly into the embryonic cerebral ventricle at the time of corticogenesis is incorporated into only a few cells and clonally-related cells follow a common migratory pathway and remain radially oriented in the developing cortical plate (Luskin et al., 1988). The labeled clones of neurons form interrupted columns, interspersed with unlabeled neurons that are presumably generated in the same proliferative unit from different progenitors just as could be predicted from the radial unit hypothesis.

Proliferative activity in the ventricular zone that leads to the establishment of the cortex can be divided into two broad phases (Rakic, 1988): the phase of unit formation which proceeds by symmetrical division and occurs mostly before E40 and the phase of ontogenetic column formation which begins around E40 and lasts until the completion of corticogenesis. I suggested that these two phases may be controlled by different regulatory genes - the first one controlling the size and duplication of cytoarchitectonic areas, the second controlling the formation and differentiation of neuronal phenotypes within each ontogenetic column (Rakic, 1988). There are several lines of evidence that these two phases can be separately affected by experimental manipulations or genetic perturbations. A deficit occurring during unit formation produces a cortex with a small surface area but normal thickness, while a defect during the phase of ontogenetic column formation produces a thin cortex with a large surface (Rakic, 1988).

Relationship between ontogenetic and functional columns

At later stages of development and in adults the initially simple radial organization of the cortical plate is distorted due to neuronal growth, development of their processes, introduction of afferent fibers, formation of synapses, myelination, proliferation of glial cells, and ingrowth of blood vessels. A severalfold increase in the cortical surface between completion of the migratory phase and maturity causes horizontal displacement of neurons which initially had a simple radial relationship. The anatomical and functional relationship between clonally-related cells in mature cortex remains a challenge to researchers in this field.

At present it is not possible to establish a direct relationship between ontogenetic columns and functional columns because we do not know their precise boundaries. For example, receptive field columns present in sensory areas are defined as a group of radially deployed neurons that all have the same receptive field (Mountcastle, 1957, 1979). On the other hand some "columns" that are defined by the terminal fields of afferent input such as ocular dominance columns (Hubel and Wiesel, 1977) or callosal columns (Goldman and Nauta, 1977; Goldman-Rakic and Schwartz, 1982) are in reality stripes. Other types of "columns", such as orientation columns in the visual cortex, are morphologically less precisely defined and gradually merge from one to another (Hubel and Wiesel, 1977). In most instances, more than one ontogenetic column must be involved in a given functional column or, in the case of stripes, rows of columns (Rakic, 1988). The important common feature among the cellular compartments or modules described in the mature cortex is the principle of radial organization (Mountcastle, 1979). It has been repeatedly shown that the neurons in the upper and lower layers of the given cortical segment have a close anatomical and functional relationship in the adult cortex (e.g., Szenthagothai, 1983; Eccles, 1984). Although the relationship between ontogenetic and functional columns is probably not simple, available data reveal a principle of radial organization that has its origin in developmental events explicable in the context of the radial unit hypothesis. Embryological and lineage studies reviewed above show that neurons in a given radial compartment arise from the same sector of the ventricular zone. It remains to be shown directly, by double labeling techniques, whether clonally-related cells are preferentially interconnected.

The concept of a cortical protomap

Perhaps the most relevant aspect of the radial unit hypothesis for understanding cortical phylogeny is that the larger the number of proliferative units in an individual or in a given species, the larger the surface of the cortex (Rakic, 1988). Theoretically, all units and columns of neurons could be initially identical or pluripotential. In this extreme case, differentiation of cortex into cytoarchitectonic fields must be imprinted on the developing cortical plate exclusively by the information derived from the afferents arriving from the periphery via thalamo-cortical projections (e.g., Creutzfeldt, 1970) or monoamine fibers from the brain stem (e.g., Ebersole, et al., 1981). Although there is little doubt that all cortical afferents must play an important role in shaping the size of cytoarchitectonic areas (Rakic, 1988) there are several lines of experimental evidence that suggest the existence of a basic blueprint, a provisionary cortical map, or at least a molecular heterogeneity of the cerebral wall. Since, as discussed below, the basic blueprint can be modified by competitive cell interactions, I termed it a *protomap* to underscore its primordial, malleable character (Rakic, 1988).

The first line of evidence for the protomap in the developing telencephalon comes from studies of normal development. A relatively well-preserved register between proliferative units in the ventricular zone and ontogenetic columns in the cortical plate suggests a definite spatial relationship between the two structures. Furthermore, ^{3}H-thymidine autoradiographic analyses show that a portion of the ventricular zone subjacent to one area produces a different number of neurons than an equivalent size portion in the other (Rakic, 1976, 1982, 1988). Thus, a portion of the ventricular zone in the V1 area which contains a larger number of neurons in each ontogenetic column *anticipates* becoming visual by producing more cells (Rakic, 1988). The existence of some sort of map in the cortical plate prior to the arrival of axons is also supported by the common finding that specific thalamic afferents are attracted only to specific areas.

Another line of evidence for the heterogeneity of the ventricular and intermediate zones comes from the recent histochemical and immunocytochemical studies of the embryonic cerebral vesicles using a variety of molecular markers which show that various macro molecules are expressed in a spatially restricted manner. Heterogeneous classes of molecules, including some oncogenes, glycoconjugates, vimentin, or several types of adhesion molecules may be distributed transiently in the cerebral wall in broad radial "patches", "columns", or barrel fields before afferents from the periphery had an impact (Cooper and Steindler, 1986; Johnson and Van der Kooy, 1989; Hutchins and Casagrande, 1989; Steindler et al., 1989).

Our recent study of the distribution of a mitochondial oxidative enzyme - cytochrome oxidase (CO) - in the visual cortex of mature monkeys that were subject to prenatal binocular enucleation before neurons destined for these layers are generated provides experimental evidence that at least certain molecular features of the cytoarchitectonic areas may develop independently of the information from photoreceptors at the periphery (Kuljis and Rakic, 1990). In the normal adult monkey CO is distributed in layers II and III of V1 in the form of "blobs" or "puffs" that are interspersed with CO-free interpuff areas (Wong-Riley and Carroll, 1984; Livingstone and Hubel, 1988). In addition, neurons containing neuroactive peptide Y (NPY) are distribute predominantly in the interpuff regions (Kuljis and Rakic, 1989). It is thought that neurons in the puff and interpuff regions subserve predominantly color and non-color vision (Livingston and

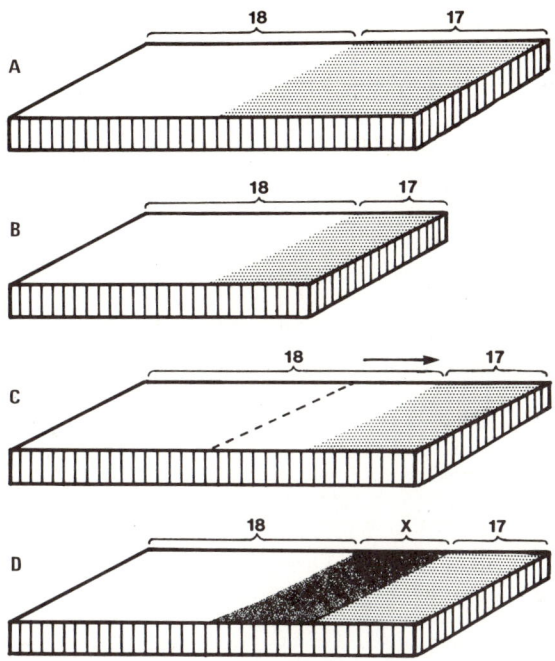

Figure 4. Schematic representation of the possible modes of decrease in the size of V1 caused by experimental reduction of thalamic input. (A) Relation between areas V1 and V2 in a normal animal; (B) differential cell death; (C) encroachment of adjacent V2 into the territory of V1; (D) formation of an abnormal cytoarchitectonic area (X) that consists of neurons genetically destined for area 17 but which receive input characteristic for area 18 (From Rakic, 1988).

Hubel, 1988). In a cortex that has developed in the absence of the retina from an early embryonic age, before photoreceptor cones and rods that mediate these two visual functions have been generated or connected to the central pathways (Rakic, 1977; Nishimura and Rakic, 1987), these cytochemical features nevertheless develop normally (Kuljis and Rakic, 1990). Thus, at least some structural and chemical characteristics of neocortical modules are specified in the absence of input from photoreceptors at the periphery.

Modifiability of the protomap in ontogeny

The existence of a cortical protomap does not imply that every detail or the final size of each cytoarchitectonic area is rigidly prespecified in the ventricular zone. Recently, I provided experimental evidence of the role that thalamic input plays in regulating the size of a given cytoarchitectonic area in monkeys (Rakic, 1988). Due to transneuronal degeneration, early prenatal bilateral enucleation diminished drastically the number of geniculocortical afferents which in turn caused a proportional reduction in the size of visual cortex. Nevertheless, V1 in enucleates was remarkably normal in thickness, layering pattern, synaptic density, and pattern of cytochrome oxidase reactivity, but had a considerably smaller surface (Rakic, 1988).

In spite of the smaller cortical size in early enucleates, CO-rich puffs retain the same size and similar packing density per unit area as in normal monkeys (Kuljis and Rakic, 1990). Our results, therefore, indicate that the reduction in the size of V1 cortex in enucleates is accomplished by deletion of a number of cellular modules rather than by their shrinkage. This suggests that the protomap which exists at the stage of unit formation is subject to modification through interactions with afferent inputs. The final number of ontogenetic columns devoted to a given area, and therefore its final size can be regulated by the number of specific thalamic afferents present at the critical developmental stage (Rakic, 1988). This finding is in full harmony with the results obtained by the manipulation of the somatosensory cortex where deletion of the vibrisae at the skin leads to a loss of the corresponding "barrels" in the developing rodent cerebrum (Van der Loos and Dorfl, 1978).

The cellular mechanism of the reduction of V1 in early enucleates is not fully understood, but some possibilities are illustrated in Figure 4: area V1 can simply lose a number of ontogenetic columns by diminishing the total size of the cortex (Figure 4B). Alternatively, V2, which normally receives input from the adjacent thalamic nucleus (pulvinar) and from the other parietal and temporal cortices, could take over some of the territory from V1 (Figure 4C). Finally, a number of columns that were specified for V1 (X in Figure 4D) could, in the absence of normal afferents, receive input from the pulvinar and become a "hybrid" cortex that is genetically V1 and connectionally V2. It is difficult to test these possibilities experimentally. However, we have some preliminary evidence that enucleation during the midgestational period can result in the formation of a novel cytoarchitectonic area situated at the periphery or as an "island" in the middle of an otherwise normal-looking V1 (Rakic et al., 1990). The experimentally induced "hybrid" area, which we termed V_x, may provide a useful model for the study of how novel cytoarchtectonic fields can be introduced during evolution. In our experiment we made specific input relatively smaller while the cortical area destined to become V1 remains the same. In the application of this model to evolution, however, one has to assume the opposite: an initial increase in the number of ontogenetic columns which then could serve as a substrate for competition among incoming afferent axons resulting in establishment of new synaptic relationships. This is further elaborated below.

Implications for cortical evolution

Examination of the endocasts indicate that between a common ancestor of Mesozoic mammals some 200 million years ago and living primates there must have lived a now-extinct species with numerous combinations of cortical cytoarchitectonic maps that eventually led to the present forms (Fleagle, 1988; Passingham, 1982). According to the theory of natural selection, a species gradually acquires a more elaborate cortex because individuals with higher mental skills are more likely to survive longer and, therefore, have a higher rate of reproduction thereby transmitting their characteristics to their offspring. However, when it comes to providing the genetic or cellular basis of how the expansion of neocortex could have been achieved during the process of evolution, we have much less knowledge. There are several attempts to relate data from comparative neuroanatomy and contemporary research in neuronal plasticity to the cellular issues of brain phylogeny. (e.g., Armstrong and Falk, 1980; Ebbeson, 1984; Allman, 1989; Kaas, 1988; Finlay et al., 1987). The major drawbacks of such theories is that they can not be substantiated since experimental methods can not be used on fossils. The basic strategy of our approach is to extrapolate cellular mechanisms and principles of corticogenesis derived from embryonic development of living primate species to the possible cellular events that may take place during the evolution of cortical parcellation. Although the macaque species under study are not in the same phylogenetic line as humans, we do share with them a common ancestor. I will discuss below several aspects of cortical evolution that are explicable in terms of the radial unit hypothesis and our present knowledge of corticogenesis in normal, genetically or experimentally perturbed rhesus monkeys.

The neocortex in living primitive placental mammals that are thought to be close to early primates such as the hedgehog is one thousand times smaller in surface area than in humans. Yet the thickness of the cortex in the same period of evolution increases only about two times. I have argued elsewhere (Rakic, 1988) that this can be explained by the enormous increase in the number of proliferative units produced during the first phase (of unit formation) and only a moderate increase in production of neurons by each unit during the second phase (of formation of ontogenetic columns). Thus, during the first phase each symmetric division doubles the number of progenitors during each cycle (Figure 2A). While, during the second phase, when asymmetrical divisions predominate, an additional cycle adds only a single neuron to a given ontogenetic column (Figure 2B). As a result, an extra round of cell cycle at the end of corticogenesis produces only about 1% of the total cortical population. The difference between the linear increase which prevails at later stages of corticogenesis stands in contrast to the exponential increase that occurs during the earlier phase. This distinction is important. For example, three divisions at the early phase would produce eight times more progenitor cells and as a consequence an eight times larger number of radial units (Figure 2A). Therefore, only slightly more than three rounds of cell divisions at the end of the first phase of unit formation can account for the difference in cortical size between rhesus monkey and homo sapiens. Theoretically, substantial changes in the total surface of the cortex can be attributed to a single or only a few heterochronic processes, but this hypothesis is not sufficient to explain the uneven growth of various cytoarchitectonic areas.

Remarkable variation in the number and size of cytoarchitectonic areas devoted to vision in living primates suggests that they are introduced during evolution sequentially and expand or retract independently (Allman, 1989). Accordingly, this area represents about 15% of the total cerebral surface in monkey and only 3% in human because other cytoarchitectonic areas differ even more in relative size, particularly association cortex which is in man expanded more than in any other species (Blinkov and Glezer, 1968). Considering that about 22 million years ago the common ancestor of rhesus monkey and *homo sapiens* had a smaller surface of the cortex, how is this species variation achieved? What are the underlying cellular mechanisms?

I suggest that the differential expansion of the cortex can be explained in the context of the radial unit hypothesis (Rakic, 1988). According to this hypothesis, changes in the cytoarchitectonic map could be explained by the heterochronic process or modification of timing and proliferation kinetics in the ventricular zone coupled with competitive interactions with extracortical neurons. For example, extra rounds of mitotic divisions before E40 could produce a larger number of radial units. After additional ontogenetic columns of postmitotic neurons are added to the existing areas, this expanded cellular mass could provide a new synaptic territory with an opportunity for the spread, new combinations, and competitive interactions among incoming afferents. The last statement underscores that the enlarged cortex, without new input-output relationships, is not sufficient to provide extra functional benefits to the individual.

Although the radial unit hypothesis may be compelling in terms of what we presently know about cortical ontogeny, it is obviously difficult to prove its validity for phylogeny. Let's suppose that the entire, or only part of the ventricular zone undergoes an additional round of cell division during the first phase of unit formation. Additional columns of cortical cells resulting from this event then could compete for the various afferent inputs and form novel input-output affiliations that can be adverse, neutral, or beneficial to an organism. This extra round of cell divisions, which can occur either due to genetic mutation or normal variation, must be inheritable so that new cell relationships could be enhanced through the process of natural selection. If additional columns turn out to present an advantage for the survival of the organism it may be selected by preferential propagation of individuals with this trait. This hypothesis may not be directly testable, but some ongoing experiments in our laboratory are relevant to this issue.

Recently, we designed a model in which a larger than normal amount of cortex is devoted to a given function. In an initial study of the visual system in animals with prenatal monocular enucleation the remaining eye projects to the lateral geniculate nucleus which retains a normal number of neurons (Rakic, 1981). Geniculate cells in turn project to the visual cortex, which also retains a normal number of neurons but does not have ocular dominance columns. Since synapses seem to be ultrastructurally and biochemically normal in such cases, information from a single eye in these animals can be analyzed by a two times larger number of cortical neurons and synapses than in the normal monkeys (Rakic, 1981). It has been theorized that the computational ability of V1 for the analysis of certain visual features, such as Vernier hyperacuity, depends on the number of local circuit neurons in the primary visual cortex in relation to the number of photo receptors (Barlow, 1975; Crick, et al., 1981). To test this hypothesis we examined whether the change in numerical relationship in monocular enucleates is beneficial, neutral, or deleterious to the animals visual cortical capacity. We first performed a 2-deoxyglucose experiment in behaving animals which showed that all cortical neurons are active (Rakic, Friedman and Goldman-Rakic, unpublished). That is, a single remaining eye apparently drives a two times larger number of cortical neurons. After that we performed a psychophysical test of Vernier acuity, a specific function of visual cortex. Our preliminary evidence shows statistically better Vernier hyperacuity in monocular animals than in controls with both eyes and the difference is even larger when the results are compared to a single eye in the control animals (MacAvoy, et al., 1987). One can, therefore, speculate that thalamic input which is experimentally spread to a larger than normal area of the cortex devoted to a given function may provide synaptic relationships that improve performance. In the context of an evolutionary hypothesis one can speculate that when new inheritable thalamocortical affiliations are formed, they also could provide the conditions for better performance with a presumed survival advantage for the individual that is fostered through selective breeding.

The purpose and limit of the present article precludes further elaboration of the radial unit model of cortical evolution and does not allow for provision of additional examples of supporting evidence. However, I hope that the examples selected demonstrate how studies of normal ontogeny and experimental manipulation of cortical development can be used to suggest possible cellular and genetic mechanisms that could underlie phylogenetic development. I believe that this type of research will not only explain the pathogenesis of certain genetic and developmental disorders of higher cortical function, but could also provide insight into the evolution of cerebral cortex and the development of human mental capacity.

REFERENCES

Allman, J. (1989) Evolution of neocortex. In: *Cerebral Cortex*. Peters, A., Jones, E.G. (eds.), Academic Press, New York, Vol. 8, in press.
Alberch, P., Gould, S.J., Oster, G.F., and Wake, D.B. (1979) Size and shape in ontogeny and phylogeny. *Phyleobiology*, 5: 296-317.
Armstrong, E., and Falk, D. (eds.) (1982) *Primate Brain Evolution*. Plenum, New York, pp 332.
Barlow, H.B. (1975) Visual experience and cortical development. *Nature,* 258: 199-204.
Blinkov, S.M., and Glezer, I.J. (1968) *The Human Brain in Figures and Tables: A Quantitative Handbook.* Plenum Press, New York, pp 298.
Burkhalter, A., and Bernardo, A.L. (1989) Organization of cortico-cortical connections in human visual cortex. *Proc. Nat. Acad. Sci.*, in press.
Cooper, N.G.F., and Steindler, D. (1986) Lectins demarcate the barrel subfield in the somatosensory cortex of the early postnatal mouse. *J. Comp. Neurol.,* 249: 157-168.
Creutzfeldt, O.D. (1977) Generality of the functional structure of the neocortex. *Naturwissenschaften,* 64: 507-517.
Crick, F.H., Marr, D., and Poggio, T. (1981) An information-processing approach to understanding the visual cortex. In: *The Cerebral Cortex.* Schmitt, F.V., Worden, F.G. (eds.), MIT Press, Cambridge, MA, pp. 505-533.
Easter, S.S., Jr, Purves, D., Rakic, P., and Spitzer, N.C. (1985) The changing views of neuronal specificity. *Science,* 230: 507-511.
Ebbeson, S.O.E. (1984) Evolution and ontogeny of neural circuits. *Behav. Brain Sci.*, 7: 332-350.
Ebersole, P., Parnavelas, J.G., and Blue, M.E. (1981) Development of visual cortex of rats treated with 6-Hydroxydopamine in early life. *Anat. Embryol.,* 162: 489-492.
Eccles, J.C. (1984) The cerebral neocortex. A theory of its operation. In Jones, E.G., Peters, A. *Cerebral Cortex*, (eds.): Vol. 2. Plenum, New York, pp. 1-36.
Edelman, G.M. (1983) Cell adhesion molecules. *Science* 219: 450-457.
Edelman, G.M. (1988) *Topobiology. An Introduction in Molecular Embryology.* Basic Books, Inc., New York, pp. 240.
Edmondson, J.C., Liem, R.K., Kurster, J.E., and Hatten, M.E. (1988) Astrotactin: A novel cell surface antigen that mediates neuron-astroglial interactions in cerebellar microcultures. *J. Cell Biol.,* 106: 505-517.
Finlay, B.L., Wikler, K.C., and Sengelaub, D.R. (1987) Regressive events in brain development and scenarios for vertebrate brain evolution. *Brain, Behavior and Evolution,* 30: 102-117.
Fleagle, J.G. (1988) *Primate Adaptation and Evolution.* Academic Press, New York, pp 486.
Goldman, P.S. and Nauta, W.J.H. (1977) Columnar organization of association and motor cortex: Autoradiographic evidence for cortico-cortical and commissural columns in the frontal lobe of the newborn rhesus monkey. *Brain Res.,* 122: 369-385.
Goldman-Rakic, P.S., Schwartz, M.L. (1982) Interdigitation of contralateral and ipsilateral columnar projections to frontal association cortex in primates. *Science,* 216: 755-757.
Gould, S.J. (1977) *Ontogeny and Phylogeny.* The Belknap Press, Cambridge, MA, pp. 501
Hatten, M..E and Mason, C.A. (1986) Neuron-astroglia interactions *in vitro* and *in vivo.* TINS, 9: 168-174.
Hubel, D.H. and Wiesel, T.N. (1977) Ferrier lecture. Functional architecture of macaque monkey visual cortex. *Proc. R. Soc. Lond. B.,* 198: 1-59.
Hutchins, J.B. and Casagrande, V.A. (1989) Vimentin: Changes in distribution during brain development. *Glia,* in press.
Johnson, J. and Van der Kooy, D. (1989) Protooncogene expression identifies a columnar organization of the ventricular zone. *Proc. Nat. Acad. Sci.(USA),* in press.
Kaas, J.H. (1988) Development of Cortical Sensory Maps. In: *Neurobiology of Neocortex.* Rakic, P., Singer, W. (eds.), Wiley and Sons, New York, pp 101-113.

Kuljis, R.O., Rakic, P. (1989) Neuropeptide Y-containing neurons are situated predominantly outside cytochrome oxidase puffs in macaque visual cortex. *Visual Neurosci.,* 2: 57-62.

Kuljis, R.O., Rakic, P. (1990) Hypercolumns in primate visual cortex can develop in the absence of cues from photoreceptors. *Proc. Nat. Acad. Sci.,* in press.

Levitt, P., Cooper, M.L. and Rakic, P. (1981) Coexistence of neuronal and glial precursor cells in the cerebral ventricular zone of the fetal monkey: An ultrastructural immunoperoxidase analysis. *J. Neurosci.* ,1: 27-39.

Levitt, P., Cooper, M.L. and Rakic, P. (1983) Early divergence and changing proportions of neuronal and glial precursor cells in the primate cerebral ventricular zone. *Dev. Biol.,* 96: 474-484.

Levitt, P. and Rakic, P. (1980) Immunoperoxidase localization of glial fibrillary acid protein in radial glial cells and astrocytes of the developing rhesus monkey brain. *J. Comp. Neurol.,* 193: 815-840.

Livingstone, M.S. and Hubel, D.H. (1988) Segregation of form, color, movement, and depth: anatomy, physiology and perception. *Science,* 240: 740-749.

Luskin, M.B., Pearlman, A.L. and Sanes, J.R. (1988) Cell lineage in the cerebral cortex of the monkey studied *in vivo* and *in vitro* with a recombinant retrovirus. *Neuron,* 1: 635-647.

MacAvoy, M.G., Bruce, C.J. and Rakic, P. (1987) Effect of prenatal monocular enucleation on Vernier hyperacuity in rhesus monkeys. *Abstr. Soc. Neuroci.,* 13: 1244.

McConnell, S.K. (1988) Development and decision-making in the mammalian cerebral cortex. *Brain Res. Review* ,13: 1-23.

Mountcastle, V.B. (1957) Modality and topographic properties of single neurons of cat's somatic sensory cortex. *J. Neurophysiol.,* 20: 408-434.

Mountcastle, V.B. (1979) An organizing principle for cerebral function: The unit module and the distributed system. In F.O. Schmitt and Worden, F.G. (eds.): *The Neurosciences: Fourth Study Program.* MIT Press, Cambridge, MA.

Nishimura, Y. and Rakic, P. (1987) Development in the rhesus monkey retina: II. Three-dimensional analyzing of the sequences of synaptic combinations in the inner plexiform layer. *J. Comp. Neurol.,* 262: 190-313.

Passingham, R. (1982) *The Human Primate.* Freeman, Oxford, San Francisco, pp 390.

Purves, D. (1988) *Body and Brain.* A Trophic Theory of Neural Connections. Harvard University Press, Cambridge, MA, pp. 230.

Rakic, P. (1971) Guidance of neurons migrating to the fetal monkey neocortex. *Brain Res.,* 33: 471-476.

Rakic, P. (1972) Mode of cell migration to the superficial layers of fetal monkey neocortex. *J. Comp. Neurol.,* 145: 61-84.

Rakic, P. (1974) Neurons in rhesus monkey visual cortex: Systematic relation between time of origin and eventual disposition. *Science,* 183: 425-427.

Rakic, P. (1976) Differences in the time of origin and in eventual distribution of neurons in areas 17 and 18 of visual cortex in rhesus monkey. *Exp. Brain Res. Suppl.,* 1: 244-248.

Rakic, P. (1977) Prenatal development of the visual system in the rhesus monkey. *Phil. Trans. Roy. Soc. Lond. Ser. B* , 278: 245-260.

Rakic, P. (1978) Neuronal migration and contact guidance in primate telencephalon. *Postgrad. Med. J.,* 54: 25-40.

Rakic, P. (1981) Development of visual centers in primate brain depends on binocular competition before birth. *Science,* 214: 928-931.

Rakic, P. (1982) Early developmental events: Cell lineages, acquisition of neuronal positions, and areal and laminar development. *Neurosci. Res. Program Bull.,* 20: 439-451.

Rakic, P. (1985) Limits of neurogenesis in primates. *Science,* 227: 154-156.

Rakic, P. (1988) Specification of cerebral cortical areas. *Science,* 241: 170-176.

Rakic, P. and Singer, W. (1988) *Neurobiology of Neocortex.* Wiley & Sons, New York.

Rakic, P., Stensaas, L.J., Sayre, E.P. and Sidman, R.L. (1974) Computer aided three-dimensional reconstruction and quantitative analysis of cells from serial electron microscopic montages of fetal monkey brain. *Nature,* 250: 31-34.

Rakic, P., Suner, I. and Williams, R.W. (1990) A novel cytoarchitectonic area induced experimentally within the primate visual cortex. (Submitted).

Sanes, J.R. (1988) Analysing cell lineages with a recombinant retrovirus. *TINS* , 12: 21-28.

Schachner, M., Faissner, A., Fischer, G., Keilhauer, G., Kruse, J., Kunemund, V., Lindner, J., Wernecke, H. (1985) Functional and structural aspects of the cell surface in mammalian nervous system development. In: *The Cell in Contact: Adhesions and Junctions as Morphogenetic Determinants.* Wiley & Sons, New York, pp. 257-276.

Schmechel, D.E., Rakic, P. (1979a) A Golgi study of radial glial cells in developing monkey telencephalon: Morphogenesis and transformation into astrocytes. *Anat. Embryol.,* 156: 115-152.

Schmechel, D.E., Rakic, P. (1979b) Arrested proliferation of radial glial cells during midgestation in rhesus monkey. *Nature,* 277: 303-305.

Shkol'nik-Yarros, E.G. (1971) *Neurons and International Connections of the Central Visual System.* Plenum, New York.

Szenthagothai, J. (1983) The modular architectonic principle of neural centers. *Rev. Physiol. Biochem. Pharmacol.,* 98: 11-61.

VanEssen, D.C. (1985) Functional organization of primate visual cortex. In: *Cerebral Cortex.* Peters, A., Jones, E.G. (eds.), Plenum, New York, pp 259-329.

Van der Loos, H. and Dorfl, J. (1978) Does the skin tell the somatosensory cortex how to construct a map of the periphery? *Neurosci. Lett.,* 7: 23-30.

Wong-Riley, M., Carroll, E.W. (1984) Effect of inputs blockage on cytochrome oxidase activity in monkey visual system. *Nature,* 307: 262-264.

CONTROL OF CELL NUMBER AND TYPE IN THE DEVELOPING AND EVOLVING NEOCORTEX

Barbara L. Finlay

Department of Psychology
Uris Hall
Cornell University
Ithaca, NY 14853 USA

INTRODUCTION

Large mammalian brains show "encephalization": that is, in larger brains, the neocortex claims a disproportionately high percentage of their volume (Jerison, 1973; Jerison, this volume). The magnitude of encephalization is impressive, and its manner is strikingly regular across multiple mammalian radiations (Hofman, 1989). I would like to take advantage of this forum to speculate on the relationship of encephalization to current research on the ontogenetic regulation of neuron number and type in the cortex. The discussion is organized around three questions:

1. Can any persistent feature of development account for the disproportionately large volume of the cortex in large brains?

2. Why do developmental stabilizing mechanisms permit cortical hypertrophy? For example, in the spinal cord there is some trophic relationship between the number of neurons and peripheral muscle and sensory mass, regulated by normally-occurring cell death (Hamburger and Levi-Montalcini, 1949; reviewed in Hamburger and Oppenheim 1982; Oppenheim, 1981). Why is relative cortical volume allowed to grow so large with respect to the volume of its input and terminal zones?

3. Can development give us any clue as to how local areas of cortex get wired for their particular functions? Two views of the brain have competed for decades, whether the brain is best understood as a generalized computing device, or as an accretion of specialized capacities. While some of the functional change in larger brains might be described as faster, more powerful, and more general computing, the most prominent functional changes are the addition of particular, specialized skills. These skills, like echolocation, language, predictive prey tracking and the like, are apparently represented in a modular way in the brain. How can a generalized and regular neocortical hypertrophy be reconciled with the development of modularly-organized special functions?

DEVELOPMENTAL CONSTRAINTS ON THE PATHS OF NEOCORTICAL EVOLUTION

What about development could result in predictable cortical hypertrophy?

Hofman (1989) has recently summarized important work on the allometric relationship of cortical volume to brain volume. The central message of this analysis is that across mammalian radiations, there is an exceedingly regular relationship of cortex area or volume to total brain volume. (An example from this paper is reproduced in Figure 1). As total brain size enlarges, the cortex subsumes an increasingly greater percentage of its total volume. This analysis includes primates, and particularly the human brain: the human cortex is the expected size for our overall brain size (primates and humans particularly do have an unusually large brain size per body size;

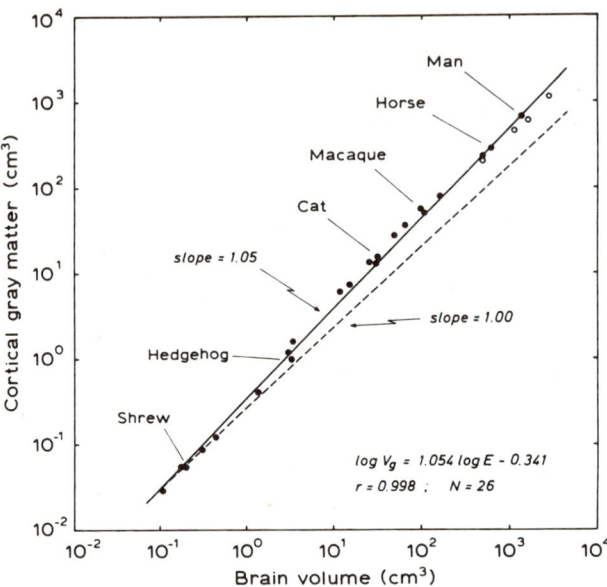

Figure 1. Volume of the neocortex (gray matter only) as a function of whole brain volume for a number of mammals. Logarithmic scale. The dashed line represents scaling of cortical volume by the first power, geometric similarity, and shows that the amount of cortex increases at a greater rate than that expected by geometric scaling alone. Dolphins and whales are indicated by open circles. (Reprinted with permission from M. A. Hofman (1989), *Progress in Neurobiology, 32:* 137-158)

given the brain size, however, the cortex is typical). The increasing contribution of the cortex to total brain volume at larger brain sizes is not a simple geometrical consequence of enlargement of a sphere. To take one of Hofman's examples, if the tree shrew brain were enlarged to the size of a human brain without altering its conformation, its cortical surface area would measure 425 cm^2. The actual area of the human cortex is 2,430 cm^2. All reasonable metrics of cortical area and volume, including total surface area, grey matter volume, white matter volume and total volume, demonstrate this regularity, appropriately scaled. Thus, when the brain enlarges in evolutionary time it does so in a regular and highly constrained way. This relationship has arisen independently over a number of mammalian radiations. While it is remotely conceivable that some feature of behavioral adaptation required for many different behavioral niches would require a disproportionately large cortex in large brains, in horses, primates and whales alike (for example, Pagel and Harvey, 1990), it seems much more likely that there is a developmental regularity or constraint that produces this predictable relationship of cortex volume to total brain volume.

To get to development directly, another allometric regularity can be used, and that is the relationship of brain size to gestational length (as reviewed in Eisenberg, 1981; Pagel and Harvey, 1990). Although the lack of variability is not as striking as in the case of the relationship of brain and cortex size, various investigations all support the basically unsurprising conclusion that it generally takes longer to make a bigger brain.

What feature of neurogenesis might routinely favor the neocortex, producing not a linear change in cortex volume with a longer gestational length, but rather an exponential change? How are neurogenetic events packaged into gestational periods that may differ almost by an order of magnitude, from tree shrew to elephant? Two types of observations seem central here, first, the order in which cells in different brain structures exit the symmetric divisions of the cell cycle, or are "born", and second, the relative timing of onset and offset of neurogenesis in species with varying gestational lengths and brain sizes.

Shown in Figure 2 are the approximate dates of last cell division in a number of brain areas in the rat (taken from Rodier, 1980) plotted against the increasing DNA content of its brain in the

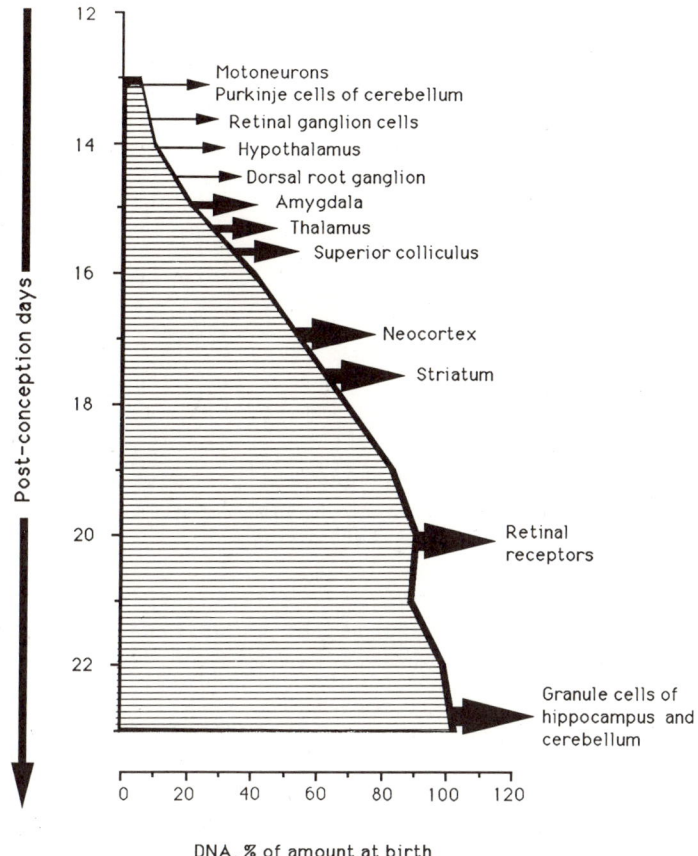

Figure 2. Day of last neurogenesis in a variety of brain areas in the rat, plotted over the increase in total brain DNA in the same period (Brain DNA data redrawn from Zamenhof and Van Marthens, 1979; events in neurogenesis as reviewed in Rodier, 1980).

same time period (redrawn from Zamenhof and Van Marthens, 1979). The intent of this dual plot is to give an indication of the size of the brain, as measured by cell contents, at the time particular structures are withdrawn from the mitotic cycle. The neocortex and striatum, retinal photoreceptors, and the granules cells of the cerebellum and hippocampus are the last to withdraw, of the structures plotted here. This relative order is generally conserved across species (see also Dreher and Robinson, 1988): although the number of species on which systematic birthdating has been done is not large, there is an extensive body of knowledge on several rodents (Sidman, 1961; Angevine and Sidman, 1962; Shimada and Langman, 1970; Bruckner et. al., 1976; Mustari et al, 1979; Crossland and Uchwat, 1982; Kane et al., 1984; Sengelaub et al., 1988; review, rat in Rodier, 1980), the cat (Hickey and Hitchcock, 1984; Walsh et al., 1983; Luskin and Shatz, 1985b) and the macaque (Rakic, 1974; Hendrickson and Rakic, 1977; Cooper and Rakic, 1981).

The accommodation of cortical neurogenesis (grey bars) into gestational periods of varying lengths is shown in Figure 3 for the golden hamster, domestic cat and macaque monkey. In this representation, time to eye opening is used to equilibrate gestational periods, which occurs two weeks after birth in hamsters, about a week after birth in the cat, and *in utero* in the macaque; this physical measure correlates well with a number of indicators of functional maturity, while birth itself is quite variably related to maturity (Dreher and Robinson, 1988). The time of birth (arrows), as well as the duration of neurogenesis in the lateral geniculate nucleus of the thalamus (white bars) is also presented for reference. Two observations can be drawn from this graph.

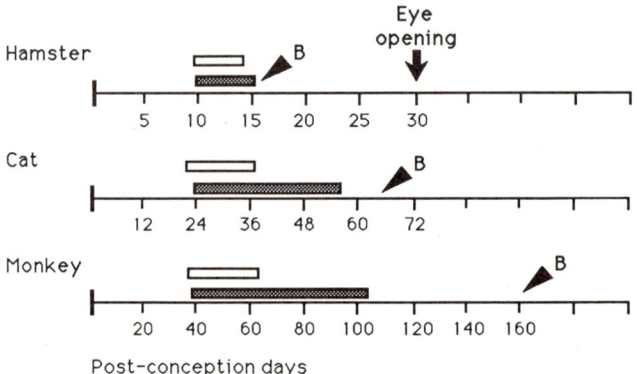

Figure 3. Duration of neurogenesis in the lateral geniculate nucleus (white bar) and neocortex, including subplate (grey bar) in the golden hamster (Crossland and Uchwat, 1982; unpublished results, this laboratory), domestic cat (Hickey and Hitchcock, 1984; Luskin and Shatz, 1985b) and rhesus monkey (Rakic, 1974; Hendrickson and Rakic, 1977). The X-axis has been equilibrated in the three species for the time to conception to eye opening. Eye opening information from Blakemore and Cummings (1977) and Dreher and Robinson (1989). The time of birth is marked with an arrowhead.

First, future cortical neurons begin to exit from the cell cycle at very much the same percentage of the closed eye period (approximately 30%), even though the absolute amount of this time varies widely, from 10 days post conception in the hamster to 40 days postconception in the monkey. Second, the duration of cortical neurogenesis occupies a proportionately greater percentage of the closed eye period in longer gestations. By contrast, generation of the lateral geniculate body appears to remain at a fixed percentage of the closed eye period.

The principal factor in expanding cortex volume is the expanding cortex surface area, though non-negligible change in cortical depth also occurs (Hofman, 1989). Since expansion of cortical area is caused by an increase in the number of radial units that comprise it, which are related in turn to proliferative units in the ventricular zone and their associated radial glia (Rakic, this volume; Caviness et al., this volume), the factor that might produce a greater number of proliferative units is the critical issue. In this case, therefore, the duration over which cells are undergoing symmetric divisions in the ventricular zone is critical, before cells exit the cell cycle, migrate and differentiate as neurons. Linear extensions of gestational length, permit doubling and redoubling of the cortical germinal zone, an exponential change in the number of "radial units". Because the cortex is one of the late structures exiting the cell cycle, it would experience doublings and redoublings for a longer period and show a exponential enlargement compared to other brain areas after a linear change in gestational length. Additional support for this hypothesis can be found in the numerical stability of late versus early generated structures across mammals. The cerebellum is late of generation, and may also show proportional hypertrophy in large brains, though this has not been quantified. Interestingly, retinal ganglion cells, early to start terminal divisions, are rather stable in number across mammals, while the numbers of late-generated photoreceptors differ by nearly an order of magnitude more (Finlay and Pallas, 1989). The hippocampus may present a significant challenge to this hypothesis, in that there is no evidence as yet for a regular relationship of hippocampal volume to whole brain volume.

A number of assumptions that are reasonable, but that nevertheless require demonstration, are required by the hypothesis that extended neurogenesis regularly produces disproportionately large cortex. These are that the onset of cell division in the forebrain is not disproportionately retarded in larger brains; that the cell cycle time is not markedly slower; nor are the proportions of cells in the germinal zone entering the cell cycle markedly less. Many more species need to be investigated, particularly those that show any unusual relationships in the duration of maturation to brain size.

The significance of the extended period of cortical terminal divisions in large brains is unclear (Figure 3). When cells undergo their last division and differentiate as neurons, the consequence of a cell division is a linear addition of one cell, not a doubling of precursors. Cells undergo their last division and migrate to the cortex over a period of approximately six days in the hamster, but sixty days in the monkey. While the monkey cortex is somewhat thicker as measured in cells per unit column than the hamster cortex, it is not ten times thicker. The duration of the cell cycle, and the tangential spread of newly-generated cells might be avenues to investigate.

Why does the rest of the brain permit the cortex to hypertrophy?

In many other neural systems, a linear relationship between numbers of neurons and their targets is defended during development. Cell-death-mediated trophic relationships were first demonstrated in the classic experiments of Hamburger and Levi-Montalcini (1949) and explored further by many others (reviewed in Oppenheim, 1981; but see Sperry, 1990). Target dependence can be shown for retinal ganglion cells (Finlay et al., 1986; Wikler et al., 1986), components of the the auditory system, parasympathetic nervous system, features of cerebellar connectivity and many others (reviewed in: Cowan, 1973; Oppenheim, 1981; Williams and Herrup, 1988; Finlay and Pallas, 1989). Trophic relationships of neuronal populations is a central concept in developmental neurobiology (Purves, 1988). If cortical neurons behaved like spinal cord motoneurons, the consequence of "overproduction" of axons from neurons in the subcortical output layers V and VI would be failure of those axons to find sufficient downstream targets, and the subsequent neuronal death in those layers. Nothing like this has ever been observed: there is little indication of any substantial amount of cell death in layers 5 and 6, and no indication of phylogenetic variability of the appropriate type: that is, excessive cell death in "encephalized" mammals (Finlay and Slattery, 1983; Heumann and Leuba, 1983; Finlay and Pallas, 1989).

Developmental manipulations, however, might reveal trophic dependencies that could be obscured in a phylogenetic analysis (Ebbeson, 1980). In a series of studies, the trophic dependence of cortical cells on subcortical and intracortical targets during development was investigated in my laboratory. In a first experiment, a major subcortical target of the visual cortex was removed, the superior colliculus, early in development before the entry of cortical fibers. The effect on Layer V cells of the visual cortex was examined, during development and at maturity (Pallas et al., 1988). Removal of the superior colliculus in this fashion produced no measurable effect on evidence of cell degeneration or neuron number during development, nor on total cortex size, number of cells or density of cells in a "unit column," or in the number or density of layer V cells per column. A similar result was found by Ramirez and Kalil (1985) for the motor cortex after prevention of pyramidal tract development.

In a second experiment, the corpus callosum was sectioned in early development, transecting some fibers and prohibiting the remainder from growing (Windrem et al., 1988). While this experiment is more difficult to interpret due to the simultaneous interruption of efference and afference by the callosal transection, there was no hint of any deviation in cell number in those areas and layers of cortex that contribute most heavily to callosal innervation, for example, layers II and III of the 17-18 border. We concluded from these two experiments and from other results from other laboratories that cortical neurons are relatively insensitive to the deletion of particular target populations. In both of these cases, a lamina-specific loss of cells would have been easily detectable. While cells projecting to the colliculus or through the callosum are a small minority of all cortical cells (Lamantia and Rakic, 1990), they are a substantial fraction of their particular, identifiable laminae (Sefton et al., 1981; Pallas et al., 1988). The fact that afferent-induced alterations in cortical number are easily detectable, as described in the third section of this paper, reduces the probability that we have failed to detect a numerical change.

Not at issue here is whether cortical neurons might have basic trophic dependence on some target tissue. There are a number of explanations for these results that preserve some measure of target dependency in cortical neurons. Cortical efferent neurons branch widely early in development (Innocenti and Caminiti, 1980; Innocenti, 1981; Ivy and Killackey, 1982; Stanfield et al., 1982; reviewed in O'Leary, 1989), and there was ample time for readjustment in connectivity. Since the majority of synapses on a cortical neuron appear to derive from intrinsic cortical connectivity (Elhanany and White, 1990), it may be impossible in principle to deprive cortical neurons of a sufficient amount of connectivity by any manipulation extrinsic to a local cortical area. Even with this qualification it is important to remember that, in these experiments, a major extrinsic target of the cortex was removed, with no apparent numerical readjustment in the cortex. Cortical neurons cannot thus be selective for trophic support from any particular target and must be able to accept a range of potential targets, including possibly wholly intrinsic targets. The amount of axon permitted to a viable cortical cell must be variable. This type of target independence in the cortex might be a permissive factor for cortical hypertrophy.

How does the cortex get its local character and modularity?

Little is known how functional circuits for particular skills arise in the cortex over phylogeny. The cortex in adulthood is not a uniform layered sheet, but shows local differences in cell number and constituents that are related to local processing demands: greater specialization of layer IV in primary sensory cortices; Betz cells in the motor cortex and the like. While we cannot address issues of the design of complex functional circuitry, we do have some information on how some of these local differences arise through developmental interaction with thalamic afference.

There is much indirect evidence that the amount of thalamic innervation may control the number of thalamorecipient cells in layer IV. First, that the overall "granularity" of cortex is related to the volume of its thalamic innervation is widely assumed: primary sensory areas of the cortex, associated with dense thalamic projections, have a highly developed layer IV (von Economo, 1929). In both cat and monkey striate cortex, the number of cells in layer IV is greater in parts of the cortex receiving binocular innervation from the thalamus than in monocular zones (O'Kusky and Collonier, 1982; Beaulieu and Collonier, 1983). Early enucleation in monkeys has been shown to reduce thalamic volume and secondarily the extent of striate cortex (Rakic, 1988; Rakic, this volume). A peculiarity of the early pattern of thalamic innervation of the cortex may be explained if cortical cells are differentiated or selected as thalamorecipient by their thalamic afference. Prior to innervation of the cortex, thalamic fibers reside under the cortical plate for an extended period contacting a transitory population of cells, while the cells they will eventually innervate migrate through them (Luskin and Shatz, 1985a; Shatz and Luskin, 1986).

Throughout the cortex, there are marked local differences in the numbers of degenerating cells in early development, and their distribution is correlated, though imperfectly, with natural variation in the volume of thalamic afference to particular cortical zones (Finlay and Slattery, 1983). Martha Windrem, in my laboratory, undertook to make thalamic lesions, and look at the alteration in cell degeneration and cell number in the cortex (Windrem and Finlay, 1985; Windrem et al., 1986). Thalamic deaffererentation of the cortex produced an increase in cell degeneration in the supragranular layers, and a particular loss of cells in Layer IV.

We argue that this developmental manipulation may mirror normal variation in cortical innervation. If the local differentiation of cortex in part arises epigenetically, as has also been argued persuasively by O'Leary (1989) for the organization of cortical efferent projections, this may in part account for the extreme species variability in the total amount and differentiation pattern of the neocortex. Relatively few simultaneous genetic changes would be required for the generation of a functional thalamocortical module.

If the thalamus in part confers local specificity on the cortex, thalamic projections themselves must be relatively specific. This is the case: in contrast to early callosal projections or subcortical projections which may be quite diffuse (Innocenti and Caminiti, 1980; Innocenti, 1981; Ivy and Killackey, 1982; Stanfield et al., 1982), or in contrast to projections to the thalamus itself (Frost, 1984; Frost, this volume; Pallas, this volume), the early pattern of tangential innervation in the cortex is quite similar to the adult (Crandall and Caviness, 1984). Evidence for within-area remodeling of thalamocortical projections is extensive, from topographical sharpening (Naegele et al., 1988), barrel formation (Van der Loos et al., this volume) and ocular dominance columns (Hubel et al. 1977). Between-area remodeling has been reported markedly less, with local retraction of branched axons to adjoining somatosensory areas one of the few examples (Carlson et al., 1987). But if new cortex arises phylogenetically, how is it to be innervated by new thalamic afference?

Again, we looked to developmental manipulations for some insight on how the cortex might acquire new or reorganized thalamic input. In the same thalamic ablation procedure mentioned above, we investigated whether the cortex denervated by an early thalamic lesion would be reinnervated by the remaining thalamus (Miller et al., 1987). We made a thalamic lesion in neonates, in either the posterolateral visual thalamus, or the ventrobasal thalamic area. At adulthood in these animals, we made HRP injections into the cortex associated with the intended thalamic lesion, and looked for innervation of this cortex by the remaining thalamus. We found no evidence for wholesale invasion of denervated cortical areas by the remaining thalamus: in 7 of 9 cases, no reorganization was seen, and in the remaining two cases, the anomalous innervation was relatively small. In contrast, callosal connectivity in these same animals became substantially more diffuse (Howard et al., 1989) Assessment of relative thalamocortical and callosal specificity by lesion-induced plasticity thus mirrored the early connectional anatomy of these two systems.

Since the thalamic damage in these animals occurred before thalamic invasion of the cortical plate, but well after substantial interaction of thalamocortical axons with the subplate, these results may indicate that we may need to look much earlier in development for the factors that produce thalamocortical specificity. The mechanisms by which the pattern of thalamocortical innervation may be altered over development thus remain unknown.

SUMMARY

In this paper, I have attempted to use information gained from normal and perturbed cortical development to get some sense of the paths cortical evolution might take, taking much the same strategy originally described by Ebbesson (1980) in his account of neural evolution by parcellation. I have argued that regular cortical hypertrophy might be a result of predictable heterochronic and homochronic changes in the pattern of neurogenesis in animals with varying gestational lengths. Because of an unusual lack of trophic reliance on external synaptic targets, the cortex is able to enlarge at a rate proportionately greater than the rest of the brain. By contrast, thalamic afference is required for the maintenance of Layer IV, thalamorecipient cells, and this developmental dependency allows the cortex to exhibit local functional specialization. However, the nature of the tangential specificity of thalamocortical innervation and how it might be altered over phylogeny remains a mystery, a question that will be taken up by further chapters in this volume.

ACKNOWLEDGEMENTS

This work was supported by NIH grants KO4 NS00783 and RO1 NS19245. I thank Lael Hinds and Brad Miller for their helpful comments on the manuscript and Kim Stockton for her excellent secretarial help.

REFERENCES

Angevine, J.B. Jr. and Sidman, R.L. (1962) Autoradiographic study of histogenesis in the cerebral cortex of the mouse. *Anat. Rec.*, 142: 210.
Beaulieu, C. and Collonier, M. (1983) The number of neurons in the different laminae of the binocular and monocular regions of area 17 in the cat. *J. Comp. Neurol.*, 217: 337-344.
Blakemore, C., and Cummings, R.M. (1975) Eye opening in kittens. *Vision Res.* 15: 1417-1418.
Bruckner, G., Mares, V. and Biesold, D. (1976) Neurogenesis in the visual system of the rat: an autoradiographic investigation. *J. Comp. Neurol.*, 166: 245-276.
Carlson, M., O'Leary, D.D.M. and Burton, H. (1987) Potential role of thalamocortical connections in recovery of tactile function following somatic sensory cortex lesions in infant primates. *Soc. Neurosci. Abs. 13*: 75.
Cooper, M.L. and Rakic, P. (1981) Neurogenetic gradients in the superior and inferior colliculi of the rhesus monkey. *J. Comp. Neurol.*, 202: 309-334.
Cowan, W.M. (1973) Neuronal death as a regulative mechanism in the control of cell number in the nervous system. In: *Development and Aging in the Nervous System*, pp.19-41. Ed. M. Rockstein. New York: Academic Press.
Crandall, J.E. and Caviness, V.S. (1984) Thalamocortical connections in newborn mice. *J. Comp. Neurol.*, 228: 542-556.
Crossland, W.J., and Uchwat, C.J. (1982) Neurogenesis in the central visual pathways of the golden hamster. *Devel. Brain Res.*, 5: 99-103.
Dreher, B. and Robinson, S.R. (1988) Development of the retinofugal pathway in birds and mammals: evidence for a common timetable. *Brain Beh. Evol.,* 31:369-390.
Ebbeson, S.O.E. (1980) The parcellation theory and its relationship to interspecific variability in brain organization, evolutionary and ontogenetic development, and neuronal plasticity. *Cell Tissue Res.,* 213: 179-212.
Eisenberg, J.F. (1981) *The Mammalian Radiations: An Analysis of Trends in Evolution, Adaptation and Behavior.* Chicago: The University of Chicago Press.
Elhanany, E. and White, E.L. (1990) Intrinsic circuitry: synapses involving local axon collaterals of corticocortical projection neurons in the mouse primary somatosensory cortex. *J. Comp. Neurol.*, 291, 43-54.

Finlay, B.L., and Pallas, S.L. (1989) Control of cell number in the developing visual system. *Progress in Neurobiol., 32* : 207-234.

Finlay, B.L., Sengelaub, D.R. and Berian, C.A. (1986) Control of cell number in the developing visual system. I. Effects of monocular enucleation. *Dev. Brain Res.,* 28: 1-10.

Finlay, B.L. and Slattery, M. (1983) Local differences in amount of early cell death in neocortex predict adult local specializations. *Science,* 219: 1349-1351.

Frost, D.O. (1984) Axonal growth and target selection during development: retinal projections to the ventrobasal complex and other "nonvisual" structures in neonatal Syrian hamsters. *J. Comp. Neurol., 230:* 576-592.

Hamburger, V., and Levi-Montalcini, R. (1949) Proliferation, differentiation and degeneration in the spinal ganglia of the chick embryo under normal and experimental conditions. *J. Exp. Zool.,* 111: 457-502.

Hamburger, V. and Oppenheim, R.W. (1982) Naturally occurring neuronal death in vertebrates. *Neuroscience Commentaries,* 1: 39-55.

Hendrickson, A., and Rakic, P. (1977) Histogenesis and synaptogenesis in the dorsal lateral geniculate nucleus (LGd) of the fetal monkey brain. *Anat. Rec.,* 187: 602.

Heumann, D. and Leuba, G. (1983) Neuronal death in the development and aging of the cerebral cortex of the mouse. *Neuropath. & Applied Neurobio.,* 9: 297-311.

Hickey, T.L. and Hitchcock, P.F. (1984) Genesis of neurons in the dorsal lateral geniculate nucleus of the cat. *J. Comp. Neurol.,* 228: 186-199.

Hofman, M.A. (1989) On the evolution and geometry of the brain in mammals. *Progress in Neurobiol.* 32: 137-158.

Howard, B., Miller, B. and Finlay, B.L. (1989) Reorganization of visual callosal projections after early thalamic lesions in the golden hamster. *Soc. Neurosci. Abst.,* 15: 1339.

Hubel, D.H., Wiesel, T.S. and LeVay, S. (1977) Plasticity of ocular dominance columns in monkey striate cortex. *Phil. Trans. R. Soc. Lond.,* 278: 377-409.

Innocenti, G.M. (1981) Growth and reshaping of axons in the establishment of visual callosal connections. *Science,* 212: 824-827.

Innocenti, G.M. and Caminiti, R. (1980) Postnatal shaping of callosal connections from sensory areas. *Exp. Brain Res.* 38: 381-394.

Ivy, G.O. and Killackey, H. P. (1982) Ontogenetic changes in the projections of neocortical neurons. *J. Neurosci., 2* : 735-743.

Jerison, H.J. (1973) *Evolution of the Brain and Intelligence.* New York: Academic Press.

Kane, M.H., Sengelaub, D.R. and Finlay, B.L. (1984) An autoradiographic analysis of the role of cell death in regulation of neocortical cell number. *Soc. Neurosci. Abs., 10:* 462.

Lamantia, A.-S. and Rakic, P. (1990) Cytological and quantitative characteristics of four cerebral commissures in the rehsus monkey *J. Comp. Neurol.,* 291: 520-537.

Luskin, M.B. and Shatz, C.J. (1985a) Studies of the earliest generated cells of the cat's visual cortex: cogeneration of subplate and marginal zones. *J. Neurosci., 5:* 1062-1075.

Luskin, M.B. and Shatz, C.J. (1985b) Neurogenesis in the cat's primary visual cortex. *J. Comp. Neurol.,* 242: 611-631.

Miller, B., Windrem, M.S., Anllo-Vento, L. and Finlay, B.L. (1987) Minor reorganization of thalamocortical projections following large neonatal thalamic lesions in the golden hamster. *Soc. Neurosci. Abs.,* 13: 1419.

Mustari, M.J., Lund, R.D. and Graubard, K. (1979) Histogenesis of the superior colliculus of the albino rat: a tritiate thymidine study. *Brain Res.,* 164: 39-52.

Naegele, J.R., Jhaveri, S., and Schneider, G.E. (1988) Sharpening of topographical projections and maturation of geniculocortical axon arbors in the hamster. *J. Comp. Neurol.,* 281: 1-12.

O'Kusky, J., and Colonnier, M. (1982) A laminar analysis of the number of neurons, glia and synapses in the visual cortex (area 17) of adult macaque monkeys. *J. Comp. Neurol.,* 210: 278-290.

O'Leary, D.D.M. (1989) Do cortical areas emerge from a protocortex? *Trends in Neurosi.,* 12: 400-406.

Oppenheim, R.W. (1981) Neuronal death and some related regressive phenomena during neurogenesis: a selective historical review and progress report. In *Studies in Developmental Neurobiology,* pp. 74-133. Ed. W.M. Cowan. New York: Oxford University Press.

Pagel, M.D. and Harvey, P.H. (1990) Diversity in the brain sizes of newborn mammals. *Bio Science,* 40: 116-122.

Pallas, S.L., Gilmour, S., and Finlay, B.L. (1988) Control of cell number in the developing neocortex: I. Effects of early tectal ablation. *Devel. Brain Res.,* 43: 1-11.

Purves, D. (1988) *Body and Brain: A Trophic Theory of Neural Connections.* Harvard University Press: Cambridge, Ma.

Rakic, P. (1974) Neurons in the rhesus monkey visual cortex: systematic relation between time of origin and eventual disposition. *Science,* 183: 425-427.
Rakic, P. (1988) Specification of cerebral cortical areas. *Science,* 241: 170-176.
Ramirez, L.F. and Kalil, K. (1985) Critical stages for growth in the development of cortical neurons. *J. Comp. Neurol.,* 237: 506-518.
Rodier, P.M. (1980) Chronology of neuron development: animal studies and their clinical implications. *Develop. Med. and Child Neurol.,* 22: 525-545.
Sefton, A.J., MacKay-Sim, A., Baur, L.A. and Cottee, L.J. (1981) Cortical projections to visual centers in the rat: an HRP study. *Brain Res.,* 215:1-11.
Sengelaub, D.R, Dolan, R.P. and Finlay, B.L. (1988) Cell generation, death and retinal growth in the development of the hamster retinal ganglion cell layer: *J. Comp. Neurol.,* 204: 311-317.
Shatz, C.J. and Luskin, M.B. (1986) The relationship between the geniculocortical aferents and their cortical target cells during the development of the cat's primary visual cortex. *J. Neurosci.,* 6: 3655-3668.
Shimada, M. and Langman, J. (1970) Cell proliferation, migration and differentaiton on the cerebral cortex of the golden hamster. *J. Comp. Neurol.,* 139: 227-244.
Sidman, R.L. (1961) Histogenesis of the mouse retina studies with thymidine 3-H. In: *The Structure of the Eye*, G.K. Smelser, ed. Academic Press, New York: 487-506.
Sperry, D.G. (1990) Variation and symmetry in the lumbar and thoracic dorsal root ganglion cell populations of newly metamorphosed Xenopus laevis. *J. Comp. Neurol.,* 292: 54-64.
Stanfield, B.B., O'Leary, D.D.M. and Fricks, C. (1982) Selective collateral elimination in early postnatal development restricts cortical distribution of rat pyramidal tract neurones. *Nature, (Lond.)* 298: 371-373.
von Economo, C.F. (1929) *The Cytoarchitecture of the Human Cerebral Cortex*. Oxford University Press: London
Walsh, C., Polley, E.H., Hickey, T.L. and Guillery, R.W. (1983) Generation of cat retinal ganglion cells in relation to central pathways. *Nature, (Lond)* P302: 611-614.
Wikler, K.C., Kirn, J., Windrem, M.S. and Finlay, B.L. (1986) Control of cell number in the developing visual system: III. Partial tectal ablation. *Devel. Brain Res.,* 28: 23-32.
Williams, R.W. and Herrup, K. (1988) The control of neurons number. *Ann. Rev. Neurosci.*, 11: 423-454.
Windrem, M.S., and Finlay, B.L. (1985) Early thalamic lesions increase neonatal cell death and alter adult cytoarchitecture in the neocortex. *Soc. Neurosci. Abstr.,* 1: 991.
Windrem, M.S., Jan de Beur, S.M., and Finlay, B.L. (1986) Effects of early callosal and thalamic lesions on differentiation of cortical cytoarchitecture. *Soc. Neurosci. Abst.,* 12: 867.
Windrem, M.S., Jan de Beur, S. and Finlay, B.L. (1988) Control of cell number in the developing neocortex: II. Effects of corpus callosum transection. *Devel. Brain Res.,* 43: 13-22.
Zamenhof, S. and Van Marthens, E. (1979) Brain weight, brain chemical contents and their early manipulation. In: Hahn, M.E., Jensen, C. and Dudek, B.C. eds. *Development and Evolution of Brain Size: Behavioral Implications*. New York: Academic Press.

PATHWAYS BETWEEN DEVELOPMENT AND EVOLUTION

Giorgio M. Innocenti
Institute of Anatomy
Universite de Lausanne
9 rue du Bugnon
1005 Lausanne, Switzerland

At a recent round table discussion on evolution and development, a friend developmental neurobiologist whose opinion I cherish said more or less: "What is the use of thinking about evolution? Did it ever generate an experiment?" The ensuing silence seemed to settle the issue. The first part of this paper readdresses these questions. The second part deals with the possibility that developmental concepts may allow one to model, or even predict aspects of evolution. Although evolution is undoubtedly the most powerful theoretical concept in biology it cannot really be tested experimentally. An evolutionary scenario can, however, be represented by a model; known developmental mechanisms or rules can provide criteria for testing the coherence of the model.

PREDICTING BRAIN DEVELOPMENT

Evolution can be handy to discard as evolutionary oddity aspects of brain function, structure and development one fails to understand. Furthermore, it can profitably be used to settle controversies on results by invoking species differences.

A different usage of evolution involves across species comparisons of developmental events and their temporal relations as a way to probe their robustness and possible causal relations. Sometimes this approach allows for guesses about phenomena which for some reason cannot be directly analyzed.

This approach is exemplified in recent studies on the development of cortical projections (Innocenti, 1986; 1990). Transient corticocortical projections, callosal projections in particular, but also intrahemispheric and intra-areal projections are generated in development and are subsequently eliminated in the cat and in a number of other species, some of which are separated, according to Johnson (1980) by 65 million years or more of independent evolution. In cat and rat, the elimination of these projections, as studied with retrograde transport of HRP occurs during the beginning of the period of fast synaptogenesis and precedes myelination of the corpus callosum. This appears also to be the case in the monkey (compare Dehay et al., 1988 and Killackey and Chalupa, 1986 with Rakic et al., 1986). The robustness of these temporal correlations across species could indicate that i) the elimination of transient projections is triggered by synaptogenesis or by the same events which trigger the latter, and ii) the axons to be eliminated are unmyelinated; myelination may become possible only after the axon has escaped developmental elimination.

In the cat, the elimination of the transient projections coincides with a decrease in the number of callosal axons, which it probably causes (Berbel and Innocenti, 1988). The massive elimination of callosal axons in turn causes the growth of the cross sectional area of the corpus callosum to stop during the first three postnatal weeks as first noticed by Innocenti and Caminiti (1980) in a study by Fleischhauer and Schlüter (1970). In man, counts of callosal axons and tracer studies are difficult to perform. However, recently Clarke et al. (1989) discovered that the corpus callosum of man undergoes a growth pause similar to that of the cat. This growth pause coincides, as in the cat, with the beginning of the fast phase of synaptogenesis and precedes myelination. Thus it probably corresponds to a period of massive elimination of callosal axons in man.

PREDICTING BRAIN EVOLUTION

Insofar as evolution involves the appearance of new phenotypes, it implies modifications of development, and these have to be heritable. However, those developmental rules and/or mechanisms which do not change in parallel to evolution must act to constrain and possibly channel the latter. This concept has gained wide acceptance among evolutionary biologists, and is illustrated by specific models (Alberch, 1982; Maderson, 1982; Kauffman, 1983). This view implies that a robust knowledge of developmental rules should allow one to guess the likelihood of a specific phenotypic evolution and could therefore illuminate our evolutionary past as well as future, the latter on the condition that evolution is not achieved yet (Salk, 1983).

Is our knowledge of development robust enough to allow for evolutionary guesses?

In a general sense, the answer is obviously negative. In particular, we miss a general theory of development.

Figures 1 and 2 illustrate the possible format of such a theory by two examples (Weiss, 1955; Arthur, 1988). The common ground of the two theories is the assumption that development can be represented by causal links i.e. by propositions of the sort: "A causes B". In Arthur's formulation, the overall structure of developmental causations can be modeled as a tree, illustrating the hierarchical nature of development. Each branching in the tree leads to an increase in "heterogeneity". Arthur's model allows the classification of at least three types of evolutionary changes. In Weiss's formulation, instead, the overall structure of developmental causation is a horribly complicated network. I do not believe that Weiss meant this to be a joke, although this is certainly how some developmental neurobiologists take it.

Developmental causation, in the nervous system, does indeed show network properties, in the sense that each step in a morphogenetic sequence depends on several interacting factors. These factors may not be independent of each other and can be back-regulated by the outcomes of the morphogenetic process. Causal interactions in such a network are insufficiently described by the proposition: "A causes B". A more adequate proposition is "B changes as a function of A" (and of C, D etc.), where each "function" should be mathematically defined.

The multifactorial origin of brain structure is well exemplified by the development of ocular dominance. It has been known for some time that this aspect of the connectivity of visual cortical neurons is determined by the degree of synchronism in the visual input from the two eyes (Wiesel, 1982; Miller et al., 1989). More recently it was found that the development of ocular dominance could also be affected by ocular proprioception (Maffei and Bisti, 1976; Buisseret and Singer, 1983), catecholamines (Kasamatsu, 1987), acetylcholine (Bear and Singer, 1986), the past visual experience of the animal (Cynader and Mitchell, 1980), and centrencephalic inputs related to motivation or possibly attention (Singer, 1982; Singer and Rauscheker, 1982). To what extent these factors interact with each other and whether they do so in a hierarchical fashion is unknown. The possibility exists that they interact at some molecular common step whereby the stabilization or possibly the gain of some geniculocortical synapse can also be regulated. Singer and collaborators (this volume) have gone a long way towards the identification of this molecular mechanism.

The fate of juvenile corticocortical connections seems similarly decided by multiple factors. The stabilization or elimination of juvenile visual callosal connections for example, seems to depend not only on normal binocular vision during the second postnatal month, but also on integrity of the eyes, and the geniculocortical projections, and possibly on selective affinity between the presynaptic and postsynaptic elements. For other corticocortical connections (e.g. the transient intra- and interhemispheric projections from auditory to visual areas) stabilization or elimination seem also to depend on factors related to the target of the projection (for references and discussion see Innocenti, 1986; 1990). Unfortunately, one does not know if the different factors listed above influence the same or different subclasses of neurons. Furthermore, as in the case of binocular properties, it is unknown if the different controls which lead to selection of certain juvenile projections and the elimination of others interact with each other or are hierarchically organized.

The different factors which control the fate of juvenile corticocortical projections may operate through a common molecular step. In early postnatal development, cortical axons appear to

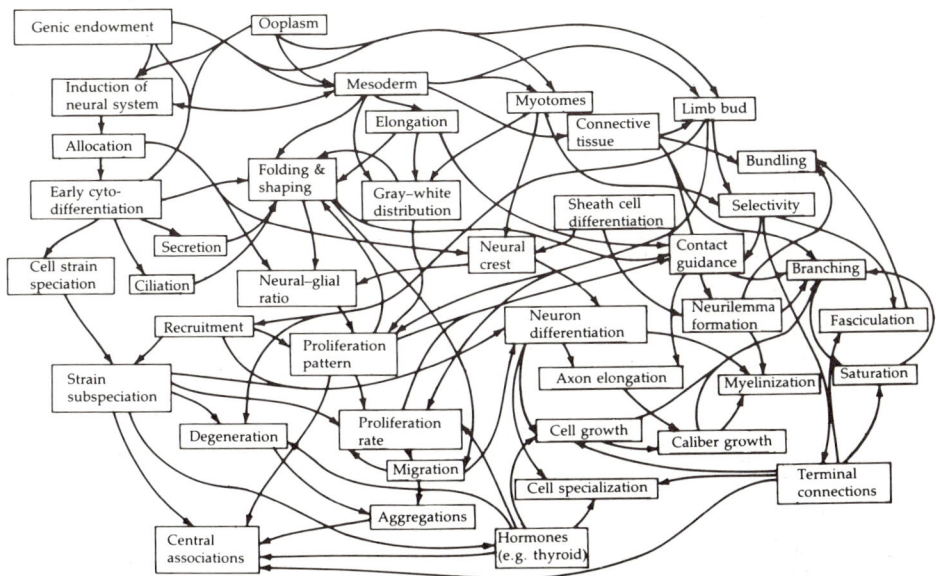

Figure 1. Causal relations in neurogenesis. An early attempt at a synthesis of neural development (Weiss, 1955; reproduced with permission). This diagram could be greatly complicated were the present knowledge of neural development added to it. Although the diagram identifies the overall network structure of causal relations in development it is singularly "undifferentiated" in the sense that it appears devoid of any hierarchical structure and that all the pathways appear to have the same importance.

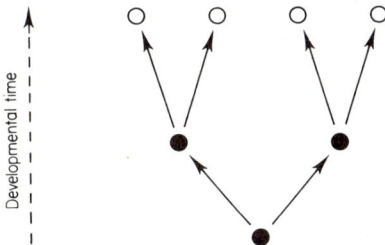

Figure 2. A recent conceptualization of development (Arthur, 1988; reproduced with permission). Causal links hierarchically organized in a tree lead to increasing morphogenetic heterogeneity (filled circles; open circles: terminal heterogeneity) along developmental time. One disadvantage of this model is that single morphogenetic events appear to occur in relative isolation from each other.

be in a labile state from which they undergo a transition either to a stable state or to elimination. This state transition presumably coincides with the period of axonal elimination as revealed by tracer studies or electronmicroscopic counts. In fact, the period during which the fate of a callosal axon can be modified seems to correspond roughly with the period of massive axonal elimination. During the same time, the cytoskeletal proteins of callosal axons appear to undergo several changes. These include the appearance of the heavy subunit of neurofilaments (Figlewicz et al., 1988), the partial dephosphorylation of this protein and perhaps also of the medium subunit of neurofilaments (Guadano-Ferraz et al., 1990), the dephosphorylation of microtubule associated protein MAP5 (Riederer et al., submitted) and the transformation of juvenile Tau proteins into their adult counterparts (Innocenti, 1990).

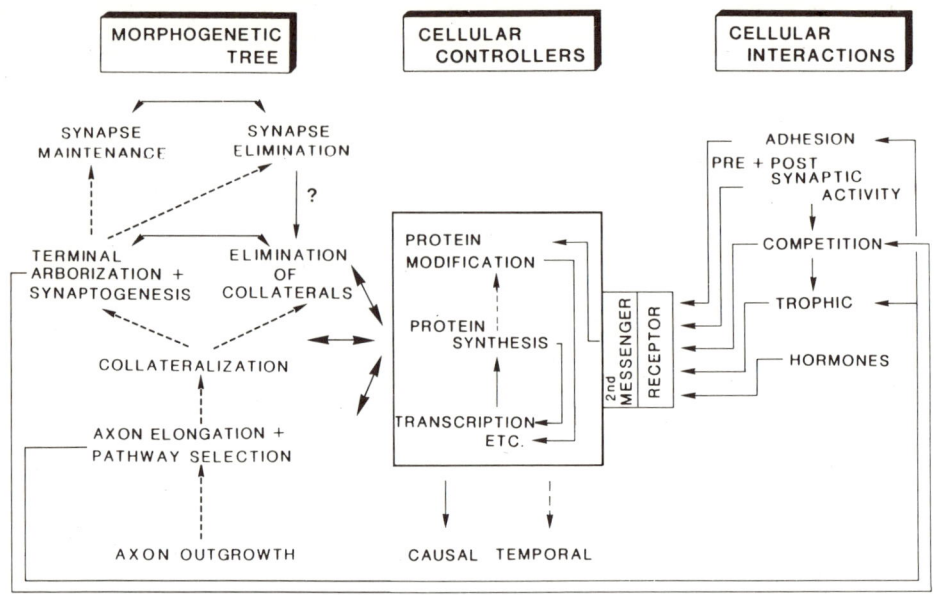

Figure 3. Proposed structure of causal interactions underlying a specific aspect of neural development: the formation of neural connections. In the left-hand panel, the formation of connections is represented by a tree (morphogenetic tree) of temporally (interrupted arrows) or causally (continuous arrows) related morphogenetic events; double headed arrows indicate events which can reciprocally influence each other, e.g. whether an axon collateral is eliminated can depend on whether another collateral of the same neuron is maintained and arborizes. The central panel represents the cellular mechanisms which control (primary controllers) the formation of connections; a chain of causally related events leading to production and modification (e.g. by phosphorylation) of one or more specific proteins presumably regulate each step in the formation of connections. The outcomes of a given morphogenetic event could, however, back-regulate the cellular controllers, e.g. axonal arborization presumably depends on the production of specific cytoskeletal proteins but may in turn regulate the production of these proteins or possibly that of synaptic proteins and/or neurotransmitters. The right-hand panel lists classes of cellular interactions (secondary controllers) which can regulate the primary controllers but can be back-regulated by the morphogenetic events; many cellular interactions operate via receptor-second messenger systems; some cellular interactions may be causally related with each other, e.g. activity may regulate competition which in turn may regulate the uptake of trophic substances. The factual basis for this simplified view of the development of connections is partially discussed in the text.

Which of these biochemical changes may be critical for the stabilization of juvenile axons is unknown. They do, however, occur simultaneously to changes in axonal size, and in the number and distribution of neurofilaments and microtubules (Berbel et al., 1989), both of which strongly suggest axonal maturation.

Figure 3 summarizes in a model the general structure of causal interactions responsible for the formation of connections in the central nervous system. The left panel is a tree representing, as in Arthur's model, the temporal causal sequence of events. Each step in this sequence is probably controlled by a chain of molecular interactions leading to protein production and their post-translational modification. Notice that, in principle, the outcome of each developmental step can back-regulate the cellular/molecular controllers and that the fate of an axonal branch or synapse can, in addition, be influenced by what happens to other branches or synapses of the same neuron. The molecular/cellular events can be considered the "primary controllers" of the process. However, they are in turn regulated by "secondary controllers", i.e. cell-cell interactions of different sorts. There can be direct interactions between a step in the hierarchical tree and a secondary controller. For example, the competition could be regulated by the amount of terminals an axon produces and the type of adhesion molecules or trophic substances an axon encounters may be determined by the pathway it selected. Notice that given the network structure of the CNS practically each neuron can be influenced by several others even though they are not directly linked to it. Furthermore two neurons can interact in more than one way.

One may argue that a general theory of development is not really useful to understand development. Nevertheless, two general properties of complex networks (Hinton and Sejnowski, 1986) may be directly relevant in understanding development and its relations with evolution. One of these properties is some degree of "redundancy" or "degeneracy": the network could be partially deleted without its output being dramatically altered. This property could explain the "invisibility" of microevolutionary changes and their necessity to accumulate before significant phenotypic change occurs. On the other hand, certain complex networks tend to assume only few stable configurations (Hopfield and Tank, 1986) and this could explain the channeling of evolution by developmental mechanisms, already noticed by Waddington (1957). Finally, the understanding of complex networks requires modeling and this implies quantitative rather than qualitative observations. A general theory of neural development may therefore have normative consequences at the data acquisition level.

Cerebral cortex appears to be a good system on which to test to what extent current "local" knowledge (in contrast to the "general" knowledge discussed above) of neural development, may contribute to understanding for evolutionary predictions. Neocortex underwent dramatic evolution in mammals and at least some of its basic developmental mechanisms are understood.

Let us first consider what I will call an evolutionary trajectory for neocortex, i.e. a line connecting mammalian species in the direction of increased cortical volume with respect to total brain volume (Hofman, 1988). This evolutionary trajectory does not reflect or imply phylogenetic relations among the extant mammals. Although the variable determining this evolutionary trajectory is arbitrary, it has some advantages. First, it can be expressed numerically. Second, it puts man at the top of an evolutionary ranking system, or scala naturae which is reassuring although not necessarily justified.

The subsequent questions are: what changes in cortical organization parallel with this evolutionary trajectory? and what developmental changes can account for the evolutionary changes?

Although not all aspects of cortical organization and/or function have been analyzed equally in depth, in all the orders connected by the evolutionary trajectory, at least two generalizations seem to be possible. First, the increase in cortical volume seems to reflect an increase in cortical thickness but even more in cortical surface, at least in gyrencephalic brains (Hofman, 1985). Second, cortical areas, defined as cytoarchitectonically distinguishable fields, and/or distinct representations of sensory modalities, tend to increase in number (Kaas, 1980), with increased cortical surface.

What change in developmental mechanisms could allow the cortex to become thicker? In species with more evolved neocortex, and in phylogenetically more recent areas, supragranular layers are relatively more developed (Sanides, 1969). Given that later generated neurons acquire progressively more superficial positions (Angevine and Sidman, 1961; Rakic, 1974; Luskin and Shatz, 1985), a prolongation of the time of generation could result in more supragranular layer neurons. Although additional mechanisms, in particular a differential rate of neuronal death in supra- and infragranular layers could also contribute, a heterochronic change in the time of neurogenesis could be involved in this aspect of cortical evolution. Interestingly, in the monkey, the time of neurogenesis is shorter for the limbic cortex, presumably a more ancient cortex than area 17 (Rakic, 1988).

What could be the consequences of changes in cortical thickness? These cannot be predicted by the current knowledge of cortical development. A series of studies in the cat demonstrated that when the cortex was drastically reduced in thickness by perinatal deletion of the deep cortical layers, several aspects of cortical function and functional organization are preserved although specific aspects of cortical connectivity are profoundly modified (Innocenti et al., 1987, 1988; Assal et al., 1989; Hornung et al., 1989).

How can neocortex increase tangentially? One mechanism could be to increase the surface of the germinative epithelium from which neocortical neurons originate. This may be obtained by multiplication of the "proliferative units" in which the epithelium may be organized (Rakic, 1988). Alternatively, the size of the epithelium may remain the same but the channels of radial glia could increase in number leading to dispersion of cortical neurons over a tangentially more extended cortical plate.

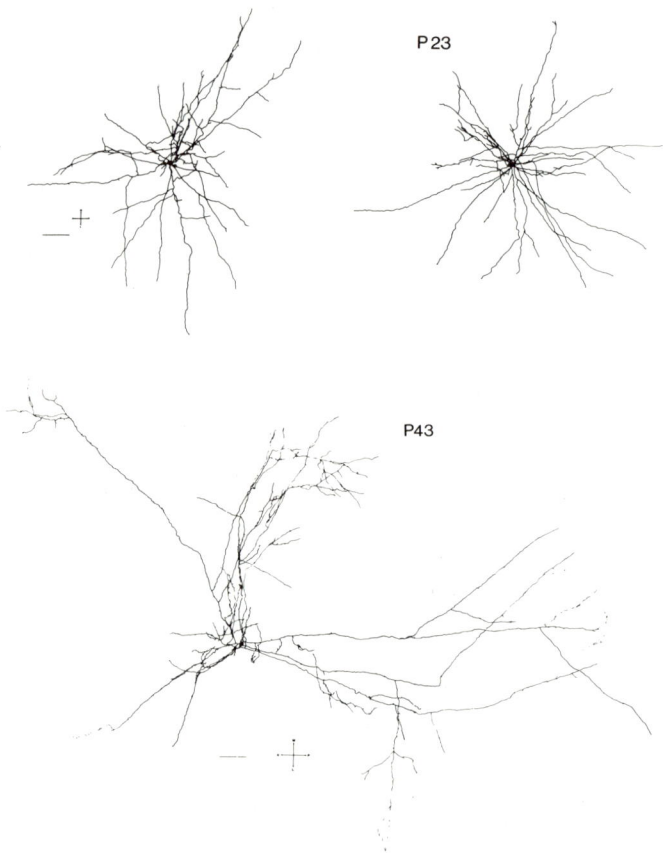

Figure 4. Layer 2/3 pyramidal cells intracellularly injected with Lucifer yellow in 400 um thick tangential slices of visual area 17 of P23 and P43 day kittens. In the P23 cells, many long, unbranched horizontal initial axon collaterals are visible. Notice that at P23 the axon collaterals radiate more or less symmetrically in all directions; at P43, they are denser in certain sectors around the cell body. The transition from the P23 to the P43 morphology involves selective elimination of some collaterals, and the selective growth of others (from Callaway and Katz, unpublished).

When does the tangential expansion of neocortex result in the formation of a new cortical area rather than the enlargement of those which were already there? This question cannot yet be answered, however, the size of a given cortical area may not be rigidly regulated in development but instead, at least in the case of primary sensory areas, depend on some relation with the periphery early in development (Rakic, 1988; Dehay et al., 1989), possibly because the periphery controls the size of the thalamic nuclei. Another unanswered question is whether the tangential increase produced by phylogenesis implies addition of neurons (or "modules", Sawagouchi and Kubota, 1986) at the periphery of the phylogenetically older cortex or throughout it.

The emergence of a new cortical area necessitates that this area establish connections with preexisting areas. At first glance this appears to be possible without violation of major developmental rules (Innocenti, 1988). In development, each cortical area sends and receives projections in excess to those which are found in the adult (see Innocenti, 1990, for review). Which projections an area receives and sends is, however, constrained. For example, in a young kitten, areas 17 and 18 receive projections from a characteristic elongated territory in the contralateral hemisphere extending mediolaterally across most visual areas but not restricted to them (Innocenti and Clarke, 1984). A similar territory was labeled from injections in the presplenium in a recent study in which the topographical order of callosal axons was investigated by direct applications

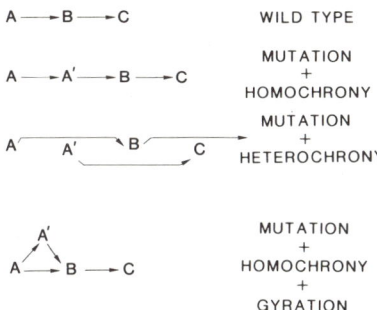

Figure 5. Consequences of the rules dictate the development of connections in the case of emergence of a "new" cortical area. Explained in text. Notice that heterochrony produces exuberance i.e. projections from B overshoot their target C.

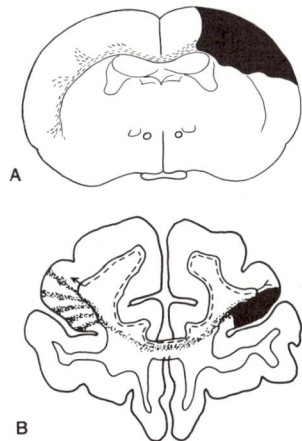

Figure 6. Two rather similar experiments showing callosal connections with different tracer techniques, one from Heimer et al. (1967), in the rat (A) the other by Goldman-Rakic (1982) in the rhesus monkey (B). They illustrate how gyration, and reorganization of the white matter can shorten the trajectory of cortico-cortical axons. The interrupted arrow (added by the author), in the bottom drawing indicates the trajectory that axons would have taken from their origin to their destination had the monkey cortex not been gyrated and therefore had callosal axons travelled as in the rat.

of HRP to the corpus callosum (Nakamura and Kanaseki, 1989) in adult cats. Thus, it was argued (Innocenti, 1990) that rules of axonal growth and order in the callosum may determine the initial pattern of connections between the hemispheres. These rules could constrain connectivity in such a way that, for example, a new area arising in "supercat" in the dorso-frontal cortex would have no possibility of ever becoming connected to the contralateral visual cortex, unless the rules of axonal order in the corpus callosum, i.e. the mechanisms which produce this order, are also changed. Interestingly, the same tracer injection which labels a characteristically restricted region of the contralateral hemisphere labels a different, territory in the ipsilateral hemisphere, more or less a circle centered around the injection (Clarke and Innocenti, 1986). Thus, axonal growth of intrahemispheric projections may be unconstrained or relatively unconstrained. Recent work by Callaway and Katz (Fig. 4) supports this conclusion: they see that the initial axonal growth around a cortical neuron is more or less uniformly radial and unselective, in the tangential dimension albeit rather specific in the choice of the layer through which growth occurs. The directionally patterned and clustered distribution of axon collaterals seems to be acquired secondarily by elimination of some of the initial branches and growth of the remaining ones. The first developmental constraint on evolution of corticocortical connectivity thus may be the pathways the axons can chose and follow and the ordering rules within the pathways. The nature of the pathways, extracellular matrix, glia or other axons, is undefined. Elsewhere I have speculated that the pathways may be more conserved in phylogenesis than the number of axons of different origin coursing through them (Innocenti, 1990).

Another probable constraint is the total length that axons can grow in the phase of exuberant development, which is presumably determined by growth speed times total growth time. Probably a few, if any of the exuberant axons grow to the very end of the pathway which is available to them and there may be species differences in the extent of growth (O'Leary and Stanfield, 1986). Fig. 5 illustrates the consequences of this constraint in the case that a new area appears in cortical evolution. If a new area (A') appears between A and B and the total distance that the axons grow remains the same because the total time allowed for growth and speed of growth do not change (homochrony), the result is that A becomes disconnected from B. If, on the other hand, the total distance travelled by axons in the exuberance phase increases because either the speed of growth or the total time of growth have increased (heterochrony), then B (as well as other cortical areas) could overshoot its target and have access to more distal areas than before the evolutionary change had occurred. This condition produces exuberance in the juvenile projections if the part of the projection which has overshot the target has to be eliminated. Thus, an increase in the number of cortical areas, unparalleled by an increase in the number of pathways (discussed in Innocenti, 1990) or accompanied by heterochronic changes in axonal growth, can be responsible for the appearance of connectional exuberance in development.

SUMMARY AND CONCLUSIONS

The absence of a general theory of neural development is obviously a severe limitation for the understanding of evolution. However, in the case of neocortex, the present knowledge of development is sufficient to at least guess which specific developmental change could result in phenotypic changes that probably occurred in evolution. Undoubtedly, the predictive power of developmental neurobiology will become increasingly relevant for a general theory of the living matter, and hopefully for the diagnosis and treatment of early brain pathology.

ACKNOWLEDGEMENTS

Supported by Swiss National Science Foundation Grant 3.359-0.86. I am grateful to Marc Weisskopf for his helpful criticism of this text and to the authors who allowed me to reproduce part of their work.

REFERENCES

Alberch, P. (1982) Developmental constraints in evolutionary processes, in: "Evolution and Development." J. T. Bonner, ed., Dahlem Konferenzen, Springer-Verlag, Berlin.
Arthur W. (1988) "A Theory of the Evolution of Development." John Wiley & Sons, Chichester.
Assal, F., Melzer, P., and Innocenti, G.M. (1989) Functional analysis of a visual cortical circuit resembling human microgyria. *Europ. J. Neurosci.*, Suppl. 2: 256.
Angevine, J.B., and Sidman, R.L. (1961) Autoradiographic study of cell migration during histogenesis of cerebral cortex in the mouse. *Nature*, 192: 766.
Bear, M.F., and Singer, W. (1986) Modulation of visual cortical plasticity by acetylcholine and noradrenaline. *Nature*, 320: 172.
Berbel, P., and Innocenti, G.M. (1988) The development of the corpus callosum in cats: A light- and electron-microscopic study. *J. Comp. Neurol.*, 276: 132.
Berbel, P., Innocenti, G.M., Prieto, J.J., and Kraftsik, R. (1989) A quantitative study on the development of the cytoskeleton of callosal axons in cats. *Europ. J. Neurosci.*, Suppl. 2:47.
Buisseret, P., and Singer, W. (1983) Proprioceptive signals from extraocular muscles gate experience-dependent modifications of receptive fields in the kitten visual cortex. *Exp. Brain Res.*, 51: 443.
Clarke, S., and Innocenti, G.M. (1986) Organization of immature intrahemispheric connections. *J. Comp. Neurol.*, 251: 1.
Clarke, S., Kraftsik, R., Van der Loos, H., and Innocenti, G.M. (1989) Forms and measures of adult and developing human corpus callosum: is there sexual dimorphism? *J. Comp. Neurol.*, 280: 213.
Cynader, M., and Mitchell, D.E. (1980) Prolonged sensitivity to monocular deprivation in dark-reared cats. *J. Neurophysiol.*, 43: 1026.
Dehay, C., Kennedy, H., Bullier, J., and Berland, M. (1988) Absence of interhemispheric connections of area 17 during development in the monkey. *Nature*, 331: 348.

Dehay, C., Horsburgh, G., Berland, M., Killakey, H., and Kennedy, H. (1989) Maturation and connectivity of the visual cortex in monkey is altered by prenatal removal of retinal input. *Nature*, 337: 265.

Figlewicz, D.A., Gremo, F., and Innocenti, G.M. (1988) Differential expression of neurofilament subunits in the developing corpus callosum. *Dev. Brain Res.*, 42: 181.

Fleischhauer, K., and Schlüter, G. (1970) Ueber das postnatale Wachstum des Corpus callosum der Katze (Felis domestica). *Z. Anat. Entwick-Gesch.*, 132: 228.

Goldman-Rakic, P.S. (1982) Neuronal development and plasticity of association cortex in primates. *Neurosciences Res. Prog. Bull.*, 20: 520.

Guadano-Ferraz, A., Riederer, B., and Innocenti, G.M. (1990) Developmental changes in the heavy subunit of neurofilaments in the corpus callosum of the cat. *Dev. Brain Res.*, submitted.

Heimer, L., Ebner, F.F., and Nauta, W.J.H. (1967) A note on the termination of commissural fibers in the neocortex. *Brain Res.*, 5: 171.

Hinton, G.E., and Seijnowski, T.J. (1986) Learning and relearning in Boltzmann machines. In: "Parallel Distributed Processing. Explorations in the Microstructure of Cognition". Volume 1: Foundations, D.E. Rumelhart, J.L. McClelland, and the PDP Research Group, eds., The MIT Press, Cambridge.

Hofman, M.A. (1985) Size and shape of the cerebral cortex in mammals. I. The cortical surface. *Brain Behav. Evol.*, 27: 28.

Hofman, M.A. (1988) Size and shape of the cerebral cortex in mammals. II. The cortical volume. *Brain Behav. Evol.*, 32: 17.

Hopfield, J.J., and Tank, D.W. (1986) Computing with neural circuits: a model. *Science*, 233: 625.

Hornung, J.P., Assal, F., and Innocenti, G.M. (1989) Distribution of diffuse afferents and interneurons in experimentally induced microcortex in cat visual cortex. *Europ. J. Neurosci.*, Suppl. 2: 105.

Innocenti, G.M., (1986) General organization of callosal connections in the cerebral cortex. In: "Cerebral Cortex", vol. 5, E.G. Jones, and A. Peters, eds., Plenum Publishing Corporation.

Innocenti, G.M., (1988) Loss of axonal projections in the development of the mammalian brain, In: "The Making of the Nervous System", J. G. Parnavelas, C.D. Stern, and R.V. Sterling eds., Oxford University Press.

Innocenti, G.M. (1990) The development of projections from cerebral cortex. *Progress in Sensory Physiol.*, in press.

Innocenti, G.M., Berbel, P., and Melzer, P. (1987) Stabilization of transitory cortico-cortical projections following lesions provoked by neonatal ibotenic acid injections. *Neuroscience*, 22: S227.

Innocenti, G.M., Berbel, P., Aschoff, A., and Melzer, P. (1988) Connections and functional properties of an experimental cortical network. *Europ. J. Neurosci.*, Suppl. 1: 334.

Innocenti, G.M., and Caminiti, R. (1980) Postnatal shaping of callosal connections from sensory areas. *Exp. Brain Res.*, 38: 381.

Innocenti, G.M., and Clarke, S. (1984) The organization of immature callosal connections. *J. Comp. Neurol.*, 230: 287.

Johnson, J.I. (1980) Morphological correlates of specialized elaborations in somatic sensory cerebral neocortex. In: "Comparative Neurology of the Telencephalon," S.O.E. Ebbesson, ed., Plenum Press, New York.

Kaas, J.H. (1980) A comparative survey of visual cortex organization in mammals. In: "Comparative Neurology of the Telencephalon," S.O. Ebbesson, ed., Plenum Press.

Kasamatsu, T. (1987) Norepinephrine hypothesis for visual cortical plasticity: thesis, antithesis, and recent development. *Current Topics in Developmental Biology*, 21: 367.

Kauffman, S.A. (1983) Developmental constraints: internal factors in evolution. In: "Development and Evolution," B.C. Goodwin, N. Holder, and C.C. Wylie, eds., Cambridge University Press, Cambridge.

Killackey, H.P., and Chalupa, L.M. (1986) Ontogenetic change in the distribution of callosal projection neurons in the postcentral gyrus of the fetal rhesus monkey. *J. Comp. Neurol.*, 244: 331.

Luskin, M.B., and Shatz, C.J. (1985) Neurogenesis of the cat's primary visual cortex. *J. Comp. Neurol.*, 242: 611.

Maderson, P.F.A. (1982) The role of development in macroevolutionary change. Group report. In: "Evolution and Development," J.T. Bonner, ed., Springer-Verlag, Berlin.

Maffei, L., and Bisti, S. (1976) Binocular interaction in strabismic kittens deprived of vision. *Science*, 191: 579.

Miller, K.D., Keller, J.B., and Stryker, M.P. (1989) Ocular dominance column development: analysis and simulation. *Science*, 245: 605.

Nakamura, H., and Kanaseki, T. (1989) Topography of the corpus callosum in the cat. *Brain Res.* 485: 171.
O'Leary, D.D.M., and Stanfield, B.B. (1986) A transient pyramidal tract projection from the visual cortex in the hamster and its removal by selective collateral elimination. *Dev. Brain Res.*, 27: 87.
Rakic, P. (1974) Neurons in rhesus monkey visual cortex: systematic relation between time of origin and eventual disposition. *Science*, 183: 425.
Rakic, P. (1988) Specification of cerebral cortical areas. *Science*, 241: 170-176.
Rakic, P., Bourgeois, J.-P., Eckenhoff, M.F., Zecevic, N., and Goldman-Rakic, P.S. (1986) Concurrent overproduction of synapses in diverse regions of the primate cerebral cortex. *Science*, 232: 232.
Riederer, B., Guadano-Ferraz, A., and Innocenti, G.M., Difference in distribution of microtubule-associated proteins 5a and 5b during the development of cerebral cortex and corpus callosum in cats: dependence on phosphorylation, submitted.
Salk, J. (1983) "Anatomy of Reality. Merging of Intuition and Reason." Columbia University Press, New York.
Sanides, F. (1969) Comparative architectonics of the neocortex of mammals and their evolutionary interpretation. *Ann. N.Y. Acad. Sci.*, 167: 404.
Sawaguchi, T., and Kubota, K. (1986) A hypothesis on the primate neocortex evolution: column-multiplication hypothesis. *Intern. J. Neuroscience*, 30: 57.
Singer, W. (1982) Central core control of developmental plasticity in the kitten visual cortex: I. Diencephalic lesions. *Exp. Brain Res.*, 47: 209.
Singer, W., and Rauschecker, J.P. (1982) Central core control of developmental plasticity in the kitten visual cortex: II. Electrical activation of mesencephalic and diencephalic projections. *Exp. Brain Res.*, 47: 223.
Waddington, C.H. (1957) The Strategy of the Genes", George Allen and Unwin Ltd, London.
Weiss, P. (1955, 1971) Nervous system (neurogenesis). In: "Analysis of Development," B.H. Willier, P.A. Weiss, and V. Hamburger, eds., Hafner Publishing Company, New York.
Wiesel, T.N. (1982) Postnatal development of the visual cortex and the influence of environment. *Nature*, 299: 583.

INTRODUCTION TO SECTION 2 and 3

2: COMPARATIVE ASPECTS OF FOREBRAIN ORGANIZATION
3: NEUROEMBRYOLOGY OF THE NEOCORTEX

Section 2: "Comparative aspects of forebrain organization" contains chapters about the organizational properties of the telencephalon and associated structures for major vertebrate lineages, at many levels of analysis. Section 3, "Neuroembryology of the neocortex" contains parallel chapters about the developmental mechanisms that produce the neocortex. While these sections can be read independently, it was the intent of the conference to find explicit relationships between these two ways of comprehending the brain. In the many discussions at the Angelo Mosso Institute, some clear relationships did emerge, and as many questions. The following summary reflects both these discussions, and the editors' ideas of the relationships of these papers. The authors themselves need bear no responsibility for heedless overinterpretation of their papers!

What is the relationship of cell masses and laminae of the forebrain of different vertebrates?

Cell groups in the nervous system (i.e. nuclei, cortical areas or cell laminae) can be homologized on a number of attributes, from peptide expression to behavioral function. The choice of which attributes are central to an understanding of cell group identity has led to persistent controversies in the assignment of homologies (and which persist in this volume, as can be seen by comparing the conclusions of Shimizu and Karten versus Lohmann and Smeets on homologizing the avian, reptilian and mammalian forebrain).

The advantage of using a developmental approach to understand relationships between cell groups in the telencephalon was reaffirmed in the discussions. When comparative anatomists attempt to establish homologies, they describe a list of overlapping attributes of two cell groups. However, much of the message of developmental neurobiology of the last decade is that the notion of "cell type" as defined by a list of particular attributes is quite mutable. In vertebrates, clonally related cells can give rise to a number of morphological and functional cell classes. The expression of particular cell type is often dependent on the particular environment of the cell; connectivity and topology may similarly be changed in a large number of ways depending on the cellular and behavioral environment.

What developmental neurobiologists must establish for the forebrain is a "cladistics" of neuronal groups, rather than a sorting of cell groups dependent on their mature attributes. As in evolutionary systematics, the only way to establish relationships between neuronal classes is by examination of each neuron's developmental history.

Can differences in cell migration and disposition account for evolutionary variations in the vertebrate forebrain?

The papers of Shimizu and Karten, and Lohmann and Smeets point the desirability of extending the elegant descriptive work of Caviness and colleagues to a number of nonmammalian species. Caviness et al. demonstrate that the orchestration of neuronal migration on radial glia is clearly central to understanding the organization of the mammalian neocortex; it is equally clear in the comparative papers that there must be striking differences in the patterns of early neuronal migration across species. While something is known about the patterns of neuron settling and migration in nonmammalian brains, the relationship of these patterns to the radial glia scaffolding is unknown.

A new structure for classifying cell migration and dispostion in the forebrain is outlined in the chapter by Soriano, Del Rio and Ferrer. Soriano and colleagues propose the interesting idea that a form of cell disposition after migration they term "sandwich gradients" of late-generated cells nested between early-generated cells is a form that may be basic to the forebrain and is expressed both in the mammalian hippocampus and neocortex. That this was characteristic of neocortical migration had not been widely emphasized due to the specialized, transitory nature of the early generated subplate cells. It would clearly be of interest to investigate the dispositional relationships of early and late generated cells in the Wulst, dorsal ventricular ridge and the dorsal telencephalon of the avian brain, as well. Migration, and termination of migration is again shown to be central to the understanding of relationships of cell groups in the telencephalon, with the further proviso that transitory cell groups like the subplate must also be considered in the understanding of relationships and homologies.

How stable is the neocortex itself in its connectivity and cell classes?

Valverde describes the species variability in cell types in the cortex that belies the notion that in mammals thalamocortical axons contact a specialized class of cells termed "stellate cells". Cells accepting thalamocortical input show a wide range of specialization for this purpose: in the hedgehog, thalamocortical axons contact the basal and apical dendrites of pyramidal cells of layers II-V, while in primates, thalamocortical fibers contact the specialized spiny stellate cell with no extrinsic axon. All forms of intermediate specialization are present. Valverde raises the interesting possibility that these variations represent "the capability (of the cortex) to adapt cell types and patterns of synaptic input to whatever is most convenient for its specific function". Developmental manipulations might thus be employed in any one species to elicit a range of variability in thalamorecipient neurons that might normally only be seen across species.

In the context of plasticity, Cynader et al. describe the changing developmental expression of various neurotransmitters in the kitten cortex. Could species differences in cortex be related to alterations in the developmental expression of these agents?

How are new areas specified? What is the evidence for the phylogenetic "invasion" of terminal fields by new afferents?

The introductory chapters of Rakic, Finlay and Innocenti all converge on the question of what is required to specify a new cortical area as a central one in understanding the paths of cortical evolution, and all view the problem of how the cortex acquires new afference as a central part of this question. The papers of Fritzsch, Pallas, and Metin and Frost investigate the question of invasion of a terminal zone by new afference, and the resultant creation of a new area (or sensory system) but from different perspectives: Fritzsch describes a likely case of invasion in phylogeny, and Pallas and Metin and Frost both describe a developmentally-induced form of neocortical invasion. The likelihood that new functions can reside in old structures is the suggestion of all of these papers, and all propose or describe developmental mechanisms in the brain generally and the cortex specifically that tend to stabilize functional representations. Pallas show how topographic ordering is preserved when retinal afferents are induced to innervate auditory targets; Frost shows that many features of visual receptive field properties are preserved when visual input innervates somatosensory cortex. The source of this preserved order, either cortical, from the afferents themselves, or deriving from the interaction of cortical mechanisms with environmental regularities was the subject of continuing debate.

Van der Loos, Welker, Dorfl and Hoogland raise the related issue of the altered organization of cortical maps in the face of altered afference. This paper and that of Pallas' suggests that cortical areas contain some sort of intrinsic polarity cue, perhaps of the type described by Walter and Bonhoeffer in the retinotectal system. Within these topographic constraints, however, the sensory periphery or thalamic afference may produce substantial reordering.

Is there any advantage conferred by cortical lamination? How do the response properties of single units differ in "homologous" cortical and nuclear structures?

The impressive burgeoning of cortical volume in our own species (Jerison), coupled with its most impressive structural feature, lamination, raises the questions of the functional advantage conferred by lamination.

Three of the chapters in this volume provide information on the functional consequences of neocortical lamination. Pettigrew contrasts visual information processing in the visual Wulst of owls with the visual cortex of cats; Scheich contrasts auditory representations in Field L of chickens with auditory cortex in gerbils. In these two cases, the functional advantages remain elusive. In neither case is there any obvious behavioral advantage conferred by lamination, and in both cases, functional homologies are more striking than functional differences. Diamond and Ebner describe an emergent property of modular organization in the rodent somatosensory cortex not found in the turtle cortex, which does possess a laminar organization of fewer cell types. However, the functional advantage conferred by columns or modularity in the mammalian cortex is unknown. Whether the absence of a striking degree of encephalization in the non-mammalian telencephalon results from some functional limitation of a nonlaminar structure, unrelated embryological constraints, or evolutionary accident remains unresolved.

Radial and modular units: a central feature in neocortical organization?

In development migrating cortical neurons may be organized into radially oriented "Ontogentic columns" (Rakic). A modular organization of the neocortex in adulthood is something that appears in cases where the periphery has an obvious segmentation that impresses itself on the cortex, as in the whisker-to barrel organization studied by Van der Loos, but that also appears when the periphery has no obvious segmentation, as described by Diamond and Ebner. Is there a necessary relationship of ontogentic columns and the adult modularity of the cortex? Does this modularity in some way confer selective advantage on the cortex? Swindale shows how such a structure can accommodate the parallel mapping of a number of computed dimensions, with equivalent "coverage" of all dimensions in every location. Would it be possible to realize the formal properties of Swindale's model in a nonmodular structure?

Is cortical plasticity central to cortical diversity? What are its mechanisms?

The ability of the cortex to conform to alterations in the volume and structure of its input would seem to be central to understanding the possibilities for functional divergence of the forebrain, but there has been very little comparative study of plasticity. Cynader, Shaw, van Huizen and Prusky, and Welker and van der Loos both describe possible biochemical candidates for synaptic plasticity in the visual cortex and barrel cortex respectively. Perhaps the "suspicious coincidence" of expressions of neurotransmitters with developmental plasticity can be extended to a comparative analysis.

2. COMPARATIVE ASPECTS OF FOREBRAIN ORGANIZATION

THE DORSAL VENTRICULAR RIDGE AND CORTEX OF REPTILES IN HISTORICAL AND PHYLOGENETIC PERSPECTIVE

Anthony H.M. Lohman and Wilhelmus J.A.J. Smeets

Department of Anatomy and Embryology
Vrije Universiteit. P.O. Box 7161
1007 MC Amsterdam, The Netherlands

INTRODUCTION

From a comparative point of view, reptiles are of particular interest, since they are believed to be ancestral to both birds and mammals. It is, therefore, not surprising that in many laboratories species of the reptilian class have been and are still being used in search for basic features of the central nervous system of amniotes, i.e. reptiles, birds, and mammals. One of the most intriguing questions is whether a structure homologous to the mammalian neocortex is already present in the forebrain of reptiles. Before dealing with this question, some introductory comments will be made on the classification of the reptilian species mentioned in this chapter and on the anatomy of the reptilian forebrain.

Living reptiles comprise four orders (Fig. 1): (1) the Chelonia (turtles); (2) the Rhynchocephalia (tuataras); (3) the Squamata (lizards and snakes); and (4) the Crocodilia (alligators and crocodiles). Crocodilians and turtles have received special attention, since the former are believed to be most closely related to the reptilian stem from which birds evolved, whereas turtles probably have the closest relation to the stem reptiles that gave rise to mammals. In our laboratory, a research program is carried out of which the main objective is to study the forebrain organization of reptiles. Early studies were performed in the Tegu lizard, *Tupinambis nigropunctatus*, but currently the Tokay gekko, *Gekko gecko*, is used as the core species in cytoarchitectonical, hodological, and immunohistochemical studies.

Frontal sections at several levels through the forebrain of *Gekko* (Fig. 2) show that the pallium (roof) of the telencephalon contains cell plates that, according to the definition of Kuhlenbeck (1929), are properly referred to as cortical plates. In *Gekko*, as in reptiles in general, three separate plates can be identified. Following the nomenclature used by Kuhlenbeck, these plates are named, according to their relative mediolateral position, medial, dorsal, and lateral cortex. The medial cortex can be further subdivided into small-celled and large-celled portions. The basal part of the hemisphere is divided by the lateral ventricle into medial and lateral portions. The medial portion contains the septal area and the nucleus accumbens, whereas the lateral portion comprises the dorsal ventricular ridge (DVR), the striatum, and the amygdaloid complex.

In frontal sections through the gekkonid diencephalon (Fig. 3), eight nuclei are identified in the dorsal thalamus: nucleus dorsomedialis thalami, nucleus dorsolateralis thalami, nucleus rotundus, nucleus medialis thalami, corpus geniculatum laterale pars dorsalis, nucleus intercalatus, nucleus medialis posterior and nucleus posterocentralis. The nucleus rotundus, located at mid-diencephalic level, is the most conspicuous cell group. The dorsolateral thalamic nucleus consists of a dorsomedial small-celled part and a ventrolateral large-celled part. For further details of the anatomy of the forebrain in *Gekko*, reference is made to Smeets et al. (1986a).

The search for a putative reptilian homologue of the mammalian neocortex has focussed in particular on two telencephalic structures, viz. the dorsal ventricular ridge and the dorsal cortex (Nauta and Karten, 1970; Kirsche, 1972). Therefore, we will first discuss the structure and organization of the DVR. Next, we will deal with the structure and organization of the dorsal

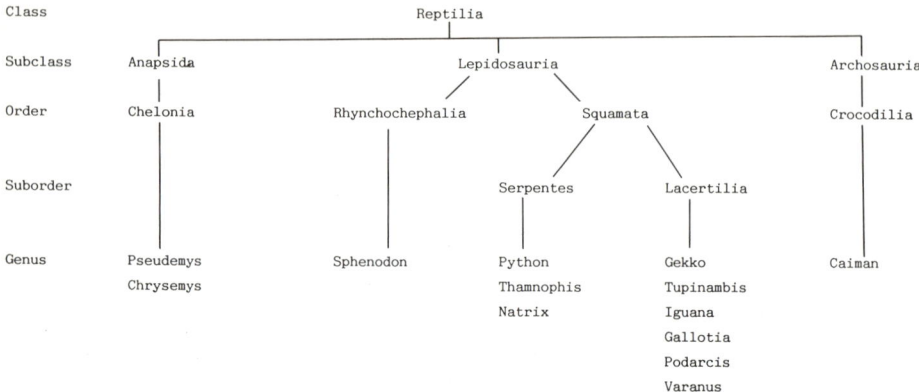

Figure 1. Classification of the living reptiles (based on Romer and Parsons, 1977).

cortex in relation to the other cortical regions. Finally, it will be discussed whether or not a reptilian homologue of the mammalian neocortex exists.

DORSAL VENTRICULAR RIDGE

Structure

The prominent eminence in the lateral ventricle of reptiles was for the first time described in 1861 by the Englishman John Hunter and in some publications is still called Hunter's eminence. Because in the early decades of this century most authorities on comparative neuroanatomy viewed Hunter's eminence as homologous to the basal ganglia of the mammalian brain, it became general use to apply the term "striatum" to this structure. As shown in Table 1, the striatum was subdivided into a dorsal part, the neostriatum, and a ventral part, the paleostriatum. A cell-free zone, which marks the course of the lateral striatal artery, constitutes the boundary between the two subdivisions. At caudal telencephalic levels, the neostriatum is still present, but the paleostriatum is replaced here by the nuclei of the amygdaloid complex, also named the archistriatum.

Not all students of the reptilian brain have used this nomenclature which originally was introduced by Ariëns Kappers in 1928 (see Ariëns Kappers et al., 1936). For instance, Elliot Smith (1919) called the part of the eminence located above the lateral striate artery the hypopallium, because he was of the opinion that it was due to an ingrowth of cells from the pallial part of the hemisphere (Table 1). In the present study, we will use the term "dorsal ventricular ridge" which was originally introduced by the American comparative neuroanatomist Judson B. Johnston (1915) as a purely descriptive label. The major rostral part of the ridge is by most authors called the anterior dorsal ventricular ridge (ADVR), whereas the small caudal part is termed the posterior dorsal ventricular ridge (PDVR). The PDVR is laterally and, more caudally, also ventrally bordered by the nuclei of the amygdaloid complex (Fig. 2). Nearly all experimental studies, carried out on the dorsal ventricular ridge, deal with the ADVR.

Thalamic afferents

Many investigations by means of anterograde and retrograde tracing techniques have attempted to determine the afferent connections of the ADVR. They have revealed that at least three major sensory modalities reach this brain structure: visual, auditory, and somatosensory.

Visual information in reptiles reaches the telencephalon via two pathways: a retino-tecto-thalamo-telencephalic pathway and a retino-thalamo-telencephalic pathway. The nucleus of the dorsal thalamus that relays visual information in the former pathway is nucleus rotundus. Bilateral projections from the midbrain tectum to the nucleus rotundus have been described for *Gekko* (Butler, 1978) as well as for other lizard species (Butler and Northcutt, 1971; Foster and Hall, 1975; Hoogland, 1982). Similar projections have been reported for snakes (Dacey and Ulinski,

Table 1

1. The three subdivision of Hunter's eminence

Ariëns Kappers ('28)	Elliot Smit ('19)	Curwen ('38)
Neostriatum	Hypopallium	Neostriatum
Paleostriatum	Paleostriatum	Paleostriatum
Archistriatum	Amygdaloid complex	Amygdala
Goldby and Gamble ('57)	Hoogland ('77)	Present Usage
Striatum, pars dorsalis	Dorsal striatum	Dorsal ventricular ridge
Striatum, pars ventralis	Ventral striatum	Striatum
Amygdaloid complex	Amygdaloid complex	Amygdaloid complex

2. The three longitudinal zones of the cortex

Edinger (1896)	Unger ('06)	Goldby ('34)
Cortex mediodorsalis	Ammonsrinde	Hippocampal cortex
Cortex dorsalis	Cortex dorsalis	Dorsal cortex
Cortex lateralis	Cortex lateralis	Pyriform cortex
Curwen ('37)		Present usage
Fascia dentata		Medial cortex
Ammonsformation and general cortex		Dorsal cortex
Pyriform cortex		Lateral cortex

1983), turtles (Hall and Ebner, 1970a; Rainey and Ulinski, 1982), and crocodilians (Braford, 1972). Projections originating from nucleus rotundus and reaching the lateral part of the ADVR have been observed in lizards (Lohman and Van Woerden-Verkley, 1978; Bruce and Butler, 1984b; Fig. 4), snakes (Ulinski, 1978). Turtles (Hall and Ebner, 1970b; Kosareva, 1974; Balaban and Ulinski, 1981), and crocodilians (Pritz, 1975).

The second pathway relays visual information usually by way of the corpus geniculatum laterale pars dorsalis (dorsal lateral geniculate nucleus). Its telencephalic target area shows considerable variations in the various reptiles studied. Since this projection has played a key role in the discussion on a putative reptilian homologue of the mammalian neocortex, it will be discussed in the final section of this chapter.

Auditory information generated in the internal ear reaches the cochlear nuclei of the brainstem (crocodilians: Leake, 1974; lizards: Miller, 1975; Foster and Hall, 1978; Barbas-Henry and Lohman, 1988; turtles: Miller and Kasahara, 1979; snakes: Miller, 1980). From here, fibers project to the central nucleus of the torus semicircularis in the midbrain (crocodilians: Manley, 1971; lizards: Foster and Hall, 1978). The auditory midbrain nucleus, in turn, projects bilaterally to the caudal thalamus, i.e. the nucleus medialis thalami, sometimes referred to as the nucleus reuniens (crocodilians: Pritz, 1974a; lizards: Foster and Hall, 1978; turtles: Balaban and Ulinski, 1981). The subsequent projection of the medial thalamic nucleus terminates in the medial part of the ADVR (crocodilians: Pritz, 1974b; lizards: Foster and Hall, 1978; Bruce and Butler, 1984b;

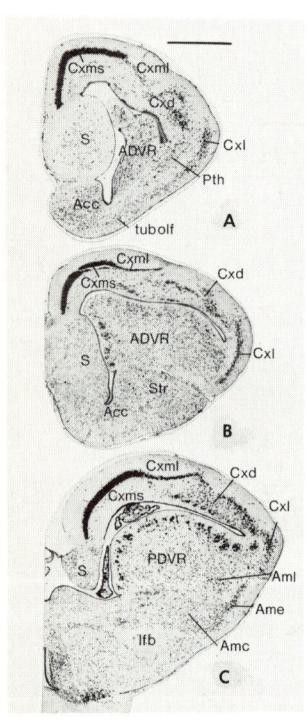

	abbreviations
Acc	nucleus accumbens
ADVR	anterior dorsal ventricular ridge
Amc	nucleus centralis amygdalae
Ame	nucleus externus amygdalae
Aml	nucleus lateralis amygdalae
Amm	nucleus marginalis amygdalae
Cgld	corpus geniculatum laterale, pars dorsalis
Cglv	corpus geniculatum laterale, pars ventralis
Cgp	corpus geniculatum pretectale
Cxd	cortex dorsalis
Cxl	cortex lateralis
Cxml	cortex medialis, large-celled part
Cxms	cortex medialis, small-celled part
Dll	nucleus dorsolateralis thalami, large-celled part
Dls	nucleus dorsolateralis thalami, small-celled part
Dm	nucleus dorsomedialis thalami
Ic	nucleus intercalatus
lfb	lateral forebrain bundle
Mp	nucleus medialis posterior
Mt	nucleus medialis thalami
Nsph	nucleus sphericus
Pc	nucleus posterocentralis
PDVR	posterior dorsal ventricular ridge
Pth	pallial thickening
Rot	nucleus rotundus
S	septum
Str	striatum
tect	tectum mesencephali
topt	tractus opticus
tub olf	tuberculum olfactorium

Figure 2. Photomicrographs of Nissl-stained transverse sections of the telencephalon of the lizard *Gekko gecko*. Bar = 1 mm.

Fig. 4). Thus, in general, it can be stated that in all reptiles studied auditory information reaches the medial part of the ADVR through a series of relays in the rhombencephalon, the mesencephalon, and the diencephalon.

Compared to the data available on the visual and auditory systems, those on the *somatosensory system* of reptiles are more limited. After spinal cord hemisections at upper cervical levels in *Gekko*, bilateral projections were observed to the nucleus medialis posterior and the nucleus posterocentralis (Bruce and Butler, 1984b), suggesting a nonfacial somatosensory input to these nuclei. The same study revealed that the medial posterior and posterocentral thalamic nuclei project to the ADVR (Fig.4). Spinothalamic projections have also been demonstrated in other lizards (Ebbesson, 1978; Hoogland, 1981), snakes (Ebbesson, 1969), turtles (Ebbesson,

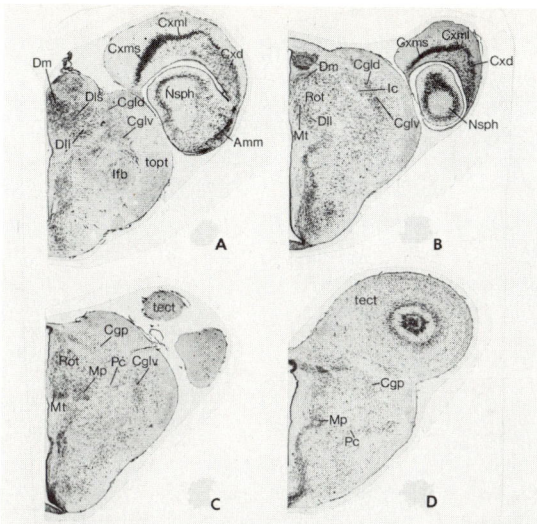

Figure 3. Photomicrographs of Nissl-stained transverse sections of the diencephalon and the pretectum of the lizard *Gekko gecko*. Bar = 1mm.

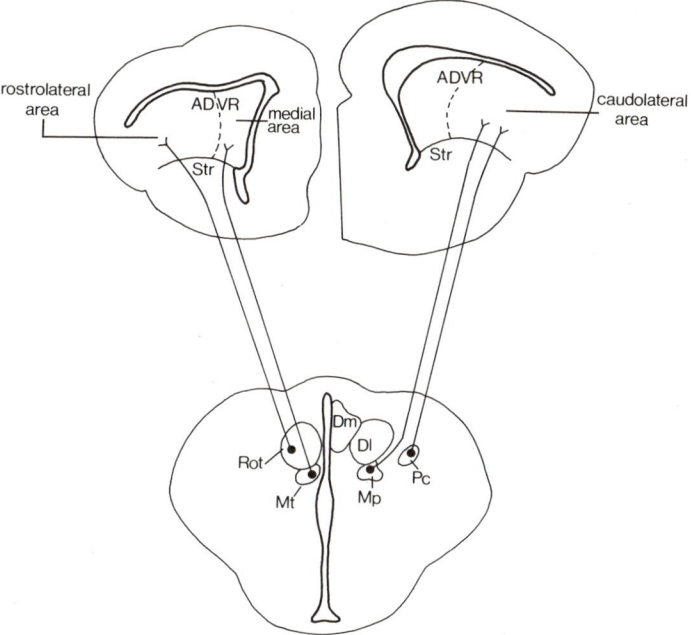

Figure 4. Diagram summarizing the ascending connections of visual, auditory, and somatosensory systems in the lizard *Gekko gecko* that synapse in the thalamus. Depicted on the left side are the visual and auditory thalamotelencephalic projections and on the right side the somatosensory projection.

1969; Künzle and Woodson, 1982), and crocodilians (Pritz and Northcutt, 1980; Ebbesson and Goodman, 1981), although there is some variation in the location and the terminology of the

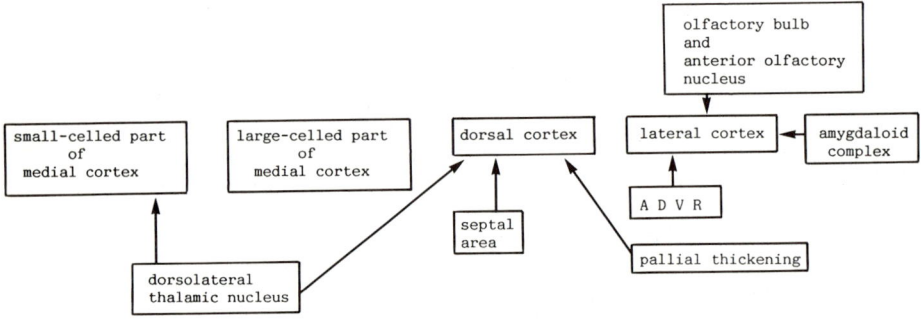

Figure 5. Diagram summarizing the afferent connections of the reptilian cortex.

spinothalamic targets among the various reptiles studied. The existence of a somatosensory thalamotelencephalic pathway was earlier identified in crocodilians (Pritz and Northcutt, 1980) and turtles (Balaban and Ulinski, 1981). A careful analysis of the available data reveals that the somatosensory input reaches the intermediate part of the ADVR in *Iguana, Pseudemys,* and *Caiman*. In *Gekko*, this input terminates in the caudolateral part or the ADVR (Bruce and Butler 1984b; Fig. 4). It thus appears that in all reptilian species studied a spino-thalamo-ADVR somatosensory pathway is present.

Apart from these segregated thalamic inputs, the ADVR is reached by diffuse projections from the dorsomedial thalamic nucleus (Lohman and Van Woerden-Verkley, 1978; Bruce and Butler, 1984b). Since this nucleus also projects to the striatum and the nucleus accumbens, it might be viewed as the reptilian counterpart of the midline paraventricular and parataenial nuclei in the brain of mammals (Gonzalez et al., 1990).

Brainstem afferents

By several authors (Lohman and Van Woerden-Verkley, 1978; Bruce and Butler, 1984b; ten Donkelaar and de Boer-van Huizen, 1988), it has been demonstrated in retrograde experiments that non-thalamic projections to the ADVR come from the brainstem. Immunohistochemical studies by means of antibodies against dopamine, noradrenaline, and serotonin in various reptilian species (Smeets et al., 1986b, 1987; Smeets, 1988b; Smeets and Steinbusch, 1988, 1989) have revealed substantial, diffusely organized, monoaminergic inputs to the ADVR. When the data of these immunohistochemical studies are combined with the results of the retrograde tracing studies, it appears that the ADVR receives input from serotonergic cells in the nucleus raphes superior, dopaminergic cells in the midbrain tegmentum, and noradrenergic cells in the locus coeruleus of the rhombencephalon. However, it should be emphasized that in the reptiles that were studied, the distribution of the monoaminergic fibers in the ADVR and the cortex is far from constant. A striking difference, for example, is the dopaminergic innervation. In various lizards and in snakes, there is a rather dense dopaminergic innervation that equals that of the striatum (Smeets, 1988a, see also Stoof et al., 1987). On the basis of this, the ADVR may be compared with the dorsal striatum of mammals which also receives a strong dopaminergic projection from the substantia nigra in the midbrain tegmentum. The ADVR of the turtle is, by contrast, practically devoid of dopaminergic fibers (Smeets et al., 1987). These observations argue against an approach in comparative neuroanatomy in which structures are compared solely on the basis of the presence or absence of a particular neurotransmitter system.

TELENCEPHALIC CORTICAL AREAS

Structure

As already mentioned, the pallial region in *Gekko* can be subdivided into three, longitudinally oriented cortical zones. From the beginning of this century, a variety of names has been given to these three zones, depending on the individual author's view on their mammalian homologies (Table 1). For example, Curwen (1937) called in the Tegu lizard the most medial zone fascia dentata, the middle zone Ammonsformation, and the lateral zone cortex piriformis. In the present study, the purely descriptive terminology of Kuhlenbeck (1929) is adopted and the cortical regions are, therefore, labeled medial, dorsal, and lateral cortex (Fig. 2). The three cortical areas are not continuous with each other: the large-celled part of the medial cortex overlaps the medial extreme of the dorsal cortex, whereas the lateral portion of the dorsal cortex is overlapped by the lateral cortex. Each cortex, except the lateral cortex, consists of five layers. From superficial to deep these are: (1) the molecular layer; (2) the cellular layer that contains most of the neurons; (3) the subcellular layer in which at most places only a few scattered cells are found; (4) the fiber layer; and (5) the periventricular neuronal layer. In the dorsal cortex, cortical, and subcortical layers cannot be recognized separately because of the dispersed position of their constituent neurons.

At rostral telencephalic levels, a cell mass can be recognized that has been termed the pallial thickening by Johnston (1915) . This cell mass is situated between the dorsal ventricular ridge and the lateral cortex (Fig. 2). By most authors the pallial thickening is considered to be a lateral extension of the dorsal cortex (Curwen, 1937; Northcutt, 1970; Davydova and Goncharova, 1979; Desan, 1984).

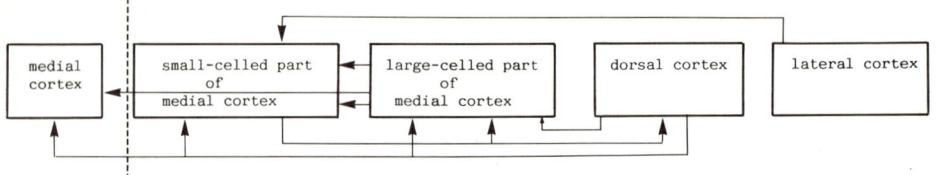

Figure 6. Diagram summarizing the intracortical and interhemispheric connections of the reptilian cortex.

Intracortical and interhemispheric connections

The intrinsic connections of the cerebral cortex have been studied mainly in lizards (Lohman and Mentink, 1972; Martinez-Garcia et al., 1986; Hoogland and Vermeulen-VanderZee, 1988; 1989) and are schematically depicted in Figure 6. The lateral cortex, which receives predominantly olfactory information, projects to the small-celled medial cortex. The fibers of this projection terminate in the outer one-third of the molecular layer. The small-celled medial cortex projects, in turn, to the large-celled medial cortex and the dorsal cortex. The large-celled medial cortex is the site which gives rise to the major intracortical and interhemispheric connections. Butler (1976) found that in *Gekko* also the dorsal cortex issues intracortical and interhemispheric fibers to the ipsilateral medial cortex and the contralateral medial and dorsal cortices. Based on recent experiments carried out in the same species, Hoogland and Vermeulen-VanderZee (1989) concluded that only the most medial part of the dorsal cortex is connected with the contralateral medial and dorsal cortices. All these projections terminate in a laminar pattern. Hoogland and Vermeulen-VanderZee (1989) further demonstrated that the dorsal cortex, at least in *Gekko*, is a very heterogeneous structure as far as its efferent connections are concerned. This observation is supported by immunohistochemical studies which have shown regional differences in the distribution of serotonin, dopamine. and noradrenaline (Smeets, 1988b).

Efferent connections of the reptilian cortex

From the schematic diagram of the efferent connections of the reptilian cortex (Fig. 7) it is clear that the dorsal cortex is the main origin of cortical efferents. In the Tegu lizard, it was observed that also the small-celled part of the medial cortex contributes to these connections (Lohman and Mentink, 1972). The cortical efferents are predominantly directed to the septal area and the medial and lateral regions of the hypothalamus and, taken together, resemble the medial corticohypothalamic tract described by Raisman et al. (1966) in the mammalian brain. Minor projections from the dorsal cortex reach the ADVR, the striatum, and the nucleus accumbens (Butler, 1976; Lohman and Van Woerden-Verkley, 1976; Hoogland and Vermeulen-VanderZee, 1989; Gonzalez et al., 1990). The lateral cortex has a sparse projection to the ADVR (Lohman and Van Woerden-Verkley, 1976).

IS THERE A REPTILIAN HOMOLOGUE OF THE MAMMALIAN NEOCORTEX?

In the previous sections, a survey was presented of the structure and the fiber connections of the ADVR and the cortical areas. Historically, it was first believed that the dorsal cortex is the sole candidate for a reptilian homologon of mammalian neocortex. More recent hodological studies have emphasized the fact that the ADVR receives visual, auditory, and somatosensory input from thalamic nuclei which makes it likely to view this structure as comparable to the sensory regions of the mammalian neocortex (Bruce and Butler, 1984b).

In this section the arguments in favor of the existence of a homologon in reptiles of the mammalian neocortex are discussed. Two lines of reasoning will be followed. One line concerns the embryology of the ADVR and the neocortex, the other deals with the organization of the reptilian ADVR and the dorsal cortex as compared to that of the mammalian neocortex.

Development of the ADVR

There are different opinions about the origin of the ADVR in reptiles. As mentioned before, Elliot Smith (1919) called this structure the "hypopallium", because he thought that it was due to an ingrowth of cells from the pallial part of the hemisphere. A pallial origin of the ADVR was

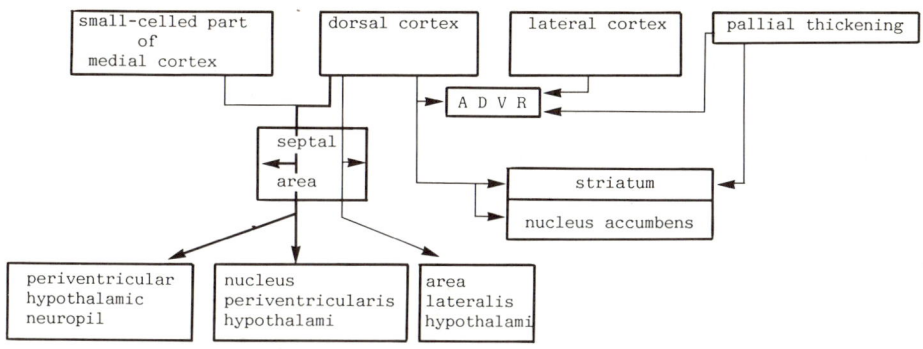

Figure 7. Diagram summarizing the major efferent connections of the reptilian cortex.

also postulated by Goldby (1934) and Northcutt (1970). By contrast, other authors (Holmgren, 1925; Källen, 1951) suggested that the ADVR is not derived from cells migrating from the pallial mantle or from the basal part of the hemisphere, but develops *in situ* from the epithelium that lines the telencephalic ventricle. Evidence that this is actually the case has recently been provided by Yanes et al. (1987) who studied the development of ADVR in embryos of the lizard *Gallotia galloti* (see Fig. 8). From their study it is evident that the ADVR develops from a zone in the dorsolateral wall of the ventricle underneath its lateral corner. Already before hatching a cell-free zone is present between the ADVR and the underlying striatum which develops from the ventral wall of the telencephalic ventricle (Yanes et al., 1989). Furthermore, there is no indication of the presence of a cortical plate in the developing ADVR. This is in contrast to the dorsal and lateral cortical zones above the telencephalic ventricle in which during development cortical plates can clearly be distinguished (Goffinet, 1983).

The part of the ventricular zone, that lines the lateral corner of the ventricle and is labeled SL in figure 8A-D, deserves special attention. It might well be, and this has already been suggested by Webster (1973), that the cells proliferating in this region in some reptiles, for instance in turtles, migrate into the pallial region and abut on or become part of the dorsal cortex, whereas in other reptiles, for instance in lizards, they are for the greater part taken up in the ADVR. When the developmental stages in the lizard, as shown in Figure 8C-D, are compared with the development of the forebrain of the rat at embryonic day 17 (Fig. 9), it is clear that the configuration of the lateral ventricle in the two species is strikingly similar and, moreover, that in the lizard the ADVR has developed from a similar part of the ventricular zone and occupies the same position with regard to the ventricular wall as does the dorsal striatum in the rat (Voorn et al., 1988). This suggests that the mammalian neocortex develops *de novo* in the pallial mantle and receives, instead of the dorsal striatum, the projections from the specific sensory thalamic nuclei. This notion is in contrast to the hypothesis, postulated in 1970 by Nauta and Karten, that the same neurons, that in reptiles compose the ADVR, in mammals come to occupy the pallial mantle and form a major proportion of the cell population of the neocortex.

The organization of the reptilian ADVR and dorsal cortex as compared to that of the neocortex of mammals

The pathway, that has played a key role in the debate whether part of the reptilian cortex is homologous to the mammalian neocortex, is the geniculotelencephalic projection. In mammals, the dorsal lateral geniculate nucleus, which is in receipt of direct retinal input, projects to the visual cortex. A retinal input to the dorsal lateral geniculate nucleus has been described in all reptiles studied (see e.g. Northcutt and Butler, 1974; Repérant et al. 1978; Halpern and Frumin, 1973; Bass and Northcutt, 1981). However, in the different species the ascending projection from this nucleus differs considerably with regard to its telencephalic target. In the turtle, Hall and Ebner (1970a) found that the dorsal lateral geniculate nucleus projects to the lateral part of the dorsal cortex, also called the general cortex, in the rostral pole of the telencephalic hemisphere. A more recent study by Heller and Ulinski (1987) in turtles of the genera *Pseudemys* and *Chrysemys* has revealed that fibers from the dorsal lateral geniculate nucleus terminate not only in the general cortex but also in the pallial thickening. According to the terminology of Desan (1988), these two fields together form area D2 of the dorsal cortex.

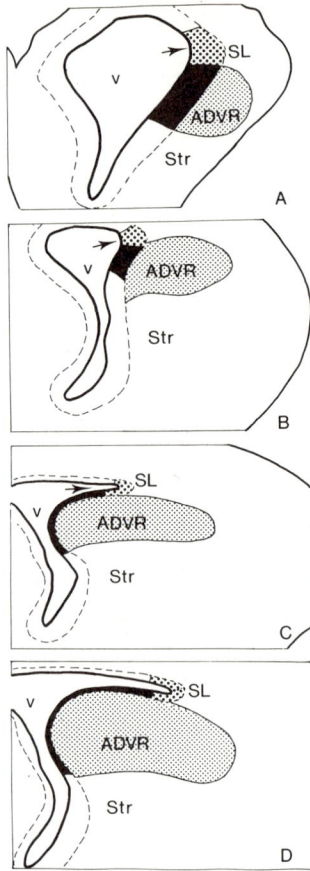

Figure 8. A-D Drawings of the topographical location of the anterior dorsal ventricular ridge (ADVR) in transverse sections at a rostral telencephalic level at stages E-32 (A), E-35 (B), E-37 (C), and hatching (D). SL, proliferative zone of sulcus lateralis (arrow); V, ventricle (Modified from Yanes et al., 1987).

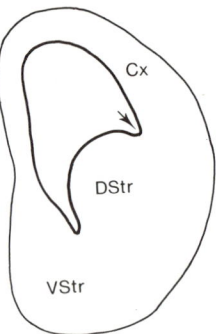

Figure 9. Drawing of the right hemisphere at rostral telencephalic level in the rat at embryonic day 17. Note the position of the cortex (CX) and the dorsal striatum (DStr) with regard to the sulcus (arrow). VStr, ventral striatum. (Modified after Voorn et al., 1988).

In snakes (Wang and Halpern, 1977) and in the Tegu lizard (Lohman and Van Woerden-Verkley 1978) it was observed that the projection from the dorsal lateral geniculate nucleus

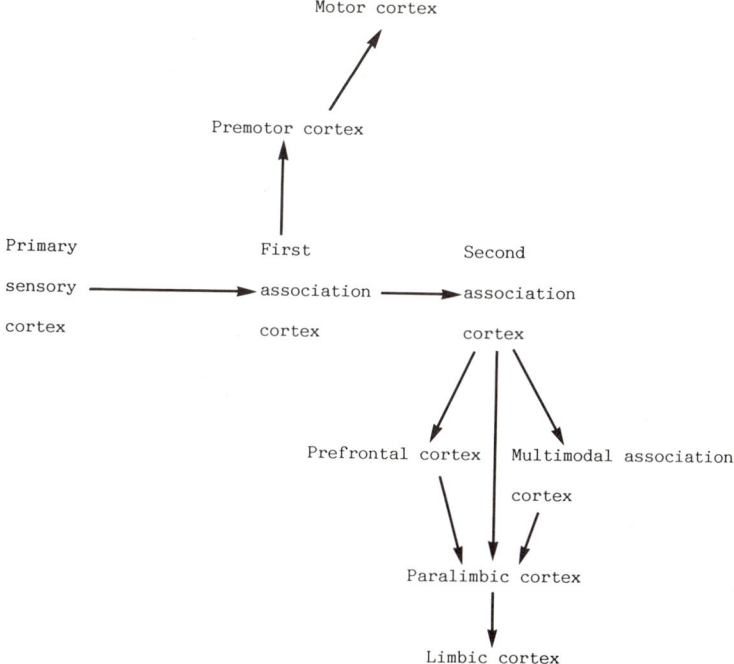

Figure 10. The processing of sensory information in the neocortex of mammals. Reciprocal connections are not indicated.

terminates not in the cortex but in the rostrolateral portion of the ADVR and perhaps also in the pallial thickening. In retrograde tracing experiments carried out in the lizards *Gekko* and *Iguana*, Bruce and Butler (1984a) found labeled cells in the retinorecipient zone of the dorsal thalamus after HRP injections in the pallial thickening. Surprisingly, these cells were not located in the dorsal lateral geniculate nucleus but in the nucleus intercalatus which also receives a direct retinal projection (Butler and Northcutt, 1978). It thus appears that in all reptiles studied there is a visual pathway from the retina via the dorsal thalamus to a dorsal and lateral region of the telencephalic hemisphere that includes either the lateral part of the dorsal cortex, the pallial thickening, or the rostrolateral portion of the ADVR.

It is, however, important to emphasize again that the inputs to the telencephalon from the other sensory nuclei of the dorsal thalamus all terminate in the ADVR. When the organization of the latter structure is compared to that of the neocortex of mammals, completely different pictures emerge. As can be concluded from the experiments by Pritz (1974b, 1975) and Pritz and Northcutt (1980) in *Caiman*, by Balaban and Ulinski (1981) in turtles and by Bruce and Butler (1984b) in *Iguana* and *Gekko* (Fig. 4), the visual (from nucleus rotundus), auditory and somatosensory modalities in the ADVR are completely segregated or show only a small overlap. There is, however, no indication of the existence of any integration between the three modalities, since the connections within the ADVR are predominantly organized radial to the telencephalic ventricle. For instance, following HRP injections in this structure retrogradely labeled cells are located mainly superficial and deep and in far smaller number medial and lateral to the injection sites (Bruce and Butler, 1984b). This concurs with the observation made by Ulinski (1978) in Golgi preparations that there is a predominance in the ADVR of neurons with radially oriented axons and dendrites. In recent experiments in *Gekko* by means of the PHAL technique it was observed that also the output of the ADVR is segregated: medial (auditory) zones in the ADVR project to medial zones in the striatum, central (somatosensory) zones in the ADVR project to central striatal zones, and lateral (visual) zones in the ADVR project to parts of the striatum not occupied by the other projections (Gonzalez et al., 1990). It thus appears that the conclusion about the functional significance of the ADVR by Ulinski (1983), namely that this structure only forms a linkage between sensory projections to the telencephalon and particular brainstem regions involved in the modulation of motor behavior, is quite justified.

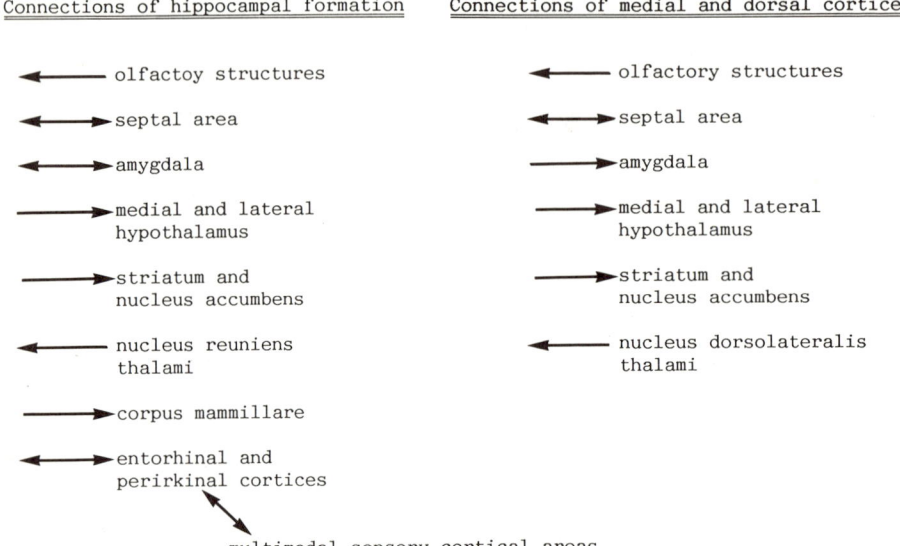

Figure 11. Comparison between the extrinsic connections of the hippocampal formation of mammals and those of the medial and dorsal cortices of reptiles. Efferent ------>: Afferent <---------.

By contrast, the sensory information, that reaches the neocortex of mammals, is processed through successive stages of unimodal and polymodal elaboration. This is shown in the diagram of Figure 10 which is based on the work by Pandya, Seltzer and Van Hoesen (for references see Pandya and Seltzer, 1982; Pandya and Yeterian, 1985). Of special importance is the convergence of the sensory information to the limbic cortex of which the core is formed by the hippocampal formation (Lopes da Silva et al., 1990).

By many authors the medial cortex and the dorsal cortex of the reptilian forebrain have been compared to the hippocampal formation of mammals (e.g. Unger, 1906; Goldby, 1934; Curwen, 1937; see Table 1). Arguments for this are the similarities in relative position, architecture and efferent connections. The latter are predominantly directed to the septal area and the hypothalamus (Fig. 11) . There are, however, also two great differences. The first is the absence in reptiles of a fornix system projecting to the mammillary bodies and the anterior thalamus (Stoll et al., 1983). The second difference is that connections between multimodal sensory association areas and the hippocampal formation, as found in mammals (Witter et al., 1989), are completely lacking in reptiles. No association areas can be distinguished in the ADVR and significant projections from the ADVR to the medial and dorsal cortices have never been observed. This leads, together with the other data on the connectional anatomy of the medial, dorsal and lateral zones presented earlier in this review, to the conclusion that the cortex of reptiles processes predominantly olfactory information and has its main output, outside the forebrain, to the hypothalamus.

CONCLUDING REMARKS

On the basis of the current data, reviewed in this chapter, the question "is there a reptilian homologue of the mammalian neocortex?" must be answered in the negative. Furthermore there is not enough evidence to view the ADVR as the structure where the sensory inputs to the central nervous system of reptiles are integrated. Alternative candidates to integrate the different types of sensory information are the tectum of the midbrain, the dorsal thalamus and, in the course of the descending central pathway, the striatum.

It is quite likely that to a certain degree integration takes place in the tectum, since in several studies (e.g. Ebbesson 1967, 1969; Welker et al., 1983; Dacey and Ulinski, 1986) it has been observed that projections to this structure come not only from the retina (visual input) but also from the torus semicircularis (auditory input), the trigeminal complex (facial somatosensory input), and the spinal cord (non-facial somatosensory input). For two reasons the dorsal thalamus appears not to be relevant for the process of sensory integration. In the first place, the various sensory inputs to this structure terminate in distinct areas or nuclei which do not show any significant overlap (Künzle and Snyder, 1983). And secondly, recent studies in *Caiman* by Pritz and Stritzel (1986, 1987, 1988) have provided evidence that the telencephalic-projecting cell groups lack intrinsic or local circuit neurons necessary for intrathalamic connections. The final candidate is the striatum. However, the pattern of its afferent connections from the ADVR does not suggest that there is a convergence of the three sensory modalities. The lack of information on the intrinsic organization of the striatum prevents us, as yet, to attribute an integrative function to this structure.

ACKNOWLEDGMENTS

The authors wish to thank Mr. D. de Jong for preparing the illustrations and Mrs. J. Hage for typing the manuscript.

REFERENCES

Ariëns Kappers, C.U., Huber, G.C., and Crosby, E.C. (1936) *The comparative anatomy of the nervous system of vertebrates, including man.* Vol. III, Hafner, New York.

Balaban, C.D., and Ulinski. P.S. (1981) Organization of thalamic afferents to anterior dorsal ventricular ridge in turtles. I. Projections of thalamic nuclei. *J. Comp. Neurol.*, 200: 95-130.

Barbas-Henry, H.A., and Lohman, A.H.M. (1988) The primary projections and efferent cells of the VIIIth cranial nerve in the monitor lizard, *Varanus exanthematicus. J. Comp. Neurol.*, 277: 234-249.

Bass, A.H., and Northcutt. R.G. (1981) Retinal recipient nuclei in the painted turtle, *Chrysemys picta*: An autoradiographic and HRP study. *J. Comp. Neurol.*, 199: 97-112.

Braford Jr., M.R. (1972) Ascending efferent tectal projections in the South American spectacled caiman. *Anat. Rec.*, 172: 275-276.

Bruce, L.L.. and Butler, A.B. (1984a) Telencephalic connections in lizards. I. Projections to cortex. *J. Comp. Neurol.*, 229: 585-601.

Bruce, L.L., and Butler, A.B. (1984b) Telencephalic connections in lizards. II. Projections to anterior dorsal ventricular ridge. *J. Comp. Neurol.*, 229: 602-615.

Butler, A.B. (1976) The telencephalon of the lizard *Gekko gecko* (Linnaeus): Some connections of the cortex and dorsal ventricular ridge. *Brain Behav. Evol.*, 13: 396-417.

Butler, A.B. (1978) Forebrain connections in lizards and the evolution of sensory systems., in: *Behavior and Neurology of Lizards*, N. Greenberg and P.D. MacLean, eds., NIMH, Rockville, Maryland, pp. 65-78.

Butler, A.B., and Northcutt, R.G. (1971) Ascending tectal efferent projections in the lizard *Iguana iguana. Brain Res.*, 35: 597-601.

Butler, A.B., and Northcutt, R.G. (1978) New thalamic visual nuclei in lizards. *Brain Res.*, 149: 469-472.

Curwen, A.O. (1937) The telencephalon of *Tupinambis nigropunctatus*. I. Medial and cortical areas. *J. Comp. Neurol.*, 66: 375-404.

Curwen, A.O. (1938) The telencephalon of *Tupinambis nigropunctatus*. II. Corpus striatum. *J. Comp. Neurol.*, 69: 229-247.

Dacey, D.M., and Ulinski, P.S. (1983) Nucleus rotundus in a snake, *Thamnophis sirtalis*: An analysis of a nonretinotopic projection. *J . Comp. Neurol.*, 216: 175-191.

Dacey, D.M., and Ulinski, P.S. (1986) Optic tectum of the Easter garter snake, *Thamnophis sirtalis*. V. Morphology of brainstem afferents and general discussion. *J. Comp. Neurol.*, 245: 423-453.

Davidova, T.V., and Goncharova (1979) Comparative characterization of the basic forebrain cortical zones in *Emys orbicularis* (Linnaeus) and *Testudo horsfieldi* (Gray). *J. Hirnforsch.*, 20: 245-262.

Desan, P.H. (1984) *The organization of the cerebral cortex of the pond turtle, Pseudemys scripta elegans*. Ph.D. Dissertation, Harvard University, Harvard.

Desan, P.H. (1988) Organization of the cerebral cortex in turtle, in: *The Forebrain or Reptiles. Current Concepts of Structure and Function*, W.K. Schwerdtfeger and W.J.A.J. Smeets. eds., Karger, Basel, pp. 1-11.

Ebbesson, S.O.E. (1967) Ascending axon degeneration following hemisection of the spinal cord in the tegu lizard, *Tupinambis nigropunctatus. Brain Res.*, 5: 178-206.
Ebbesson, S.O.E. (1969) Brain stem afferents from the spinal cord in a sample of reptilian and amphibian species. *Ann. N.Y. Acad. Sci.*, 167: 80-101.
Ebbesson, S.O.E. (1978) Somatosensory pathways in lizards. The identification of the medial lemniscus and related structures, in: *Behavior and Neurology of Lizards*, N. Greenberg and P.D. MacLean, eds., NIMH, Rockville, Maryland, pp. 91-104.
Ebbesson, S.O.E., and Goodman, D.C. (1981) Organization of ascending spinal projections in *Caiman crocodilus. Cell Tis. Res.*, 215: 383-396.
Edinger, L. (1896) Untersuchungen über die vergleichende Anatomie des Gehirns. III. Neue Studienüber das Vorderhirn der Reptilien, *Abh. Senckenb. Naturforsch. Gesch.*, 19: 313-388.
Elliot Smith, G. (1919) A preliminary note on the morphology of the corpus striatum and the origin of the neopallium. *J. Anat.*, (London), 53: 271-291.
Foster, R.E., and Hall, W.C. (1975) The connections and laminar organization of the optic tectum in a reptile (*Iguana*). *J. Comp. Neurol.*, 163:397-426.
Foster, R.E., and Hall, W.C. (1978) The organization of central auditory pathways in a reptile, *Iguana iguana. J. Comp. Neurol.*, 178: 783-832.
Gerfen, C.R., and Sawchenko, P.E. (1984) An anterograde neuroanatomical tracing method that shows the detailed morphology of neurons, their axons and terminals: immunohisto chemical localization of an axonally transported plant lectin, Phaseolus vulgaris Leucoagglutinin (PHAL). *Brain Res.*, 290: 219-238.
Goffinet, A.M. (1983) The embryonic development of the cortical plate in reptiles: A comparative analysis in *Emys orbicularis* and *Lacerta agilis. J. Comp. Neurol.*, 215: 437-452.
Goldby, F. (1934) The cerebral hemispheres of *Lacerta viridis. J. Anat.*, (London), 68: 157-215.
Goldby, F., and Gamble, H.J. (1957) The reptilian cerebral hemispheres. *Biol. Rev.*, 32: 383-420.
Gonzalez, A., Russchen, F.T., and Lohman. A.H.M. (1990) Afferent connections of the striatum and the nucleus accumbens in the lizard Gekko gecko. *Brain, Behav. Evol.* in press.
Hall, W.C., and Ebner, F.F. (1970a) Parallels in the visual afferent projections of the thalamus in the hedgehog (*Paraechinus hypomelas*) and the turtle (*Pseudemys scripta*). *Brain Behav. Evol.* 3: 135-154.
Hall, W.C., and Ebner, F.F. (1970b) Thalamotelencephalic projections in the turtle (Pseudemys scripta). *J. Comp. Neurol.*, 140: 101-122.
Halpern, M., and Frumin, N. (1973) Retinal projections in a snake. *Thamnophis sirtalis. J. Morphol.*, 141: 359-382.
Heller, S.B., and Ulinski, P.S. (1987) Morphology of geniculocortical axons in turtles of the genera *Pseudemys* and *Chrysemys. Anat. Embryol.*, 175: 505-515.
Herkenham, M. (1978) The connections of the nucleus reuniens thalami: Evidence of a direct thalamo-hippocampal pathway in the rat. *J. Comp. Neurol.*, 177: 589-610.
Holmgren, N. (1925) Points of view concerning forebrain morphology in higher vertebrates. *Acta Zool.*, 6: 415-477.
Hoogland, P.V. (1977) Efferent connections of the striatum in *Tupinambis nigropunctatus. J. Morphol.*, 152: 229-246.
Hoogland, P.V. (1981) Spinothalamic projections in a lizard, *Varanus exanthematicus*: An HRP study. *J.Comp. Neurol.*, 198: 7-12.
Hoogland, P.V. (1982) Brainstem afferents to the thalamus in a lizard, *Varanus Exanthmaticus, J. Comp. Neurol.*, 210: 152-162.
Hoogland, P.V., and Vermeulen-VanderZee. E. (1988) Intrinsic and extrinsic connections of the cerebral cortex of lizards, in: *The Forebrain of Reptiles. Current Concepts of Structure and Function*, W.K. Schwerdtfeger and W.J.A.J. Smeets, eds., Karger, Basel, pp. 20-29.
Hoogland, P.V., and Vermeulen-VanderZee, E. (1989) Efferent connections of the dorsal cortex of the lizard *Gekko gecko* studied with Phaseolus vulgaris-leucoagglutinin. *J. Comp. Neurol.*, 285: 289-303.
Hunter. J. (1861) *Essays and Observations on Natural History, Anatomy, Physiology, Psychology and Geology*, J. VanVoorst. London.
Johnston, J.B. (1915) The cell masses in the forebrain of the turtle *Cistudo carolina. J. Comp. Neurol.*, 25: 393-468.
Källén, B. (1951) On the ontogeny of the reptilian forebrain. Nuclear structures and ventricular sulci. *J. Comp. Neurol.*, 95: 307-347.
Kirsche, W. (1972) Die Entwicklung des Telencephalons derReptilien und deren Beziehung zur Hirn-Bauplanlehre. *Nova Acta Leopoldina* 204: 1-78.

Kosareva, A.A. (1974) Afferent and efferent links of nucleus rotundus of the tortoise *Emys orbicularis. Evol. Biochem. Physiol.*, 10: 354-360.

Kuhlenbeck, H. (1929) Die Grundbestandteile des Endhirns im Lichte der Bauplanlehre. *Anat. Anz.*, 67: 1-51.

Künzle, H. and Woodson, W. (1982) Meso-diencephalic and other target regions of ascending spinal projections in the turtle, *Pseudemus scripta elegans. J. Comp. Neurol.*, 212: 349-364.

Künzle, H., and Snyder, H. (1983) Do retinal and spinal projections overlap within the turtle thalamus? *Neuroscience*, 10: 161-168.

Leake, P.A. (1974) Central projections of the statoacoustic nerve in *Caiman crocodilus. Brain Behav. Evol.*, 10: 170-196.

Lohman, A.H.M., and Mentink, G.M. (1972) Some cortical connections of the tegu lizard (*Tupinambis teguixin*). *Brain Res.*, 45: 25-344.

Lohman, A.H.M., and Van Woerden-Verkley, I. (1976) Further studies on the cortical connections of the tegu lizard. *Brain Res.*, 103: 9-28.

Lohman, A.H.M., and Van Woerden-Verkley, I. (1978) Ascending connections to the forebrain in the tegu lizard. *J. Comp. Neurol.*, 182: 555-594.

Lohman, A.H.M., Hoogland, P.V., and Witjes, J.G.M. (1988) Projections from the main and accessory olfactory bulbs to the amygdaloid complex in the lizard *Gekko gecko*, in: *The Forebrain of Reptiles. Current Concepts of Structure and Function* W.K. Schwerdtfeger and W.J.A.J. Smeets, eds., Karger, Basel, pp. 41-49.

Lopes da Silva, F.H., Witter, M.P., Boeijinga, P.H.., and Lohman, A.H.M. (1990) Anatomical organization and physiology of the limbic cortex, *Physiological reviews*, 70: 453-511.

Manley, J.A. (1971) Single unit studies in the midbrain auditory area of Caiman. Z. Vergl., *Physiol.*, 71: 255-261.

Martinez-Garcia, F.M., Amiguet, M., Olucha, F., and Lopez-Garcia, C. (1986) Connections of the lateral cortex in the lizard *Podarcis hispanica. Neurosci.Lett.*, 63: 39-44.

Miller, M.R. (1975) The cochlear nuclei of lizards. *J. Comp. Neurol.* 159: 375-406.

Miller, M.R. (1980) The cochlear nuclei of snakes. *J. Comp. Neurol.*, 192: 717-736.

Miller, M.R., and Kasahara, M. (1979) The cochlear nuclei of some turtles. *J. Comp. Neurol.*, 185: 221-236.

Nauta, W.J.H., and Karten, H.J. (1970) A general profile of the vertebrate brain, with side lights on the ancestry of cerebral cortex, in: *The Neurosciences Second Study Program*, F.O. Schmitt, ed., Rockefeller Univ., New York, pp. 7-26.

Northcutt, R.G. (1970) The telencephalon of the western painted turtle (*Chrysemys picta belli*). III, Biol. Mon., No. 43, Univ. Illinois Press, Urbana.

Northcutt, R.G., and Butler, A.B. (1974) Evolution of reptilian visual systems: Retinal projections in a nocturnal lizard, *Gekko gecko* (Linnaeus). *J. Comp. Neurol.,* 157: 453-465.

Pandya, D.N., and Seltzer, B. (1982) Association areas of the cerebral cortex, *TINS*, 5: 386-392.

Pandya, D.N., and Yeterian, E.H. (1985) Architecture and connections of cortical association areas, in: *Cerebral cortex, Association and Auditory cortices*, vol. 4, pp.3-61, A. Peters and E.G. Jones, eds., Plenum Press, New York and London.

Pritz, M.B. (1974a) Ascending connections of a midbrain auditory area in a crocodile. *Caiman crocodilus. J. Comp. Neurol.*, 153: 179-19 8.

Pritz, M.B. (1974b) Ascending connections of a thalamic auditory area in a crocodile, *Caiman crocodilus. J. Comp. Neurol.*, 153: 199-21 4.

Pritz, M.B. (1975) Anatomical identification of a telencephalic visual area in crocodiles: ascending connections of nucleus rotundus in *Caiman crocodilus. J. Comp. Neurol.*, 164: 323-338.

Pritz, M.B., and Northcutt, R.G. (1980) Anatomical evidence for an ascending somatosensory pathway to the telencephalon in crocodiles, *Caiman crocodilus. Exp. Brain Res.*, 40: 342-345.

Pritz, M.B., and Stritzel. M.E. (1986) Percentage of relay and intrinsic neurons in two sensory thalamic nuclei projecting to the noncortical telencephalon in reptiles, *Caiman crocodilus. Brain Res.*, 376: 169-174.

Pritz, M.B., and Stritzel. M.E. (1987) Percentage of intrinsic and relay cells in a thalamic nucleus projecting to general cortex in reptiles, *Caiman crocodilus. Brain Res.*, 409: 146-150.

Pritz, M.B., and Stritzel, M.E. (1988) Thalamic nuclei that project to reptilian telencephalon lack GABA and GAD immunoreactive neurons and puncta. *Brain Res.*, 457: 154-159.

Rainey, W.T., and Ulinski, P.S. (1982) Organization of nucleus rotundus, a tectofugal thalamic nucleus in turtles.II. The tectorotundal projection. *J. Comp. Neurol.*, 209: 187-207.

Raisman, G., Cowan, W.M., and Powell. T.P.S. (1966) An experimental analysis of the efferent projection of the hippocampus. *Brain*, 89: 83-108.

Repérant, J., Rio, J.-P., Miceli, D., and Lemire, M. (1978) A radioautographic study of retinal projections in type I and type II lizards. *Brain Res.*, 142: 401-411.

Romer, A.S., and Parsons, T.S. (1977) *The vertebrate body*,VIII, Saunders, Philadelphia, pp. 624.

Smeets, W.J.A.J. (1988a) Distribution of dopamine immunoreactivity in the forebrain and midbrain of the snake *Python regius*, A study with antibodies against dopamine. *J. Comp. Neurol.*, 271: 115-129.

Smeets, W.J.A.J. (1988b) The monoaminergic systems of reptiles investigated with specific antibodies against serotonin, dopamine, and noradrenaline, in: *The Forebrain of Reptiles, Current Concepts of Structure and Function*, W.K. Schwerdtfeger and W.J.A.J. Smeets, eds., Karger, Basel, pp.97-109.

Smeets, W.J.A.J., Hoogland, P.V., and Lohman, A.H.M. (1986a) A forebrain atlas of the lizard *Gekko gecko*. *J. Comp.Neurol.*, 254: 1-19.

Smeets, W.J.A.J., Hoogland, P.V., and Voorn. P. (1986b) The distribution of dopamine immunoreactivity in the forebrain and the midbrain of the lizard *Gekko gecko*: An immunohistochemical study with antibodies against dopamine. *J. Comp. Neurol.*, 253: 46-60.

Smeets, W.J.A.J., Jonker, A.J., and Hoogland, P.V. (1987) Distribution of dopamine in the forebrain and midbrain of the red-eared turtle, *Pseudemys scripta elegans*, reinvestigated using antibodies against dopamine. *Brain, Behav. Evol.*, 30: 121-142.

Smeets, W.J.A.J., and Steinbusch, H.W.M. (1988) Distribution of serotonin immunoreactivity in the forebrain and midbrain of the lizard *Gekko gecko*. *J. Comp.Neurol.*, 271: 419-434.

Smeets, W.J.A.J., and Steinbusch, H.W.M. (1989) Distribution of noradrenaline immunoreactivity in the forebrain and midbrain of the lizard *Gekko gecko*. *J. Comp.Neurol.*, 285: 453-466.

Stoll, C.J., Smeets, W.J.A.J., and Hoogland, P.V. (1983) The fornix in reptiles. *Neurosci. Lett.* [Suppl.], 14: 359.

Stoof, J.C., Russchen, F.T., Verheijden, P.F.H.M., and Hoogland, P.V.J.M. (1987) A comparative study of the dopamine-acetylcholine interaction in the telencephalic structures of the rat and of a reptile, the lizard *Gekko gecko*. *Brain Res.*, 404: 273-281.

ten Donkelaar, H.J., and De Boer-Van Huizen, R. (1988) Brainstem afferents to the anterior dorsal ventricular ridge in a lizard (*Varanus exanthematicus*). *Anat. Embryol.*, 177: 465-475.

Ulinski, P.S. (1978) Organization of anterior dorsal ventricular ridge in snakes, *J. Comp. Neurol.*, 178: 411-450.

Ulinski, P.S. (1983) The dorsal ventricular ridge. A treatise on forebrain organization in reptiles and birds, in: *Wiley Series on Neurobiology*, R.G. Northcutt, ed., Wiley, New York.

Unger, L. (1906) Untersuchungen über die Morphologie und Faserung des Reptiliengehirns. *Anat. Hefte*, 31: 271-341.

Voorn, P., Kalsbeek, A., Jorritsma-Byham, B., and Groenewegen, H.J. (1988) The pre- and postnatal development of the dopaminergic cell groups in the ventral mesencephalon and the dopaminergic innervation of the striatum of the rat. *Neuroscience*, 25: 857-887.

Wang, R.T., and Halpern, M. (1977) Afferent and efferent connections of thalamic nuclei of the visual system of garter snakes. *Anat. Rec.*, 187: 741-742.

Webster, K.E. (1973) Thalamus and basal ganglia in reptiles and birds. *Symp. Zool. Soc. Lond*, 3: 169-203.

Welker, E., Hoogland, P.V., and Lohman. A.H.M. (1983) Tectal connections in *Python reticulatus*. *J. Comp. Neurol.*, 220: 347-354.

Witter, M.P., Groenewegen, H.J., Lopes da Silva, F.H., and Lohman, A.H.M. (1989) Functional organization of the extrinsic and intrinsic circuitry of the parahippocampal region. *Progress in Neurobiol.*, 33: 161-253.

Yanes, C.M., Perez-Batista, M.A., Martin-Trujillo, J.M., Monzon, M., and Marrero, A. (1987) Anterior dorsal ventricular ridge in the lizard: embryonic development. *J. Morphol.*, 194: 55-64.

Yanes, C., Perez-Batista, M.A., Martin-Trujillo, J.M., Monzon, M., and Rodriguez, A. (1989) Development of the ventral striatum in the lizard *Gallotia galloti*. *J. Anat.*, 164: 93-100.

MULTIPLE ORIGINS OF NEOCORTEX: CONTRIBUTIONS OF THE DORSAL VENTRICULAR RIDGE

Toru Shimizu and Harvey J. Karten
Department of Neurosciences, M-008
University of California, San Diego
La Jolla, CA 92093

INTRODUCTION

The uniqueness of mammalian neocortex may ultimately only be clarified with improved understanding of the evolutionary origins of cortical structure and cortical functions. Comparative studies of the organization of the nonmammalian and mammalian telencephalon may provide valuable clues for understanding the evolution of neocortex. In the nonmammalian telencephalon, there are neuronal populations which correspond to cell groups in the neocortex of mammals in terms of connections, single unit-responses, and functions. Some of these populations lying within the dorsal ventricular ridge, however, are organized in a non-laminar, rather than laminar fashion. These observations suggest that the emergence of basic "cortical" circuit and laminar organization are distinct evolutionary events that can be differentiated and studied independently in order to understand each of their respective contributions to the cognitive functions of the neocortex. Moreover, in contrast to an argument that many cortical visual areas are derived from a single area by gene duplication (Allman, 1977, in press), the origins of neocortex can be separable into at least the precursors of non-laminar and laminar regions, and thus multiple evolutionary origins of neocortex are proposed.

The neocortex of mammals plays a major role in sensory, motor, and cognitive performance. The evolutionary origins of the cortex, however, is still only poorly understood. Comparative studies of the organization of the nonmammalian and mammalian telencephalon may provide valuable clues for understanding the evolution of neocortex. The avian forebrain contains at least two major regions that appear directly comparable to the mammalian neocortex: the dorsal ventricular ridge (DVR) and the wulst. These two regions appear to have different evolutionary histories and morphologies, yet functionally, both closely resemble the mammalian neocortex (Karten & Shimizu, 89).

In the first part of this paper, we argue that the neuronal populations in the DVR which are organized in non-laminar fashion correspond to cell groups that exhibit laminar organization in the mammalian neocortex. We then discuss visual information processing in laminated and non-laminated neural circuits, and suggest that lamination is not essential for such processing. Finally, we discuss the significance of the non-laminar cortical equivalents to hypotheses concerning the evolutionary origin of the neocortex. Specifically, we present an hypothesis that neocortex arose from multiple precursors.

TWO VISUAL AREAS OF THE AVIAN TELENCEPHALON

Wulst and dorsal ventricular ridge

This paper deals mainly with the visual system of vertebrates, specifically birds and mammals, because the issues mentioned here can be best addressed in the framework of a well defined sensory system. The basic hypothesis of this paper, however, may be applied to other sensory (e.g., auditory and somatosensory), motor, and cognitive systems, as well as to other classes of vertebrates.

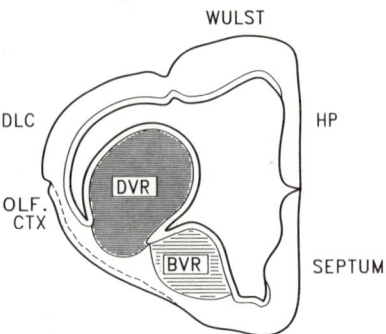

Figure 1. A schematic representation of a transverse section of the avian telencephalon. BVR = basal ventricular ridge; DLC = dorsolateral corticoid region; DVR = dorsal ventricular ridge; OLF. CTX = olfactory cortex; HP = presumptive hippocampus. (Adapted from Karten & Shimizu, 1989).

There are two major visual areas in nonmammalian telencephalon. One, located in the dorsal portion of the telencephalon, is organized in a laminar structure. The other, found in the lateral portion of the telencephalon, is organized as in a non-laminar fashion. Each area receives a major visual input (Benowitz & Karten, 1976; Karten et al., 1973; Ulinski, 1983). The dorsal portion receives the thalamofugal pathway, which corresponds to the mammalian geniculostriate pathway, whereas the lateral portion receives the tectofugal pathway, which corresponds to the mammalian colliculo-thalamo-extrastriate pathway.

In birds and reptiles, the recipient target of the thalamofugal input from the dorsal division of the visual thalamic nuclei (the equivalent of the dorsal portion of the lateral geniculate nucleus; dLGN) is *the laminated region of granule cells in the dorsal telencephalon*. In birds, these granule cells are located in a gross morphological subdivision known as the wulst; the region receiving thalamic projections has been designated as the visual wulst (Fig. 1). The visual wulst has a number of characteristics in common with the striate cortex of mammals: 1) a well defined cytoarchitectonic laminar organization (Karten et al., 1973; Pettigrew, 1979; Reiner & Karten, 1983), 2) afferent and efferent connections with similar subcortical structures (see Bagnoli & Burkhalter, 1983; Karten et al., 1973; Miceli et al., 1987), 3) retinotopic organization of the receptive fields (Denton, 1981; Miceli et al., 1979; Perisic et al., 1971; Pettigrew & Konishi, 1976; Wilson, 1980), and 4) chemical organization (Bagnoli & Casini, 1985; Yamada & Sano, 1985). Thus, although there are some structural differences (Karten et al., 1973), the wulst is considered to be the avian equivalent of the striate cortex (Bagnoli et al., 1982; Karten et al., 1973; Pettigrew, 1979).

In contrast to the laminated wulst of the thalamofugal pathway, the recipient target of the tectofugal pathway in birds and reptiles is found in *the non-laminated region of the lateral telencephalon* within DVR. As shown in figure 1, the DVR is located in the lateral wall of the telencephalon, adjacent to the olfactory cortex, the basal (or ventral) ventricular ridge (BVR, the equivalent of the mammalian basal ganglia), and a dorsolateral corticoid region (DLC). The DVR is composed of a complex multinucleate formation, rather than the familiar laminar organization of the neocortex. Within the avian DVR, some of the larger nuclear aggregates include the ectostriatum, neostriatum (including an auditory area, Field L), archistriatum, and hyperstriatum ventrale. Each aggregate has its own *relatively* uniform cytology and histochemical characteristics. *Aggregates of the DVR appear to be equivalent to individual layers of the neocortex,* as described in the detailed outline of the tectofugal visual pathway and the DVR in the next section.

Dorsal ventricular ridge and the mammalian neocortex

In birds, the region receiving the visual thalamic projections is designated as the ectostriatum of the DVR. The relationship of the ectostriatum to the extrastriate cortex is not immediately obvious because of non-laminar organization of the DVR, in contrast to that between the visual wulst and the striate cortex. The ectostriatum may be most directly compared with the specific populations of thalamic recipient neurons within layer IV of the extrastriate cortex, rather than with the complete cortex. The DVR is comparable to the extrastriate cortex in four important characteristics: 1) equivalent afferent pathways (Benowitz & Karten, 1976; Karten & Hodos, 1970); 2) similar unit-responses (Frost & DiFranco, 1976; Revzin, 1979; Kimberly et al., 1971); 3) multiple thalamic recipient areas in the DVR; and 4) descending pathways from the DVR.

1) Afferent pathways

The tectofugal pathway passes from the optic tectum to the nucleus rotundus thalami (Rt), the largest nucleus in the avian thalamus. The source of the input to Rt is the cells of the stratum griseum centrale (SGC) of the optic tectum (Benowitz & Karten, 1976). On the basis of topological position, morphology, and other afferent and efferent connections of this cell group, the SGC of birds most closely matches the laminae II of the superior colliculus of cats (Kanaseki & Sprague, 1976). Fibers from Rt pass to the ectostriatum. The ectostriatum consists of two subdivisions: a core region (Ec) and a surrounding periectostriatal belt (Ep). Only the Ec receives the direct thalamic projections (Karten & Hodos, 1970). In the tectofugal system of mammals, the superficial portion of the superior colliculus projects upon the lateral posterior nucleus (LP) and/or pulvinar (PUL) of the thalamus (see Diamond, 1973). These tectal recipient cell groups of the thalamus are in turn the source of projections to several extrastriate cortical areas, terminating mainly in layer IV (Benevento & Ebner, 1970; Benevento & Rezak, 1976; Glendenning et al., 1975; Harting et al., 1973; Kaas et al., 1972; Lin, 1977; Robson & Hall, 1977). On the basis of the connectivity, therefore, Rt corresponds to a portion of the LP or PUL, and neurons of the Ec appear to be equivalent to some neurons of layer IV of the extrastriate cortex in receipt of thalamic efferents.

2) Unit-responses

The avian optic tectum, Rt, and the ectostriatum exhibit electrophysiological properties markedly similar to those of the corresponding constituents of units in the tectofugal pathway in mammals. These units are characterized by wide receptive fields and motion sensitivity with some directional selectivity (Frost & DiFranco, 1976; Kimberly et al., 1971; Revzin, 1979).

3) Multiple recipient areas

One major difference between the ectostriatum and the extrastriate cortex is that the Ec appears to be a single thalamic recipient area, in contrast to multiple thalamic recipient areas of the extrastriate cortex. In cats, the targets of the tectofugal pathway include areas 19, 21a, the lateral and medial divisions of the Clare-Bishop complex (CBl, CBm), and the dorsal and ventral lateral suprasylvian visual areas (DLS, VLS) (Abramson & Chalupa, 1985). In macaque monkeys, the colliculo-thalamo-recipient extrastriate regions contain visual cortical areas designated as V2, V3, V4, and the middle temporal area (MT), which are uniquely selective to specific types of visual stimuli. In birds, the Ec is uniform in cytology and in histochemical sensitivity, although the Ec can be further subdivided into at least five regions, each of which receives a projection from a distinct subdivision of Rt (Benowitz & Karten, 1976). Subdivisions of Rt demonstrate different types of selective responses to specific visual stimuli, such as direction and size (Revzin, 1979). Thus, although seemingly only a single target of the tectofugal pathway, the Ec contains several distinct functional representations similar to mammalian extrastriate regions (see Fig. 2).

4) Descending pathway

Another question raised about the comparison is that the ectostriatum itself does not have the extratelencephalic projections characteristic of the extrastriate cortex (Miceli et al., 1987). Even in mammalian cortex, however, the populations of cortical cells that receive major thalamic outputs are segregated from those cortical cells that send their axons to extratelencephalic targets. Layer IV, which contains the main thalamorecipient population, does not contribute to an extrinsic descending extratelencephalic output. Thus, in the DVR, where the organization of cell populations is clustered rather than laminar, the Ec should be considered only as *a component of* the extrastriate cortex. The Ec may correspond to *only one of the laminae* of the neocortex, specifically the populations of neurons of layer IV in receipt of the thalamic outputs (Karten & Revzin, 1966; Karten & Hodos, 1970; Benowitz & Karten, 1976).

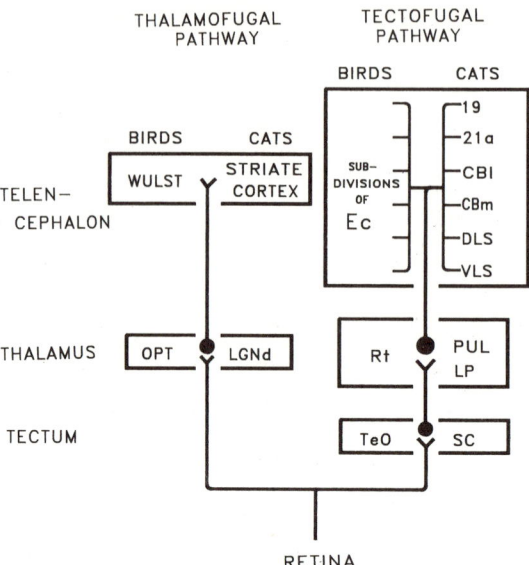

Figure 2. A schematic representation of the thalamofugal and tectofugal pathways in birds and mammals. Boxes represent main structures in the two pathways. The nomenclatures on the left side in boxes are used for birds, while those on the right side in boxes are used for cats (Abramson & Chalupa, 1985). 19 and 21a = areas 19, 21a; CBl and CBm = lateral and medial divisions of the Clare-Bishop complex; DLS and VLS = dorsal and ventral lateral suprasylvian visual areas; Ec = the core region of ectostriatum; LGNd = lateral geniculate nucleus *pars dorsalis*; LP = lateral posterior nucleus; PUL = pulvinar; OPT = opticus principalis thalami; Rt = nucleus rotundus; SC = superior colliculus; TeO = optic tectum. (Adapted from Karten & Shimizu, 1989).

The DVR also appears to contain cell groups which correspond to cell populations of layers other than layer IV. Ritchie and Cohen (1979) studied the intrinsic telencephalic conectionssubsequent to the Ec and identified the existence of neural circuits similar to the interlaminar connections of the neocortex. Thus, the Ec projects on its surrounding region (including the Ep, and the neostriatum intermedium laterale, NIL), and this region, in turn, projects on a restricted area, the archistriatum intermedium (Ai). All of these structures, (i.e., Ec, Ep, NIL, and Ai) are components of the DVR. Furthermore, Brecha, Hunt and Karten (1976) demonstrated that Ai projects on deeper laminae of the ipsilateral optic tectum. Thus, although the ectostriatum itself does not have extratelencephalic outputs, information from Ec is conveyed to the brainstem via other aggregates in the DVR. These sequential projections through the individual cell groups correspond to the interlaminar connections of neocortex; i.e., from layer IV (Ec) to layers II and III (Ep and NIL), then to layers V and VI (Ai), and then to the deep layers of the optic tectum (Karten, 1969; Nauta & Karten, 1970). Indeed, this loop of projections, that begins at and comes back to the optic tectum, is comparable to the projection pattern of regions in the extrastriate cortex, such as V2, V3, V4, MT, and the inferotemporal cortex of macaque monkeys. Analogous proposals would link other aggregates in the DVR and auditory and trigeminal cortices. The cortical circuitry in mammals and the quivalent circuitry in birds is schematized in figure 3.

FUNCTIONS OF LAMINATION

Functional capacities associated with neural circuits: nonmammals

Highly developed visual proficiency is accomplished in nonmammals, particularly in birds. In addition to a highly developed optic tectum, non-laminar neural circuits in the DVR play a major role in this ability.

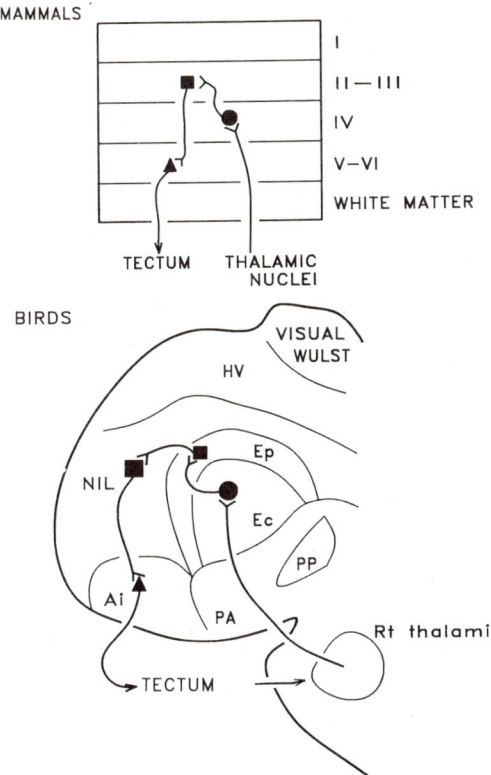

Figure 3. A comparison of visual-information circuits between the mammalian and avian brains. Circles indicate thalamic recipient neurons; squares indicate intrinsic neurons; triangles indicate neurons sending efferents to the optic tectum. Roman numerals indicate cortical layers. Ai = archistriatum intermedium; Ep = periectostriatal belt; HV = hyperstriatum ventrale; NIL = neostriatal intermedium laterale; PA = paleostriatum augmentum (avian equivalent of caudate-putamen); PP = paleostriatum primitivum (equivalent of globus pallidus). (Adapted from Karten & Shimizu, 1989).

An extensive series of studies of visual and cognitive abilities of birds (particularly pigeons) has been carried out in recent years (e.g., Blough, 1982; Herrnstein et al., 1976; Vaughn & Greene, 1984). Although the neural mechanism of visual perception and memory is still unknown, the tectofugal system and especially the DVR appear to play important roles in both visual sensory and visual cognitive behavior in birds. Lesions of any structures in the pathway (i.e., the optic tectum, Rt, or the ectostriatum) severely disrupted performance in most visual tasks, such as spatial orientation (Jarvis, 1974), and pattern and spatial-frequency discriminations (Hodos, 1976; Hodos et al., 1986; Kertzman & Hodos, 1988). Prolonged retraining was required to again achieve preoperative performance levels on various tasks. Such profound visual impairments on similar tasks, including pattern discriminations, were also found after lesions of the extrastriate cortices in mammals (Sprague et al., 1981; Antonini et al., 1985). Thus, damage to non-laminar aggregates of the DVR produce behavioral deficits directly comparable to those found after lesions in the corresponding laminar structures of the mammalian neocortex.

Functional capacities associated with neural circuits: mammals.

As the visual and cognitive abilities of birds indicate, many functional characteristics, which were often associated with lamination of the neocortex, may be attributable to neural circuits without lamination. Even in the mammalian laminated configuration, analogous observations can be found. For example, cells in each cortical layer have specific afferent and efferent projections, but lamination itself does not appear to play a critical role in specifying connections. Caviness (1977) and Dräger (1981) demonstrated that cells in the cortex of the reeler mutant mouse maintain normal connections, despite the gross malformation of cortical lamination and disposition of laminae. Furthermore, according to Dräger (1981), electrophysiological response properties of the reeler cortex are remarkably normal.

Similarly, a recent study by Murphy and Silito (1987) indicates that an interlaminar organization is less critical for visual-pattern analysis than is generally believed (reviewed by Martin, 1988). Since Hubel and Wiesel (1965) presented their model of a mechanism of pattern analysis, "feature detection" has been considered the major function of microcircuitry of the neocortex. The formulation suggested that the neural representations of visual stimuli are decomposed and analyzed into simple features, such as "orientation" or "edges." For instance, "edges" are considered to be detected by cortical "end-inhibition" cells that respond to a short stimulus, but decrease responses when the stimulus is lengthened (Bolz & Gilbert, 1986). However, Murphy and Silito (1987) reported that cells showing end-inhibition also exist in the lateral geniculate. Although removal of the cortex dramatically decreases end-inhibition in these thalamic cells, the sources of the cortico-geniculate projection are the layer VI pyramidal cells whose receptive fields show considerable length summation, not end-inhibition. Thus, Murphy and Silito argued that end-inhibition is not simply a product of the interlaminar circuits within the cortex, but rather involves cortico-geniculate loops. If such "complex" perception (edge detection) requires long cortico-geniculate loops, lamination in the neocortex and short connections between laminae are not critical for pattern analysis.

Functional capacities associated with lamination

If sophisticated visual performance can be accomplished in non-laminar neural circuits, what are the benefits of lamination in the neocortical system? The discussion above does not argue against the fact that lamination is of significance to many sensory, motor, and cognitive functions. Highly developed laminar organization is found in various brain structures other than the telencephalon, such as the retina, cerebellum, optic tectum of birds, vagal lobes of fishes, and the olfactory bulbs of rodents. These instances strongly suggest that there is some correlation between specialized functions and the degree of development of lamination. There are, however, two attributes of cortical structures, i.e. microcircuitry and lamination, that may contribute to the functional capacities of cortex. Each may have a separate evolutionary history, and each may provide unique contributions to cortical function: 1) the phylogenetically conservative pattern of the neural circuits may be responsible for the cognition and behavior observed equally in mammals and nonmammals; 2) the laminar organization more common amongst mammals may confer upon them their presumed distinctive functional capacity. Previous behavioral studies, however, have failed to distinguish the role of each of these separate aspects of cortical organization, and have generally assumed the two properties of connections and lamination were inseparable.

One of the obvious advantages of laminated organizations is the efficiency of interconnections among proximal cell groups. The interconnections between the clustered entities of non-laminated organization are presumably restricted to axo-somatic, -dendritic and -axonic synapses. The lack of laminar apposition of these clustered populations limits the extensive interlaminar dendro-dendritic and other short connections which appear to be fundamental characteristics of the neocortex. The laminar organization may be advantageous particularly when information transfers across layers perpendicularly to the surface (i.e., a columnar fashion).

Another possible advantage of lamination concerns the efficiency of control of the subjacent neural circuits. In the neocortex, each layered neuronal array represents a topographical map. Thus, the close apposition and connections between arrayed layers may provide a basis for potentially fast point-to-point operational circuitry. Such an arrangement may provide a substrate for more highly ordered control of output neurons by discrete populations of sensory neurons. Specifically, in the instance of the neocortex, this permits more direct control of output neurons of layers V and VI in relationship to sensory inputs to layer IV.

MULTIPLE ORIGINS OF THE CORTEX

Allman's hypothesis: Single origin of the neocortex

Two completely different types of configurations of neurons are present in the nonmammalian telencephalon (the laminated wulst and the non-laminated DVR). This suggests that the multiple visual cortical areas might be derived from two different origins: the precursors of striate cortex and the separate precursors of extrastriate cortices. This view contrasts with the recent hypothesis that suggests that many of the visual cortical areas have a single origin. The formulation of Allman and Kaas proposes that the multiple visual cortices might be a product of expansion of a simple primal visual area (Allman, 1977, in press; Allman & Kaas, 1974). They suggest that the striate cortex is the most ancient area of visual cortical regions, and that genetic mutation might have resulted in the replication of columnar cortical "modules" which were then "invaded" by the striate cortex to form extrastriate cortices. However, there are several obstacles to this hypothesis:

a) The tectofugal pathway apparently antedates the thalamofugal system, implying that the striate cortex is a more recent "phylogenetic development" than the extrastriate systems. Despite the widespread emphasis upon the thalamofugal or geniculostriate pathway, the tectofugal pathway represents the major visual ascending system in the majority of non-mammalian vertebrates, and is even well developed in primates.

b) There is no evidence to support the hypothesis that the cortex evolves in modular increments composed of the necessary constituent neurons of each layer that form putative discrete columns.

As described in the following sections, several lines of more recent evidence additionally suggest that the cortex may have arisen independently from two different loci and by two separate mechanisms of cell migration.

1) The birth dates of the two areas in mammals are ontogenetically different. Lateral cortical areas are elaborated earlier than the more medial regions.

2) Functions of the two areas appear to be more independent than was considered earlier. The projections from striate cortex are not necessarily critical for the function of extrastriate cortex.

3) In contrast to the current view that all neocortical areas develop by a uniform sequence of structural elaboration and radial migration, different mechanisms may contribute to the formation of lateral versus medial cortex.

Time of birth of the two cortical areas

In birds and reptiles, the embryonic pallium differentiates into two components; the dorsal portion and the lateral portion. The dorsal portion develops dorsal to the ventricle, forming the wulst, whereas the intraventricular protrusion from the lateral wall of the telencephalon develops as an intraventricular profusion, or a DVR. In birds, the DVR develops ontogenetically earlier than the wulst (Tsai et al., 1981). The neurons of Ec or Ep develop about the 4th-5th day of embryogenesis, whereas those of the wulst develops mainly after the 6th day.

In contrast to reptiles and birds, mammals do not have two clearly segregated components which contribute to the development of the embryonic pallium. Several studies, however, suggest that the birth date of the dorsal portion of the mammalian cortex is different from that of the lateral portion (Hicks & d'Amato, 1968; Smart, 1973; Smart & Smart, 1977; 1982). For example, Smart and Smart (1977) injected labelled thymidine in mice at 11 or 12 days postconception. These authors showed that cells born about the 11th day were located primarily in the lateral portion of neocortex. On the other hand, cells born about the 12th day or after tended to lie more dorsally. Although the sequence of development of the two regions need not be rigidly synchronized in relationship to their phylogeny, the earlier ontogenetic development of the extrastriate regions does not sit well with Allman's proposal that the extrastriate cortex is an evolutionary "elaboration" of the striate cortex.

Functional segregation of the two cortical areas

In mammals, the striate cortex contains units that respond to simple properties of the visual stimuli, such as line elements. On the other hand, units in the extrastriate cortex respond to more complex features, such as motion or line orientation. One notion of visual analysis presumes a hierarchy of *serial* information processing within the cortical structures (e.g., Hubel & Wiesel, 1965). In this formulation, simple features are first detected and abstracted in the striate cortex, and then the processed information is sent to the extrastriate cortex where the input is analyzed further for more complex feature recognition. In contrast to this view, comparative data strongly support another view that the extrastriate system, as well as the striate system, is one of the major structures involved in the *parallel* processing of visual information (Diamond & Hall, 1969; Stone, 1983). The functional relationship between the striate and extrastriate systems appears to be parallel rather than hierarchical, and complementary rather than dependent. Such a view is supported by many lesion studies using birds and mammals.

In nonmammals, the tectofugal ("extrastriate") system is the major structure involved in processing of visual information. The tectofugal pathway in most nonmammals is anatomically a massive projection, and its destruction causes far more substantial impairments of visual abilities than do lesions in the thalamofugal system. For example, in contrast to severe deficits after tectofugal lesions (see the earlier section), birds seem to show little or no changes after lesions of the thalamofugal route; i.e., the wulst and the nucleus opticus principalis thalami (OPT; the avian equivalent of dLGN). Pattern and intensity discrimination was disrupted only slightly and temporarily following the destruction of the OPT or the wulst (Hodos et al., 1973). Psychophysical methods revealed almost no effect on intensity-difference and visual acuity thresholds after the lesions in the thalamofugal route (Hodos & Bonbright, 1974; Hodos et al., 1984; Macko & Hodos, 1984; Pasternak & Hodos, 1977).

In mammals, Sprague and co-workers (Antonini et al., 1985; Berkley & Sprague, 1979; Sprague et al., 1981) described studies of effects on visual ability after lesions in various regions of the visual cortex and showed functional segregation of the striate and extrastriate cortex. Cats with damage in areas 17 and 18 showed perfect post-operative discrimination of coarse patterns, although visual acuity had deteriorated. On the other hand, the animals with damage in the suprasylvian gyrus showed significant impairment in visual discrimination but no changes in visual acuity. Taken together, these studies suggest that the striate system is essential for the analysis of local details, while the extrastriate system is important for perception of pattern based on its general configuration.

In primates, although lesions of the striate cortex result in severe impairment of visual functions, there are some studies showing substantial residual visual ability after such lesions. For example, Rodman and co-workers (1989) studied residual activities of the middle temporal area (MT) in the extrastriate cortex. The striate cortex and area V2 (which is completely dependent on striate cortex for visual responsiveness) send projections to MT. Single units within intact MT respond selectively to direction of motion in the visual field. After lesions of the striate cortex, the majority of MT neurons are still responsive to visual stimuli (Rodman et al., 1989). Furthermore, with residual activity weaker than before, the majority of MT neurons nevertheless maintained their direction selectivity and binocularity. This study indicates that the visual responsiveness of units in MT is not dependent exclusively on the striate input. The authors suggest that this residual ability is attributable to the other MT afferent input from the lateral and inferior portions of the pulvinar. Similar "motion detection" units are found in both avian tectum (Frost, 1982) and mammalian tectum (reviewed in Chalupa, 1984).

These studies of birds and mammals are reminiscent of a series of reports of "blindsight" in humans by Weiskrantz and co-workers (reviewed in Weiskrantz, 1986). In their reports, Weiskrantz pointed out that many patients with damage to striate cortex still respond to stimuli presented within the region of their scotomata. These people, however, are not aware of their "sight." We would suggest that "blindsight" is a manifestation of the tectofugal pathway's role that only emerges in humans after lesions of the striate cortex.

These data strongly indicate that the striate system cannot be regarded as a sensory-filter mechanism located before the extrastriate system. The extrastriate system receives input from both the striate system and the tectofugal projections and may function to varying degree without input from striate cortex. Although a hierarchical processing may be specialized, particularly in primates, in most mammals and nonmammals, the extrastriate visual areas are less dependent on striate cortex than previously appreciated.

Origins of the two cortical areas

The extrastriate and other cortices are located in the lateral portion of the neocortex, and the DVR is an intraventricular protrusion. Some developmental studies, however, suggest that cell

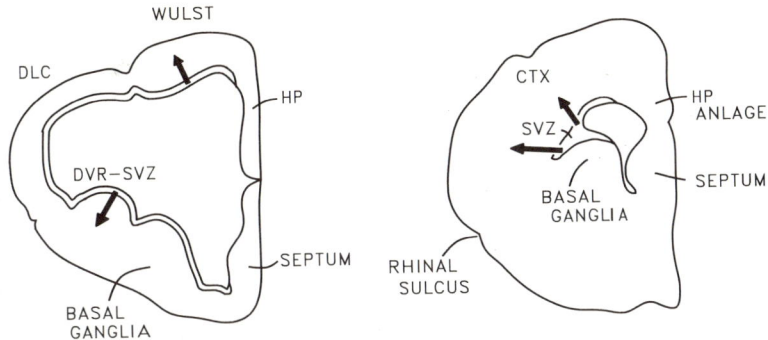

Figure 4. A schematic representation of the transverse section of avian or reptilian (left) and mammalian (right) telencephalon. Arrows indicate possible courses of migration of cells. SVZ = subventricular zone. (Adapted from Karten & Shimizu, 1989).

groups in the lateral pallium originally migrate from the mammalian equivalent of the DVR during embryogenesis. Stensaas and Gilson (1972) investigated the nature of a proliferative zone in rat and rabbit embryos. According to these authors, there is a subventricular zone (SVZ) lying at the dorsolateral angle of the basal ganglia and the ventricle (see Fig. 4). Similar observations were reported in the past, but received little attention (Källén, 1962). The SVZ contains mitotic cells that appear to consist of primitive astrocytes and migrating neuronal-like cells. These cells migrate away from the SVZ to enter the lateral pallial regions. The SVZ is prominent in embryogenesis and disappears shortly after birth. These observations suggest that ancestral reptiles and proto-mammals possessed discrete neural populations that were the precursors of cells of the SVZ of mammals and the DVR of nonmammalian amniotes. In birds and reptiles, the cells of the DVR form an intraventricular protrusion and a multinuclear complex. In mammals, on the other hand, the cells of the SVZ migrate into the lateral pallium to form the individual layers of selected cortical regions of the lateral neocortex, including the extrastriate, inferotemporal, and auditory cortices. This proposed mechanism of cortical development is consistent with that described by Berry and Rogers (1965) and Rakic (1972; 1988), who suggest that neurons migrate along a radial pathway consisting of single or multiple glial fibers. Their studies, however, were based mainly on analysis of the striate cortex. A similar pattern of radial migration presumably occurs in the dorsal pallial region that forms the wulst. Cortical maldevelopment, such as deficient laminar development and ectopias, might be related to the unusual migration course from the SVZ, instead of the uniform radial migration of cells of the pallial ependyma.

CONCLUSION

In addition to the DVR and dorsal pallium (wulst), there are several regions in the nonmammalian telencephalon of which little information is available to date on their ontogeny and functions, although their connections with other areas suggest that these areas may also be parts of the neocortex. Therefore, the evolutionary origins of the neocortex may be not only the wulst and the DVR, but multiple areas. The studies of nonlaminated cortical structures in nonmammals are increasingly important and necessary to obtain further knowledge of the origins and functions of the neocortex. The uniqueness of mammalian neocortex may ultimately only be clarified with improved understanding of the evolutionary origins of cortical structure and cortical functions.

ACKNOWLEDGEMENTS

This work was supported by NINCDS Grant NS24560, NEI Grant EY-06890-05, and ONR Contract N00014-88-K-0504 (to H. J. K.). The authors are grateful to Drs. William Hodos, Luiz R. G. Britto, James M. Sprague, Arnold Towe, and Jon H. Kaas, who made helpful comments on the early version of this manuscript.

REFERENCES

Abramson, B.P. and Chalupa, L.M. (1985) The laminar distribution of cortical connections with the tecto- and cortico-recipient zones in the cat's lateral posterior nucleus. *Neuroscience, 15:* 81-95.

Allman, J. (1977) Evolution of the visual system in the early primates. In J. M Sprague and A. N. Epstein (eds.), *Progress in Psychobiology and Physiological Psychology, 7,* (pp. 1-53). New York: Academic Press.

Allman, J. (in press) Evolution of neocortex. In A. Peters and E.G. Jones (eds.), *Cerebral cortex:* Vol.8. New York: Plenum Press.

Allman, J. and Kaas, J.H. (1974) A crescent-shaped cortical visual area surrounding the middle temporal area (MT) in the owl monkey *(Aotus trivirgatus). Brain research, 81:* 199-213.

Antonini, A., Berlucchi, G. and Sprague, J.M. (1985) Cortical systems for visual pattern discrimination in the cat as analyzed with the lesion method. In C. Chagas, R. Gattass, and C. Gross (eds.), *Pattern recognition mechanisms* (pp.153-164). New York: Springer-Verlag.

Bagnoli, P. & Burkhalter, A. (1983) Organization of the afferent projections to the Wulst in the pigeon. *Journal of Comparative Neurology, 214:* 103-113.

Bagnoli, P. and Casini, G. (1985) Regional distribution of catecholaminergic terminals in the pigeon visual system. *Brain Research, 247:* 277-286.

Bagnoli, P., Francesconi, W. and Magni, F. (1982) Visual wulst-optic tectum relationships in birds: A comparison with the mammalian corticotectal system. *Archives Italiennes de Biologie, 120:* 212-235.

Benevento, L.A. and Ebner, F.F. (1970) Pretectal, tectal, retinal and cortical projections to thalamic nuclei of the Virginia opossum in stereotaxic coordinates. *Brain Research, 18:* 171-175.

Benevento, L.A. and Rezak, M. (1976) The cortical projections of the inferior pulvinar and adjacent lateral pulvinar in the rhesus monkey *(Macaca mulatta):* An autoradiographic study. *Brain Research, 108:* 1-24.

Benowitz, L.I. and Karten, H.J. (1976) Organization of the tectofugal visual pathway in the pigeon: A retrograde transport study. *Journal of Comparative Neurology, 167:* 503-520.

Berkley, M.A. and Sprague, J.M. (1979) Striate cortex and visual acuity functions in the cat. *Journal of Comparative Neurology, 187:* 679-702.

Berry, M. and Rogers, A.W. (1965) The migration of neuroblasts in the developing cerebral cortex. *Journal of Anatomy, 99:* 691-709.

Blough, D. (1982). Pigeon perception of letters of the alphabet. *Science, 218:* 397-398.

Bolz, J. and Gilbert, C.D. (1986) Generation of end-inhibition in the visual cortex via interlaminar connections. *Nature, 320:* 362-365.

Brecha, N., Hunt, S.P. and Karten, H.J. (1976) Relations between the optic tectum and basal ganglia in the pigeon. *Society for Neuroscience Abstract, 1:* 95.

Caviness, V.S. Jr. (1977) Reeler mutant mouse: A genetic experiment in developing mammalian cortex. In W. M. Cowan & J. A. Ferrendelli, (eds.), *Approaches to the cell biology of neurons* (pp. 27-46). Bethesda, Maryland: the Society for Neuroscience.

Chalupa, L.M. (1984) Visual physiology of the mammalian superior colliculus. In H. Vanegas (ed.), *Comparative neurology of the optic tectum* (pp. 775-818). New York: Plenum Press.

Denton, C.J. (1981) Topography of the hyperstriatal visual projection area in the young domestic chicken, *Experimental Neurology, 74:* 482-498.

Diamond, I.T. (1973) The evolution of the tectal-pulvinar system in mammals: Structural and behavioral studies of the visual system. *Symposia of the Zoological Society, London, 33:* 205-233.

Diamond, I.T. and Hall, W. C. (1969) Evolution of neocortex. *Science, 164:* 251-262.

Dräger, U.C. (1981) Observations on the organization of the visual cortex in the reeler mouse. *Journal of Comparative Neurology, 201:* 555-570.

Frost, B.J. (1982) Mechanisms for discriminating objects motion from the self-induced motion in the pigeon. In. D.J. Ingle, M.A. Goodale, and R.J.W. Mansfield (eds.), *Analysis of visual behavior,* 177-196. Cambridge, MA: MIT Press.

Frost, B.J. and DiFranco, D.E. (1976) Motion specific units in the pigeon optic tectum. *Vision Research, 16:* 1229-1234.

Glendenning, K.K., Hall, J.A., Diamond, I.T., and Hall, W.C. (1975) The pulvinar nucleus of *Galago senegalensis. Journal of Comparative Neurology, 161:* 419-458.

Harting, J.K., Diamond, I.T., and Hall, W.C. (1973) Anterograde degeneration study of the cortical projections of the lateral geniculate and pulvinar nuclei in the tree shrew *(Tupaia glis). Journal of Comparative Neurology, 150:* 393-440.

Herrnstein, R.J., Lovelnad, D., and Cable, C. (1976) Natural concepts in pigeons. *Journal of Experimental Psychology: Animal Behavior Processes, 2:* 285-311.

Hicks, S.P. and D'Amato, C.J. (1968) Cell migrations to the isocortex in the rat. *Anatomical Record, 160,* 619-634.

Hodos, W. (1976) Vision and the visual system: A bird's eye-view. In J.M. Sprague and A.M. Epstein (eds.), *Progress in Psychobiology and Physiological Psychology, 6,* (pp.29-62). New York: Academic Press.

Hodos, W. and Bonbright, J.C., Jr. (1974) Intensity difference thresholds in pigeons after lesions of the tectofugal and thalamofugal visual pathway. *Journal of Comparative and Physiological Psychology, 87:* 1013-1031.

Hodos, W., Karten, H.J., and Bonbright, J.C. Jr. (1973) Visual intensity and pattern discrimination after lesions of the thalamofugal visual pathway in pigeons. *Journal of Comparative Neurology, 148:* 447-468.

Hodos, W., Macko, K.A., and Bessette, B.B. (1984) Near field acuity after visual system lesions in pigeons. II: Telencephalon. *Behavioural Brain Research, 13:* 15-30.

Hodos, W., Weiss, S.R.B. and Bessette, B.B. (1986) Size-threshold changes after lesions of the visual telencephalon in pigeons. *Behavioural Brain Research, 21:* 203-214.

Hubel, D.H. and Wiesel, T.N. (1965) Receptive fields and functional architecture in two non-striate visual areas (18 and 19) of the cat. *Journal of Neurophysiology, 28:* 229-289.

Jarvis, C.D. (1974) Visual discrimination and spatial localization deficits after lesions of the tectofugal pathway in pigeons. *Brain, Behavior and Evolution, 9:* 195-228.

Kaas, J.H., Hall, W.C., and Diamond, I.T. (1972) Visual cortex of the grey squirrel *(Sciurus carolinensis)*: Architectonic subdivisions and connections from the visual thalamus. *Journal of Comparative Neurology, 145:* 273-306.

Källén, B. (1962) Embryogenesis of brain nuclei in the chick telencephalon. *Ergebnisse der Anatomie und Entwicklungsgeschichte, 36:* 62-82.

Kanaseki, T. and Sprague, J.M. (1976) Anatomical organization of pretectal nuclei and tectal laminae in the cat. *Journal of Comaparative Neurology, 158:* 319-338.

Karten, H.J. (1969) The organization of the avian telencephalon and some speculations on the phylogeny of the amniote telencephalon. *Annals New York Academy of Sciences, 167*: 146-179.

Karten, H.J. and Hodos, W. (1970) Telencephalic projections of the nucleus rotundus in the pigeon (*Columba livia*). *Journal of Comparative Neurology, 140:* 35-52.

Karten, H.J. and Revzin, A.M. (1966) The afferent connections in the nucleus rotundus in the pigeon. *Brain Research, 2:* 368-377.

Karten, H.J. and Shimizu, T. (1989) The origins of neocortex: Connections and lamination as distinct events in evolution. *Journal of Cognitive Neuroscience,* 1: 291-301..

Karten, H.J., Hodos, W., Nauta, W.J.H., and Revzin, A.M. (1973) Neural connections of the "visual wulst" of the avian telencephalon. Experimental studies in the pigeon (*Columba livia*) and owl *(Speotyto cunicularia)*. *Journal of Comparative Neurology, 150:* 253-277.

Kertzman, C. and Hodos, W. (1988) Size-difference thresholds after lesions of thalamic visual nuclei in pigeons. *Visual Neuroscience, 1:* 83-92.

Kimberly, R.P., Holden, A.L., and Bamborough, P. (1971) Response characteristics of pigeon forebrain cells to visual stimulation. *Vision Research, 11:* 475-478.

Lin, C.S. (1977) Subdivisions of the inferior pulvinar in the owl monkey. *Anatomical Record, 187:* 637-638.

Macko, K.A. and Hodos, W. (1984) Near field acuity after visual system lesions in pigeons. I: Thalamus. *Behavioural Brain Research, 13:* 1-14.

Martin, K.A.C. (1988) The lateral geniculate nucleus strikes back. *Trends in Neurosciences, 11:* 192-194.

Miceli, D., Gioanni, H., Repérant, J. and Peyrichoux, J. (1979) The avian visual wulst: I. An anatomical study of afferent and efferent pathways. II An electrophysiological study of the functional properties of single neurons. In A.M. Granda and J.H. Maxwell (eds.), *Neural mechanisms of behavior in the pigeon* (pp. 223-254). New York: Plenum press.

Miceli, D., Repérant, J., Villalobos, J. and Dionne, L. (1987) Extratelencephalic projections of the avian Wulst. A quantitative autoradiographic study in the pigeon *Columba livia. Journal für Hirnforschung, 28:* 45-57.

Murphy, P.C. and Silito, A.M. (1987) Corticofugal feedback influences the generation of length tuning in the visual pathway. *Nature, 392:* 727-729.

Nauta, W.J.H. and Karten, H.J. (1970) A general profile of the vertebrate brain with sidelights on the ancestry of the cerebral cortex. In F. O. Schmitt (ed.), *The Neurosciences: Second Study Program.* New York: Rockefeller Press.

Pasternak, T. and Hodos, W. (1977) Intensity difference thresholds after lesions of the Visual Wulst in pigeons. *Journal of Comparative and Physiological Psychology, 91:* 485-497.

Perisic, M., Mihailovic, J., and Cuénod, M. (1971) Electrophysiology of the contralateral and ipsilateral projections to the Wulst in pigeon (*Columba livia*). *International Journal of Neuroscience, 2:* 7-14.

Pettigrew, J.D. (1979) Binocular visual processing in the owl's telencephalon. *Proceedings of the Royal Society (London), Series B, 204:* 435-454.

Pettigrew, J.D. and Konishi, M. (1976) Neurons selective to for orientation and binocular disparity in the visual Wulst of the barn owl (*Tyto alba*), *Science, 193*:: 675-678.

Rakic, P. (1972) Mode of cell migration to the superficial layers of fetal monkey neocortex. *Journal of Comparative Neurology, 145:* 61-84.

Rakic, P. (1988) Specification of cerebral cortical areas. *Science, 241:* 170-176.

Reiner, A. and Karten, H.J. (1983) The laminar source of efferent projections from the avian Wulst. *Brain Research, 275:* 349-354.

Revzin, A.M. (1979) Functional localization in the nucleus rotundus. In A. M. Granda and J. H. Maxwell (Eds.), *Neural mechanisms of behavior in the pigeon* (pp. 165-175). New York: Plenum Press.

Ritchie, T.C. and Cohen, D.H. (1979) The avian tectofugal visual pathway: Projections of its telencephalon target ectostriatal complex. *Society for Neuroscience Abstract, 2:* 119.

Robson, J.A. and Hall, W.C. (1977) The organization of the pulvinar nucleus in the grey squirrel *(Sciurus carolinensis)*. I. Cytoarchitecture and connections. *Journal of Comparative Neurology, 173:* 355-388.

Rodman, H.R., Gross, C.G., and Albright, T.D. (1989) Afferent basis of visual response properties in area MT of the macaque. I. Effects of striate cortex removal. *Journal of Neuroscience, 9:* 2033-2050.

Smart, I.H.M. (1973) Proliferative characteristics of the ependymal layer during the early development of the mouse neocortex: a pilot study based on recording the number, location and plane of cleavage of mitotic fibers. *Journal of Anatomy, 116:* 67-91.

Smart, I.H.M. and Smart, M. (1977) The location of nuclei of different labelling intensities in autoradiographs of the anterior forebrain of postnatal mice injected with [^{3}H] thymidine on the eleventh and twelfth days post-conception. *Journal of Anatomy, 123:* 515-525.

Smart, I.H.M. and Smart, M. (1982) Growth patterns in the lateral wall of the mouse telencephalon: I. Autoradiographic studies of the histogenesis of the isocortex and adjacent areas. *Journal of Anatomy, 134:* 273-298.

Sprague, J.M., Hughes, H.C. and Berlucchi, G. (1981) Cortical mechanisms in pattern and form perception. In O. Pompeiano and C.A. Marsan (eds.), *Brain mechanisms of perceptual awareness and purposeful behavior* (pp. 107-132). New York: Raven press.

Stensaas, L.J. and Gilson, B.C. (1972) Ependymal and subependymal cells of the caudate-pallial junction in the lateral ventricle of the neonatal rabbit. *Zeitschrift für Zellforschung, 132:* 297-322.

Stone, J. (1983) *Parallel processing in the visual system.* New York: Plenum Press.

Tsai, H.M., Garber, B.B., and Larramendi, L.M.H. (1981) ^{3}H-thymidine autoradiographic analysis of telencephalic histogenesis in the chick embryo: I. Neuronal birthdates of telencephalic compartments *in situ. Journal of Comparative Neurology, 198:* 275-292.

Ulinski, P.S. (1983). *Dorsal ventricular ridge,* New York: John Wiley & Sons.

Vaughan, W., Jr and Greene, S.L. (1984) Pigeon visual memory capacity. *Journal of Experimental Psychology: Animal Behavior Processes, 10:* 256-271.

Weiskrantz, L. (1986) *Blindsight: a case study and implications.* New York: Oxford University Press.

Wilson, P. (1980) The organization of the visual hyperstriatum in the domestic chick. I. Topology and topography of the avian projection. *Brain Research, 188:* 319-332.

Yamada, H. and Sano, Y. (1985) Immunohistochemical studies on the serotonin neuron system in the brain of the chicken *(Gallus domesticus).* II. The distribution of the nerve fibers. *Biogenic Amines, 2:* 21-36.

ASPECTS OF PHYLOGENETIC VARIABILITY OF NEOCORTICAL INTRINSIC ORGANIZATION

Facundo Valverde

Instituto de Neurobiología
Santiago Ramón y Cajal
Doctor Arce 37
28002 Madrid, Spain

INTRODUCTION

Our understanding of the intrinsic organization of the cerebral cortex greatly benefits from the blending of anatomy and physiology under the new concept of functional neuroanatomy. The fundamental principles of this enterprise were first conceived by Lorente de Nó (1949). He showed the existence of particular associations between a single afferent cortical fiber and groups of intrinsic neurons, advancing the concept of an *elementary unit* to designate a vertical cylinder or column of cortical tissue which has a central axis formed by a specific afferent fiber, containing all kinds of cells capable of carrying out the entire process of nerve transmission from the afferent fiber to the efferent axons. The idea was however not entirely new, for as early as 1898, Cajal suggested the existence of functional systems, or *isodynamic groups of neurons* in the visual cortex that could by activated by elementary sensory impressions. For several years, this basic unit was envisaged as a functional concept, because it explained earlier results obtained by neurophysiology in several of the primary sensory cortical areas, namely that cells having similar functional properties appear arranged along the vertical axis of the cortex from pia to the white matter. In the primary somatosensory and visual cortices, functional columns are fundamentally defined in terms of receptive field properties, while in the primary auditory cortex they are interpreted in terms of best frequency responses.

The visual cortex has been a matter of considerable interest in the last years, in part derived from a major leap exemplified in the physiological studies performed in the cat and monkey by Hubel and Wiesel (1962, 1968). They showed that the visual cortex appears subdivided into two independent systems, one for orientation preference and the other for eye dominance. The anatomical substrate of ocular dominance columns is represented by the segregation into layer IV of the geniculo-cortical fibers into alternating bands according to their origin in the lateral geniculate layers innervated by the same eye and, hence, functionally dominated by either the left or the right eye. The geometry of these bands was determined by several anatomical methods, and it was the first morphological evidence of an organization that could be correlated with physiological columns (Hubel and Wiesel, 1969, 1972). Comparable arrangements were subsequently demonstrated in almost every cortical area and newly developed methods for the tracing of pathways, and the use of enzymatic reactions and radioactive tracers provided the evidence that the cerebral cortex, as far as the thalamic input is concerned, appears organized into regularly spaced, periodic subdivisions attributable to the spatial distribution of afferent fibers in the somato-sensory (Killackey, 1973; Wise and Jones, 1976, 1978), auditory (Oliver and Hall, 1978; Andersen et al., 1980; Merzenich et al., 1982), and visual cortices (Hubel and Wiesel, 1972; Shatz et al., 1977; Ferster and LeVay, 1978; LeVay et al., 1978; Blasdel and Lund, 1983; Mower et al., 1985; Anderson et al., 1988) of various animal species, demonstrating that some of these thalamic distributions coincide with visible anatomical entities, such as the barrel field in the somato-sensory cortex of rodents (Woolsey and Van der Loos, 1970; Welker, 1976; Durham and Woolsey, 1977) and with the banding pattern of the ocular dominance system of the visual cortex (LeVay et al., 1975). This pattern of orderly partitions is not unique in cortical primary

Figure 1. Sagittal section through the visual cortex of the hedgehog showing the different neocortical layers, types of superficial and deep pyramidal cells, and various examples of large multipolar cells (**a** through **f**). Neocortical afferent fibers (**F**) ascend to ramify in the upper part of layer III-IV. Golgi method. From Valverde (1983).

sensory areas. There is now evidence that callosal and association projections also branch into alternating vertically oriented patches segregated from the thalamo-cortical ramifications (Wise and Jones, 1976; Goldman and Nauta, 1977; Jones et al., 1979; Killackey et al., 1983; Isseroff et al., 1984; Jones, 1984).

The most striking feature of the cerebral cortex is its organization into horizontal layers, but it is no less curious that the preferential mode of intrinsic connectivity is carried out in the opposite, that is, perpendicular to the layers. Each layer has its own individuality as given by their specific input and output connections, but isolated layers means nothing unless they are considered as containing characteristic elements linked to the constituents of the remaining layers. This is the philosophy underlying the module operation of the cerebral cortex, envisaged as interpenetrating spatial modules of vertically linked cell assemblies built around cortico-cortical and specific cortical afferents (Szentágothai, 1975, 1978, 1979; Eccles, 1981, 1984). Fundamental aspects of cortical organization are based on these anatomical entities which we will consider to provide a reference frame for the study of the intrinsic neocortical organization. The present contribution is based on observations obtained with the Golgi method in a large collection of brain sections from insectivora to primates. It deals with those aspects of intra-cortical connectivity that traditionally served for understanding neocortical operation from a comparative point of view.

THE MODEL OF NEOCORTICAL ORGANIZATION. THE AFFERENT FIBER

Primary cortical sensory areas receive their main afferent input from the corresponding peripheral sense organs through relays made in different thalamic nuclei. The fibers entering the cortex originating in these nuclei are known, since Lorente de Nó (1949), as "specific cortical afferents." One of the best known afferent cortical system is the geniculate input into the primary visual cortex. Early Golgi and degeneration studies showed that specific afferents terminate principally in the middle cortical layers (Valverde, 1967, 1968, 1971; Wilson and Cragg,

1967; Colonnier and Rossignol, 1969; Garey and Powell, 1971; Hubel and Wiesel, 1972; Szentágothai, 1973; Lund, 1973) demonstrating that the distribution of terminal degeneration coincides with the pattern of ramification of cortical afferents as observed in Golgi preparations in all mammals studied from insectivora to primates. More recent studies using degeneration and autoradiographic techniques (Rosenquist et al., 1974; Ribak and Peters, 1975; LeVay and Gilbert, 1976; Peters and Feldman, 1977; Ogren and Hendrickson, 1977; Gould et al., 1978; Hendrickson et al., 1978; Rezak and Benevento, 1979; Herkenham, 1980; Frost and Caviness, 1980; Winfield et al., 1982) as well as the use of bulk injections and intra-axonal filling with HRP of cortical afferent fibers (Ferster and LeVay, 1978; Gilbert and Wiesel, 1979, 1983; Blasdel and Lund, 1983; Florence et al., 1983; Conley et al., 1984; Martin, 1984, 1988; Freund et al., 1985; Humphrey et al., 1985; Lund et al., 1985; Naegele et al., 1988) have confirmed details obtained in previous Golgi observations and extended our knowledge of different characteristics related to the laminar distribution of specific afferent cortical fibers and of the functional architecture of the neocortex. We will now review the most salient features of innervation of neocortex by specific thalamic fibers.

In the hedgehog (*Erinaceus europaeus*), the pattern of ramifications of presumed specific thalamo-cortical fibers is very simple. In Golgi preparations cortical afferents to the occipital region ascend very obliquely developing conical fans of terminal fibers forming a plexus in the middle of the cortical thickness where no clear layer IV can be delimited. The fibers seem to contact large pyramidal cells of layers II-V probably representing target cells for thalamic fibers (Fig. 1). According to Gould et al., (1978) geniculo-cortical fibers in the hedgehog end in layers VI and III-IV, having some degree of topographical organization. In this animal, the same authors found a dense projection to layer I originating in the intralaminar region of the thalamus, but in contrast, using autoradiography and injections of HRP, Herkenham (1980) and Rieck and Carey (1985) found that (at least in the rat) the ventro-medial thalamic group is responsible for most heavy projections to layer I. In the hedgehog, we have described a system of thick fibers ascending in small bundles from the white matter to reach layer I, where they form extensive arborizations in the somato-sensory and visual cortices (Valverde and Facal-Valverde, 1986; Valverde et al., 1986). The existence of these projections from specific thalamic nuclei to layer I in insectivora, is reminiscent of a fundamental plan of organization found in allocortical formations, for instance in the nucleus olfactorius anterior (Valverde, 1986; Valverde et al., 1989), in which the main cortical input is made via tangential fibers running in the most superficial levels (Fig. 2).

In rodents, presumed specific thalamo-cortical fibers can be recognized by their aspect, branching pattern and distribution. They are the thickest fibers entering from the white matter and give origin to preterminal branches forming elongated clouds of terminal fibers distributed mainly in layers IV and lower part of layer III. Some, vertically ascending, isolated fibers can be followed up to layer I. Their morphology is remarkably similar to that obtained in the hamster visual cortex after single axon labeling by HRP into the optic radiation (Naegele et al., 1988) and in the somato-sensory cortex of the rat (Jensen and Killackey, 1987), but except for these characteristics, it seems that rodents show rather unspecific patterns of terminal distribution when compared to cats or monkeys (Fairén and Valverde, 1979; Valverde, 1986).

In the cat, Golgi impregnations of presumed thalamo-cortical fibers have been particularly successful in young kittens about one-month-old. In the visual cortex, the fibers follow oblique, almost horizontal courses in the lower part of layer VI and, when they could be traced down to the white matter, it was found that some divided into diverging fibers before entering the cortex indicating that the span of a single fiber may be quite large. In layer IV preterminal branches give rise, by repeated subdivisions, to interlacing bouquet plexuses, which unlike in rodents, occur as discrete entities separated by clear fields of the same size which are most apparent after computer reconstructions. The morphology of thalamo-cortical fibers as seen in Golgi preparations is similar to that obtained after injections of HRP to fill axonal arborizations in area 17 of the cat (Ferster and LeVay, 1978; Gilbert and Wiesel, 1979; Martin, 1984; Humphrey et al., 1985; Freund et al., 1985). Very fine axons entering layer I have been also described supposedly derived from the geniculate C-laminae (Ferster and LeVay, 1978). Physiological studies indicate that X- and Y-fibers do not converge on the same cortical neuron (Martin and Whitteridge, 1984), and that they may activate separate subsets of neurons (Freund et al., 1985).

In primates, the pattern of termination of specific cortical afferents differ considerably from that observed in other species. In the visual cortex the difference appears obviously related to the

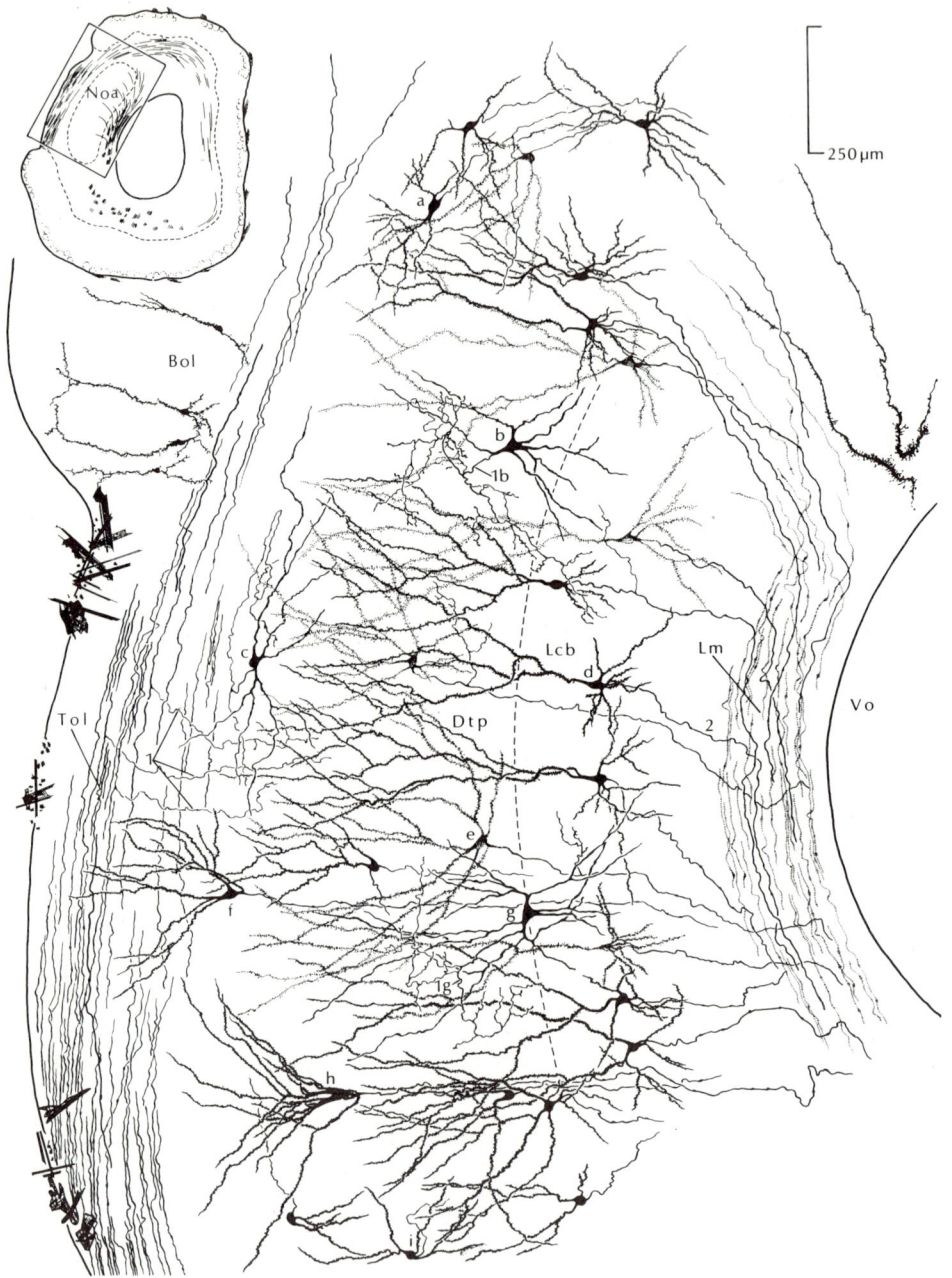

Figure 2. Transverse section through the nucleus olfactorius anterior (**Noa**) in the hedgehog. This allocortical formation represents the most primitive cortical organization composed by two distinct layers. The superficial layer (**Dtp**) contains dendritic arborizations of "primitive" pyramidal cells located in the cellular layer (**Lcb**). The figure is intended to compare the organization at a very simple level of cortical hierarchy with the most elaborate types of neocortical organization in most advanced mammals. Intercalated between the afferent system (**Tol**) and the efferent pathway (**Lm**), some intrinsic cells begin to appear (cells **b** and **g**), representing most probably the substrate of complex intrinsic circuitry found in more elaborate types of neocortical areas. Golgi method. From Valverde (1986).

dominance of the sense of vision, evidenced morphologically by the existence of one of the most elaborated types of cortical organization. In this area, layer IV is subdivided into IVa, IVb and IVc, being IVa and IVc those sublayers receiving thalamic input from the lateral geniculate nucleus. Presumed cortical afferents were described as complex terminal arbors which typically derive from thick axons ascending from lower cortical levels. They have been divided into two principal types: terminal ramifications with short-side appendages, resembling dendritic spines, and terminal axons with beaded appearance, forming spherical clumps apparently extending in single ocular dominance columns (Valverde, 1971, 1986; Lund, 1973; Blasdel and Lund, 1983). In the monkey, the different layers of the visual cortex receive distinct connections from the magno- and parvo-cellular partitions of the geniculate (Hubel and Wiesel, 1972; Hendrickson et al., 1978; Blasdel and Lund, 1983). A single afferent fiber can project to both IVa and IVc subdivisions of layer IV, and other afferent fibers can be traced to layers I and VI.

In summary, from insectivora to primates, we found that cortico-thalamic fibers first distribute in the most superficial layers including layer I, extending for a large territory occupied by superficial pyramidal cells. This stage is probably reminiscent of a primordial brain organization. Along the phylogenetic scale, the domain becomes more and more restricted until it appears circumscribed to the middle cortical layers, and even to certain subdivisions of it, as occurs in the primate brain.

THE TARGETS FOR CORTICAL AFFERENT FIBERS

Thalamo-cortical fibers contact certain specific neurons in layer IV, designated under the collective name of "stellate cells", but from the phylogenetic point of view, this is the exception to the rule. The cortex appears not to be hardwired for specific elements, but it seems to have retained along evolution the capability to adapt cell types and patterns of synaptic input in whatever is most convenient for its specific function. Target cells for specific thalamo-cortical fibers appear to vary widely in different mammals and in different cortical areas, and a common pattern is far from the rule. In lower mammals pyramidal cells are predominant, so that the number of apical dendrites and their collaterals, as well as the basal dendrites of pyramidal cells represent the principal targets for thalamic fibers. Available data, mainly based on studies carried out with the electron microscope alone or after appropriately placed lesions in the thalamic principal relay nuclei, indicate that practically all synapses identified from thalamic terminals are of the asymmetric (presumed excitatory) type (Peters and Feldman, 1976; Peters et al., 1979; White, 1979), hence it was proposed that all neurons and dendrites located in the domain of thalamic fibers in the middle cortical layers, and capable of forming asymmetric synapses, can receive thalamic input (Peters and Feldman, 1977).

It has been estimated that one out of every five axon terminals in layer IV of the visual cortex belongs to a specific geniculo-cortical fiber (LeVay and Gilbert, 1976; Davis and Sterling, 1979; White, 1979; Hornung and Garey, 1981; Winfield and Powell, 1983), and, more specifically, in area 17 of rats, cats and monkeys, about 15% of identified thalamic terminals (asymmetric synapses) contact dendritic shafts; 3% do it on neuronal somata, and the remainder, approximately 80%, synapse on dendritic spines (Garey and Powell, 1971; Peters and Feldman, 1976; Freund et al., 1985). The majority of these excitatory inputs are located on the spines of proximal dendrites, while inhibitory inputs (symmetric synapses) predominate on the cell bodies, smooth dendritic surfaces and occasionally on the spines.

Synaptic relations can not be predicted from Golgi observations alone, because only occasionally do the pre- and postsynaptic elements appear impregnated in the same specimen. However, the use of procedures combining Golgi techniques with electron microscopy, after appropriately placed lesions in the thalamus of the rat and cat, have shown that the majority of dendritic spines receiving thalamo-cortical synapses belong to dendrites of pyramidal cells whose bodies lie in layers III, IV and V, as well as varieties of intrinsic cells located in the middle cortical layers (Somogyi, 1978; Peters et al., 1979; Davis and Sterling, 1979; White, 1979; Hornung and Garey, 1981; Freund et al., 1985), constituting the population of "stellate cells" mentioned previously. However, stellate cells do not constitute a uniform population. Not only are there significant differences in their dendritic and axonal morphology in various animal species (even considering the same cortical region) and in different neocortical areas, but also their synaptic relations appear different. Several authors have used the criterion of dendritic spine density to classify stellate cells as those which lack spines (cells with smooth dendrites), have few spines (sparsely spinous), or have a number of spines approaching the density of most pyramidal cells (spinous cells). None of these cell types, however, constitute a uniform group and several other criteria,

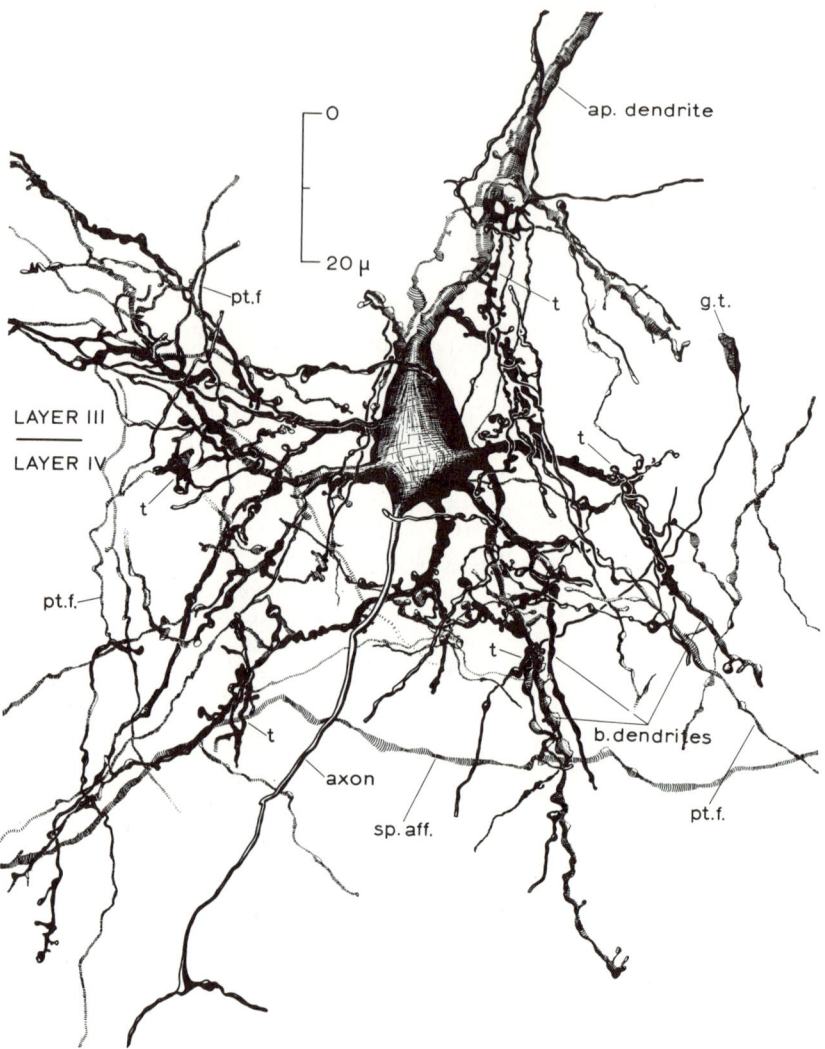

Figure 3. Pyramidal cell located between layers III and IV in the visual cortex of the mouse. The cell body and its basal and apical dendrites are immersed into a dense plexus of fine fibers. Most terminals (**t**) in contact with dendrites, and preterminal fibers (**pt.f**) could be traced to the main stems of three presumed specific cortical afferents, one of which (**sp.aff.**) courses horizontally below the cell body. Golgi method. From Ruiz-Marcos and Valverde (1970).

such as the form and distribution of the dendritic tree, cell size, and principally the axonal pattern, have been considered to achieve a comprehensive classification, from a morphological point of view (Valverde, 1976; Feldman and Peters, 1978: Fairén and Valverde, 1979; Fairén et al., 1984; Peters and Jones, 1984).

The correlation between the morphology and functional characteristics has been made possible by the intracellular HRP technique in which functionally characterized single cells are injected with the enzyme to yield its overall morphology visible. With this technique, several varieties of stellate and pyramidal cells in the primary visual cortex of the cat have been traced and their functional characteristics recorded. This technique has the disadvantage that only a limited number of cell types have thus far being recognized, but it shows that single cortical afferents only provide a small input to receiving cells and, as mentioned before, that the majority of excitatory inputs are located on the proximal dendritic segments, suggesting that thalamic fibers are more strategically located than previously thought (Freund et al., 1985).

In the zone of the somato-sensory cortex of the mouse corresponding to the barrel field, several types of cells are present in layer IV. Among these, spiny stellate cells are particularly abundant (Wooslsey et al., 1975). They have dendrites covered by numerous spines, they have no trace of an apical dendrite and their patterns of dendritic orientation depend on whether the cell body is located in the barrel wall or in the barrel hollow. Except for these highly characteristic elements in a very specific cortical area, in the visual cortex of rodents (Fig. 3), the majority of postsynaptic elements to degenerating thalamic terminals (about 83%) are represented by dendritic spines which most probably belong to the apical shafts and oblique branches of layers V and VI pyramidal cells, and to the basal dendrites of layer III pyramidal neurons (Peters and Feldman, 1977). A small proportion of postsynaptic elements (about 17%) are represented by the shafts of smooth or sparsely spinous dendrites of stellate cells (Peters et al., 1976). These cells reside in layer IV and in the immediately adjacent cortical layers (Valverde, 1968).

In the visual cortex of the cat, spiny stellate cells have been a matter of considerable interest since Kelly and Van Essen (1974) showed that some cells they recovered by intracellular dye injections after identification of their functional characteristics, were spiny stellate cells as those found in Golgi preparations (Cajal, 1911, 1921; O'Leary, 1941; LeVay, 1973; Fairén and Valverde, 1979; Lund et al., 1979; Peters and Regidor, 1981; Meyer and Ferres-Torres, 1984). These cells receive direct contact from thalamic fibers (Hornung and Garey, 1981; Martin and Whitteridge, 1984; Freund et al., 1985) and may have axons, specially those located at the 17/18 border, projecting to the contralateral hemisphere (Sanides, 1979; Innocenti, 1979; Hornung and Garey, 1980; Meyer and Albus, 1981). The fact that these cells project to the white matter had already been mentioned by Cajal (1921), and confirmed after HRP intracellular injections (Gilbert and Wiesel, 1979; Lin et al., 1979).

In primates we meet a completely different panorama; target cells for thalamic axons are intrinsic cells which do not project outside the cortex. In the primate visual cortex, sublayer IVc contains a population of spiny stellate cells with recurving axons and varieties of cells with smooth dendrites. The spiny stellate cells with recurving axons are the most distinctive element found in this sub-layer (Fig. 4). They were described by us (Valverde, 1971) as characteristic elements of the primate visual cortex, and, since then, they have been a matter of considerable interest in the analysis of the functional and anatomical organization of the visual cortex. In spite of several considerations, we always considered them to represent a very specific type of cells which probably has no counter-part in non-primate species. This type of cell, located at the very heart of thalamic terminals, represent the principal relay in cortical circuits linking neurons having simple receptive fields to complex cells in other cortical areas. The axons of these cells originates at the lower pole of the cell body, descending for a short distance and then turning upwards forming single or several characteristic loops in ascending bundles reaching layer III and sub-layers IVa and IVb where they develop into distinct elongated plexuses of terminal fibers. A second major type representing target cells for thalamic axons corresponds to certain neurons with smooth dendrites and beaded axons. The most abundant type shows a very limited axonal field formed by densely interwoven axonal collaterals and recurving dendrites, giving the ensemble the appearance of a ball of yarn. We named them "clewed cells" when we saw them for the first time (Valverde, 1971). They have been also a matter of considerable interest, not only for their specific axonal patterns, but also because they form symmetrical synaptic contacts, and appear therefore to be inhibitory interneurons. These cells are most probably identical to the small basket (clutch) cells described in the visual cortex of the cat (Kisvárday et al., 1985) and monkey (Kisvárday et al., 1986). Their synaptic relations with spiny stellate cells suggest an interesting type of local interaction (Mates and Lund, 1983; Kisvárday et al., 1986), and they have been considered important key pieces in conceptual models of neocortical operation (Eccles, 1981).

The proportion of spiny stellate cells in any cortical area varies depending on the animal. For instance, in the somatic sensory cortex of the mouse, one half of the cells in layer IV have dendrites covered by spines (Woolsey et al., 1975). The frequency of spiny stellate cells in the visual cortex of the rat (Golgi method) is 11% (Feldman and Peters, 1978), however we considered that most of these cells belongs to the category of grain or star pyramids rather than true spiny stellate cells. In the visual cortex of the cat non-pyramidal cells account for 60-80% of cells in layer IV, in which the spiny stellate cells are as common as the cells with smooth dendrites (Garey, 1971;

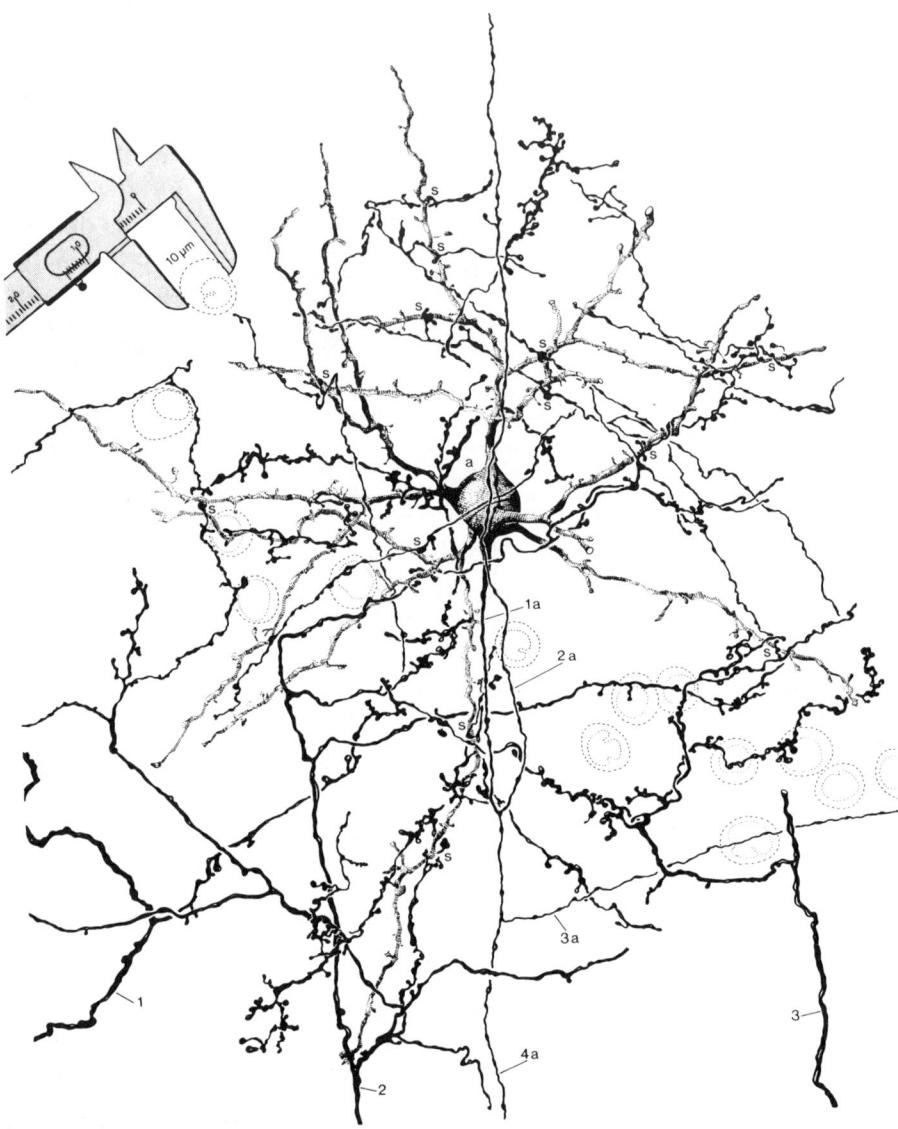

Figure 4. The spiny stellate cell of sublayer IVc of the striate area in the monkey *(Macaca mulatta)*. This cell has moderately spinous dendrites and appears surrounded by the terminal endings of afferent cortical fibers, which presumably form abundant synaptic contacts at sites marked "s". Golgi method. From Valverde (1985).

Winfield and Powell, 1976). In the visual cortex of the monkey, exclusively spiny stellate cells and cells with smooth dendrites form the population of target cells in sublayer IVc; spiny stellate cells constituting about 95% of this population (Mates and Lund, 1983). Spiny stellate cells have not been found in the visual cortex of the mouse and they are virtually absent in the rabbit's neocortex. Very occasionally, non-pyramidal spiny neurons have been observed in layers III and IV of the auditory cortex of the rabbit (McMullen and Glaser, 1982), where 87% of all impregnated neurons (Golgi-Cox) are pyramidal cells (McMullen et al., 1984). We were unable to observe them in layer III/IV of the neocortex in the insectivore hedgehog (Valverde, 1983).

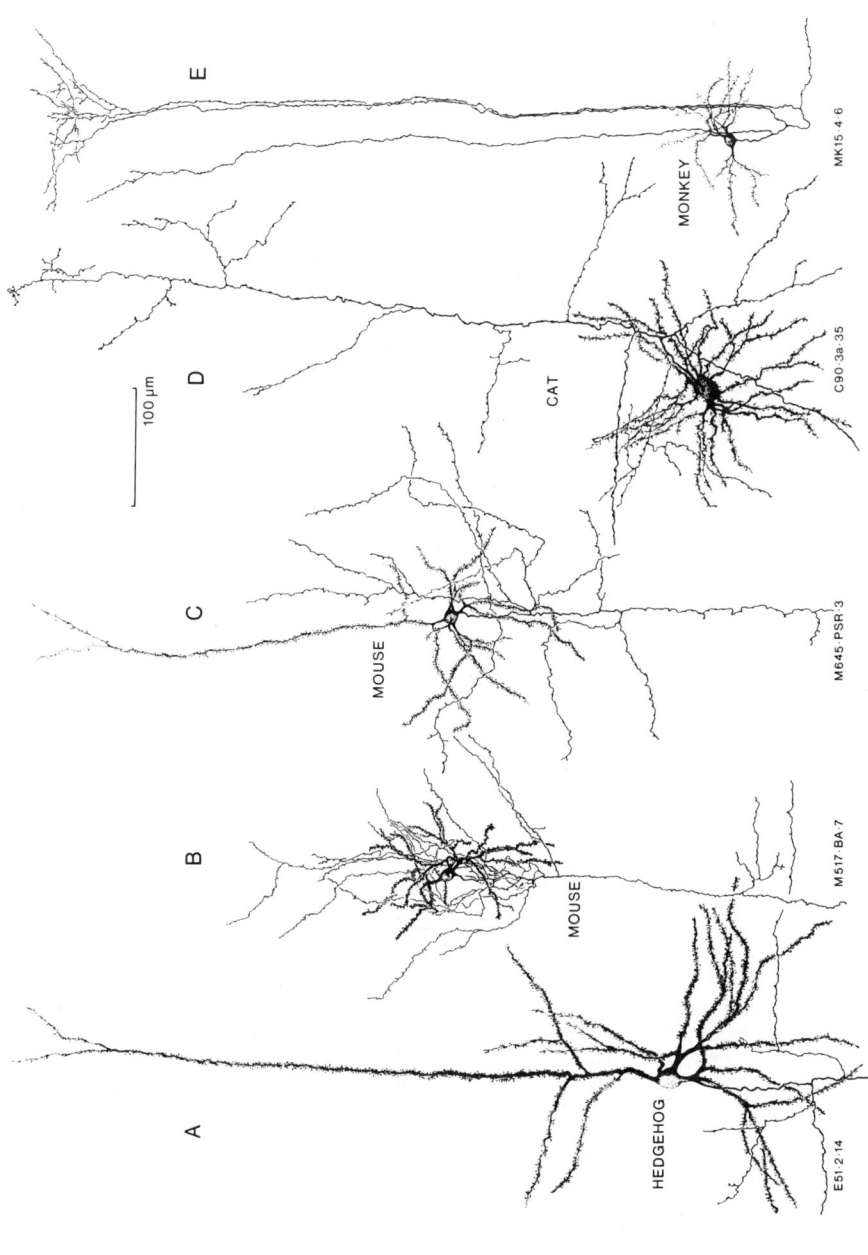

Figure 5. Varieties of cells with spinous dendrites of various mammals and cortical areas, reproduced at the same magnification. **(A)** shows a pyramid-like cell in the somato-sensory cortex of the hedgehog, **(B)** illustrates a spiny stellate cell in the barrel field of the mouse, **(C)** depicts a stellate pyramidal cell in the visual cortex of the mouse, **(D)** represents a typical spiny stellate cell in the temporal cortex of the kitten, **(E)** exhibits one example of a spiny stellate cell with recurring axon in sublayer IVC of the visual cortex in the monkey. Golgi method. From Valverde (1986).

TARGET NONSPECIFICITY FOR THALAMO-CORTICAL FIBERS

As we just reviewed, the organization of thalamo-cortical systems and the principal target cells receiving these inputs vary in different mammals. There are important considerations concerning these differences from a comparative point of view. From available data, we can state that the majority of dendritic spines receiving thalamo-cortical synapses belong to neurons other than spiny stellate cells, and that practically all involved dendritic spines belong to dendrites of pyramidal cells whose bodies lie in layers III, IV and V. These results are in agreement with our earlier observations showing the existence of presumed synaptic contacts of cortical afferent fibers on dendritic spines of apical shafts of layer V pyramidal cells (Valverde, 1967; Valverde and Ruiz-Marcos, 1969), and on the basal dendrites of layer III pyramidal cells in the visual cortex of the mouse (Ruiz-Marcos and Valverde, 1970). At that time it was clear to us that both processes might occur in direct relation to thalamo-cortical afferents because they were most sensitive to visual sensory deprivation. The visual cortex of the monkey is the exception to the rule since the thalamo-recipient zone, (sublayers IVa and IVc) are exclusively occupied by spiny stellate cells with intrinsic axons.

During evolution, these major differences appear related to changes in the laminar distribution of cortical afferents, which presumably shifted from layer I in primitive mammals to predominate in the lower part of layer III and in layer IV, which is present in all extant mammals. We believe there is no doubt that this shift modified the intrinsic neocortical organization. For instance, in the study of the forms of cells with spiny dendrites, including pyramidal cells, one gets the impression that all of them may share a common phylogenetic origin and that a continuum can be traced from lower forms to the primate brain (Fig. 5). This idea is not new as it has been put forward several times since the classical comparative neuroanatomy (Cajal, 1911) to recent times (Ramón-Moliner, 1967; Sanides, 1970; Sanides and Sanides, 1972). A similar point of view has been expressed in relation to spiny stellate cells (Lund, 1984; Saint Marie and Peters, 1985).

In the neocortex of the hedgehog, considered a direct descendant of primitive eutherians, a complete series of intermediate forms between the most extraverted pyramidal cells and fully developed ones, can always be found (Sanides and Sanides, 1972; Valverde and Facal-Valverde, 1986). During development, some pyramidal cells lose their apical terminal branches in layer I retaining a thin apical dendrite tapering at some distance from the cell body. The cell becomes stellate in form and perisomatic dendrites concentrate in more restricted volumes. This seems to be the case of stellate pyramids of the barrel field in the mouse somato-sensory cortex, and in the visual cortices of the cat and monkey in which some cells still retain unbranched apical shafts (star or stellate pyramidal cells), while others have developed into typical spiny stellate cells found in the primate brain. The stages of pyramidal cell differentiation also involve variations in the axonal patterns which change from long projecting neurons (hedgehog, rat, cat) to intrinsic cells with axons remaining inside the cortex (monkey).

CONCLUDING REMARKS

The existence of a basic plan of neocortical intrinsic organization is demonstrable throughout the mammalian scale. Layers III and IV are the major recipients of cortical afferent fibers, from here impulses are relayed mainly to the upper layers I-III, layer III is the source of long and short association fibers and intrinsic descending connections to layers V and VI, which contain the majority of cells projecting subcortically and a number of intrinsic cells with ascending axons (Martinotti cells). This suggests that the neocortex is functionally uniform at a rather fundamental level of organization. However, the study of the varieties of cells and the mode they intervene in intrinsic wiring patterns in different animals clearly shows the existence of important variations, some of which may be unique for a given species. We have commented on the differences existing in certain varieties of cells with re-curving axons seemingly unique for the primate brain, the existence of stellate pyramids peculiar of the neocortex of rodents, varieties of spiny stellate cells with axons projecting to the white matter in the visual cortex of the cat, and certain pyramidal cells in the neocortex of the hedgehog which have no counterpart in other subjects. With the exception of the visual cortex of the monkey, we have also suggested that spiny stellate cells are by no means a common target for the thalamo-cortical fibers, nor do they form a single class of neurons. The comparison in different species point out that certain pyramidal cells and portion of their dendrites, may be more directly related to cortical afferents. It is also concluded that, even though modern tracing techniques have emphasized that the major cortical afferent systems appear broken up in multiple column-like patches with remarkable constancy through the mammalian scale, the intrinsic circuitry of cortical columns may be quite different (Fig. 6).

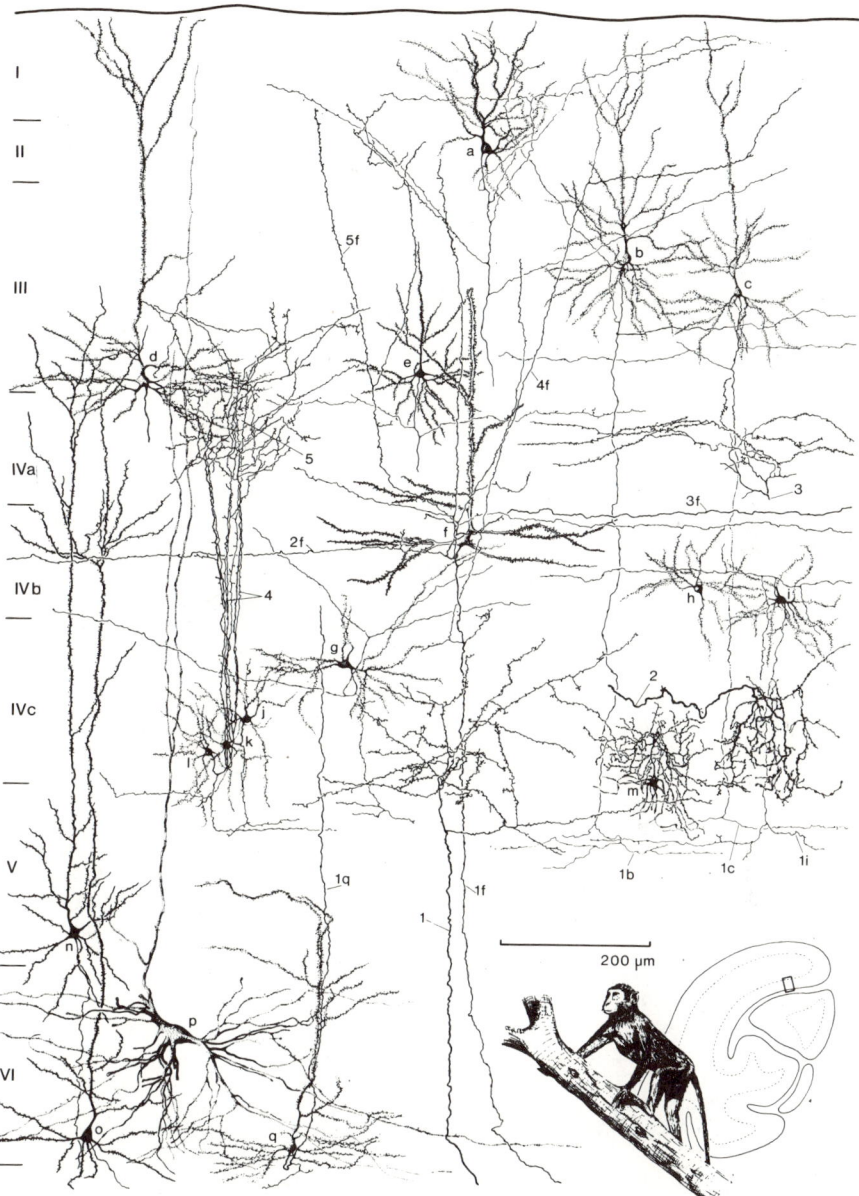

Figure 6. Composite drawing made from several adjacent sections of the visual cortex of the monkey (*Macaca irus*). Layer IV appears subdivided into three sublayers (IVa, IVb and IVc). Two types of presumed extrinsic (cortical afferent) fibers 1 and 2 can be observed. Spiny stellate cells (j,k,l) have recurving axons which ascend in small bundles (4) to develop complex terminal ramifications (5) in the lower part of layer III and in IVa. Horizontal fibers (2f, 3f) run for long distances through the stria of Gennari (sublayer IVb). Golgi method. From Valverde (1986).

The growth of neocortical afferents during ontogenetic development represents an example which might be a recapitulation of phylogeny. Cortical afferent axons first make contact with a transitory population of cells in the subplate layer, which is largely eliminated by cell death during early postnatal life (Kostovic and Rakic, 1980; Luskin and Shatz, 1985; Valverde and Facal-Valverde, 1988), suggesting that cortical afferents compete for a final target represented by newly developed cells in the middle cortical layers. Vestiges of these early connections may be represented by thalamo-cortical fibers ending in layers I and VI of the adult, known to exist in all mammals.

We have placed emphasis on the differences in intrinsic neocortical organization (Valverde, 1986, 1988) because the simple multiplications or archetypal cortical modules (Glezer et al., 1988) will not explain the diversity of cell types found in different mammals, some of which may be unique for a given species. We believe that fundamental differences can be explained by mechanisms involving the reshaping of the dendritic and axonal arbors of various categories of cortical cells, resulting in different patterns of intracortical connectivity. The differences can be minimal in closely related species, but they are substantial when the comparison is made between distant subjects such as the insectivore and the primate.

ACKNOWLEDGEMENTS

This work has been supported by grants 1329/81 and PR84-0217 (CAICYT), and PB87-0412 (GDICYT) from the Ministerio de Educatión y Ciencia.

REFERENCES

Andersen, R.A., Knight, P.L., and Merzenich, M.M. (1980) The thalamocortical and corticothalamic connections of AI, AII, and the anterior auditory field (AAF) in the cat: evidence of two largely segregated systems of connections. *J. Comp. Neurol.*, 194: 663-701.

Anderson, P.A., Olavarria, J., and Van Sluyters, R.C. (1988) The overall pattern of ocular dominance bands in cat visual cortex. *J. Neurosci.*, 8: 2183-2200.

Blasdel, G.G., and Lund, J.S. (1983) Termination of afferent axons in macaque striate cortex. *J. Neurosci.*, 3: 1389-1413.

Cajal, S.R. (1898) Estructura del kiasma óptico y teoría general de los entrecruzamientos de las vías nerviosas. *Rev. Trim. Microg.*, 4: 15-65.

Cajal, S.R. (1911) *Histologie de Systéme Nerveux de l'Homme et des Vertébrés*, Vol. 2. Maloine: Paris (Reimpress. Instituto Cajal, CSIC, Madrid, 1955).

Cajal, S.R. (1921) Textura la corteza cerebral del gato. *Trab. Lab. Invest. Biol.*, 19: 113-144.

Colonnier, M., and Rossignol, S. (1969) Heterogeneity of the cerebral cortex, in: *Basic Mechanisms of the Epilepsies*, H. Jasper, A. Ward, and A. Pope, eds., Little Brown: Boston, 29-40.

Conley, M., Fitzpatrick, D., and Diamond, I.T. (1984) The laminar organization of the lateral geniculate body and the striate cortex in the tree shrew (*Tupaia glis*). *J. Neurosci.*, 4: 171-197.

Davis, T.L., and Sterling, P. (1979) Microcircuitry of cat visual cortex: Classification of neurons in layer IV of area 17, and identification of the patterns of lateral geniculate input. *J. Comp. Neurol.*, 188: 599-628.

Durham, D., and Woolsey, T.A. (1977) Barrels and columnar organization: evidence from 2-deoxyglucose (2-DG) experiments. *Brain Res.*, 137: 169-174.

Eccles, J.C. (1981) The modular operation of the cerebral neocortex considered as the matieral basis of mental events. *Neuroscience*, 6: 1839-1856.

Eccles, J.C. (1984) The cerebral neocortex. A theory of its operation, in: *Cerebral Cortex, Functional Properties of Cortical Cells*, Vol. 2, E.G. Jones, and A. Peters, eds., Plenum Press: New York and London, 1-36.

Fairén, A., DeFelipe, J., and Regidor, J. (1984) Nonpyramidal neurons. General account, in: *Cerebral Cortex, Cellular Components of the Cerebral Cortex*, Vol. 1, A. Peters, and E.G. Jones, eds., Plenum Press: New York and London, 201-253.

Fairén, A., and Valverde, F. (1979) Specific thalamo-cortical afferents and their presumptive targets in the visual cortex. A Golgi study, in: *Progress in Brain Research. Development and Specificity of Neurons*, Vol. 51, M.Cuénod, G.W.Kreutzberg, and F.E. Bloom, eds., Elsevier, Amsterdam, 419-438.

Feldman, M., and Peters, A. (1978) The forms of non-pyramidal neurons in the visual cortex of the rat. *J. Comp. Neurol.*, 179: 761-794.

Ferster, D., and LeVay, S. (1978) The axonal arborizations of lateral geniculate neurons in the striate cortex of the rat. *J. Comp. Neurol.*, 182: 923-944.

Florence, S.L., Sesma, M.A., and Casagrande, V.A. (1983) Morphology of geniculo-striate afferents in a prosimian primate. *Brain Res.*, 270: 127-130.

Freund, T.F., Martin, K.A.C., and Whitteridge, D. (1985) Innervation of cat visual areas 17 and 18 by physiologically identified X- and Y-type thalamic afferents. I. Arborization patterns and quantitative distribution of postsynaptic elements. *J. Comp. Neurol.*, 242: 263-274.

Frost, D.O., and Caviness, V.S. (1980) Radial organization of thalamic projections to the neocortex in the mouse. *J. Comp. Neurol.*, 194: 369-393.

Garey, L.H. (1971) A light and electron microscopic study of the visual cortex of the cat and monkey. *Proc. Roy. Soc. London B*, 179: 21-40.

Garey, L.H., and Powell, T.P.S. (1971) An experimental study of the termination of the lateral geniculo-cortical pathway in the cat and monkey. *Proc. Roy. Soc. London B*, 179: 1-63.

Gilbert, C.D., and Wiesel, T.N. (1979) Morphology of intracortical projections of functionally characterised neurones in the cat visual cortex. *Nature (London)*, 280: 120-125.

Gilbert, C.D., and Wiesel, T.N. (1983) Clustered intrinsic connections in cat visual cortex. *J. Neurosci.*, 3: 1116-1133.

Glezer, I.I., Jacobs, M.S., and Morgane, P.J. (1988) Implications of the "initial brain" concept for brain evolution in cetacea. *Behav. Brain Sci.*, 11: 75-116.

Goldman, P.S., and Nauta, W.J.H. (1977) Columnar distribution of cortico-cortical fibers in the frontal association, limbic, and motor cortex of the developing rhesus monkey. *Brain Res.*, 122: 393-413.

Gould, J.J., III, Hall, W.C., and Ebner, F.F. (1978) Connections of the visual cortex in the hedgehog (*Paraechinus hypomelas*). I. Thalamocortical projections. *J. Comp. Neurol.*, 177: 445-471.

Hendrickson, A.E., Wilson, J.R., and Ogren, M.P. (1978) The neuroanatomical organization of pathways between the dorsal lateral geniculate nucleus and visual cortex in Old World and New World primates. *J. Comp. Neurol.*, 182:123-136.

Herkenham, M. (1980) Laminar organization of thalamic projections to the rat neocortex. *Science*, 207: 532-535.

Hornung, J.P., and Garey, L.J. (1980) A direct pathway from thalamus to visual callosal neurons in the cat. *Exp. Brain Res.*, 38: 121-123.

Hornung, J.P., and Garey, L.J. (1981) The thalamic projection to cat visual cortex: Ultrastructure of neurons identified by Golgi impregnation or retrograde horseradish peroxidase transport. *Neuroscience*, 6: 1053-1068.

Hubel, D.H., and Wiesel, T.N. (1962) Receptive fields, binocular interaction and functional architecture in the cat visual cortex. *J. Physiol. (Lond.)*, 160: 106-154

Hubel, D.H., and Wiesel, T.N. (1968) Receptive fields and functional architecture of monkey striate cortex. *J. Physiol. (Lond.)*, 195: 215-243.

Hubel, D.H., and Wiesel, T.N. (1969) Anatomical demonstration of columns in the monkey striate cortex. *Nature (London)*, 221: 747-750.

Hubel, D.H., and Wiesel, T.N. (1972) Laminar and columnar distribution of geniculo-cortical fibers in the macaque monkey, *J. Comp. Neurol.*, 146: 421-450.

Humphrey, A.L., Sur, M., Uhlrich, D.J., and Sherman, S.M. (1985) Projection patterns of individual X- and Y-cell axons from the lateral geniculate nucleus to cortical area 17 in the cat. *J. Comp. Neurol.*, 233: 159-189.

Innocenti, G.M. (1979) Adult and neonatal characteristics of the callosal zone at the boundary between areas 17 and 18 in the cat, in: *Structure and Function of Cerebral Commissures*, I.S. Russell, M.W. van Hof, and G. Berlucchi, eds., MacMillan Press: London, 244-28.

Isseroff, A., Schwartz, M.L., Dekker, J.J., and Goldman-Rakic, P.S. (1984) Columnar organization of callosal and association projections from rat frontal cortex. *Brain Res.*, 293: 213-223.

Jensen, K.F., and Killackey, H.P. (1987) Terminal arbors of axons projecting to the somatosensory cortex of the adult rat. I. The normal morphology of specific thalamocortical afferents. *J. Neurosci.*, 7: 3529-3543.

Jones, E.G. (1984) Laminar distribution of cortical efferent cells, in: *Cerebral Cortex, Cellular Components of the Cerebral Cortex*, Vol. 1, A. Peters and E.G. Jones, eds., Plenum Press: New York and London, 521-553.

Jones, E.G., Coulter, J.D., and Wise, S.P. (1979) Commissural columns in the sensory-motor cortex of monkeys. *J. Comp. Neurol.*, 188: 113-136.

Kelly, J.P., and Van Essen, D.C. (1974) Cell structure and function in the visual cortex of the cat. *J. Physiol.*, 238: 515-547.

Killackey, H.P. (1973) Anatomical evidence for cortical subdivisions based on vertically discrete thalamic projections from the ventral posterior nucleus to cortical barrels in the rat. *Brain Res.*, 51: 326-331.

Killackey, H.P., Gould III, H.J., Cusick, C.G., Pons, T.P., and Kaas, J.H. (1983) The relation of corpus callosum connections to architectonic fields and body surface maps in sensorimotor cortex of New and Old World monkeys. *J. Comp. Neurol.*, 219: 384-419.

Kisvárday, Z.F., Cowey, A., and Somogyi, P. (1986) Synaptic relationships of a type of GABA-immunoreactive neuron (clutch cell), spiny stellate cells and lateral geniculate nucleus afferents in layer IVc of the monkey striate cortex. *Neuroscience*, 19: 741-761.

Kisvárday, Z.F., Martin, K.A.C., Whitteridge, D., and Somogyi, P. (1985) Synaptic connections of intracellularly filled clutch cells: A type of small basket cell in the visual cortex of the cat. *J. Comp. Neurol.*, 241: 111-137.

Kostovic, I., and Rakic, P. (1980) Cytology and time of origin of interstitial neurons in the white matter in infant and adult human and monkey telencephalon. *J. Neurocytol.*, 9: 219-242.

LeVay, S., (1973) Synaptic patterns in the visual cortex of the cat and monkey: Electronmicroscopy of Golgi preparations. *J. Comp. Neurol.*, 150: 53-86.

LeVay, S., and Gilbert, C.D. (1976) Laminar patterns of geniculocortical projection in the cat. *Brain Res.*, 113: 1-19.

LeVay, S., Hubel, D.H., and Wiesel, T.N. (1975) The pattern of ocular dominance columns in macaque visual cortex revealed by a reduced silver stain. *J. Comp. Neurol.*, 159: 559-575.

LeVay, S., Stryker, M.P., and Shatz, C.J. (1978) Ocular dominance columns and their development in layer IV of the cat's visual cortex: A quantitative study. *J. Comp. Neurol.*, 179: 223-244.

Lin, C.S., Friedlander, M.J., and Sherman, S.M. (1979) Morphology of physiologically identified neurons in the visual cortex of the cat. *Brain Res.*, 172: 344-348.

Lorente de Nó, R. (1949) Cerebral cortex: Architecture, intracortical connections, motor projections, in: *Fulton's Physiology of the Nervous System*, Oxford University Press: London, 288-330.

Lund, J.S. (1973) Organization of neurons in the visual cortex, area 17, of the monkey (Macaca mulatta). *J. Comp. Neurol.*, 147: 455-496.

Lund, J.S. (1984) Spiny stellate neurons, in: *Cerebral Cortex, Cellular Components of the Cerebral Cortex*, Vol. 1, A. Peters, and E.G. Jones, eds., Plenum Press: New York and London, 255-308.

Lund, J.S., Fitzpatrick, D., and Humphrey, A.L. (1985) The striate cortex of the tree shrew, in: *Cerebral Cortex, Visual Cortex*, Vol.3, A, Peters, and E.G. Jones, eds., Plenum Press: New York and London, 157-205.

Lund, J.S., Henry, G.H., Macqueen, C.L., and Harvey, A.R. (1979) Anatomical organization of the primary visual cortex (area 17) of the cat. A comparison with area 17 of the macaque monkey. *J. Comp. Neurol.*, 184: 599-618.

Luskin, M.B., and Shatz, C.J. (1985) Studies of the earliest generated cells of the cat's visual cortex: Cogeneration of subplate and marginal zones. *J. Neurosci.*, 5: 1062-1075.

Martin, K.A.C. (1984) Neuronal circuits in cat striate cortex, in: *Cerebral Cortex, Functional Properties of Cortical Cells*, Vol. 2, E.G. Jones, and A. Peters, eds., Plenum Press: New York and London, 241-284.

Martin, K.A.C. (1988) From single cells to simple circuits in the cerebral cortex. *Quart. J. Exper. Physiol.*, 73: 637-702.

Martin, K.A.C., and Whitteridge, D. (1984) Form, function and intracortical projections of spiny neurones in the striate visual cortex of the cat. *J. Physiol. (London)*, 353: 463-504.

McMullen, T.N., Glaser, E.M., and Tagamets, M. (1984) Morphometry of spine-free nonpyramidal neurons in rabbit auditory cortex. *J. Comp. Neurol.*, 222: 383-395.

Merzenich, M.M., Colwell, S.A., and Andersen, A. (1982) Auditory forebrain organization. Thalamocortical and corticothalamic connections in the cat, in: *Cortical Sensory Organization. Multiple Auditory Areas*, Vol. 3, C.N. Woolsey, ed., Humana Press: Clifton, New Jersey, 43-57.

Meyer, G., and Albus, K. (1981) Spiny stellates as cells of origin of association fibers from area 17 to area 18 in the cat's neocortex. *Brain Res.*, 210: 335-341.

Meyer, G., and Ferres-Torres, R. (1984) Postnatal maturation of nonpyramidal neurons in the visual cortex of the cat. *J. Comp. Neurol*, 228: 226-244.

Mower, G.D., Caplan, C.J., Christen, W.G., and Duffy, F.H. (1985) Dark rearing prolongs physiological but not anatomical plasticity of the cat visual cortex. *J. Comp. Neurol.*, 235: 448-466.

Naegele, J.R., Jhaveri, S., and Schneider, G.E. (1988) Sharpening of topographical projections and maturation of geniculocortical axon arbors in the hamster. *J. Comp. Neurol.*, 277: 593-607.

Ogren, M.P., and Hendrickson, A.E. (1977) The distribution of pulvinar terminals in visual areas 17 and 18 of the monkey. *Brain Res.*, 137: 343-350.

O'Leary, J.L. (1941) Structure of the area striata of the cat. *J. Comp. Neurol.*, 75: 131-164.
Oliver, D.L., and Hall, W.C. (1978) The medial geniculate body of the tree shrew, *Tupaia glis*. II. Connections with the neocortex. *J. Comp. Neurol.*, 182: 459-494.
Peters, A., and Feldman, M.L. (1976) The projection of the lateral geniculate nucleus to area 17 of the rat cerebral cortex. I. General description. *J. Neurocytol.*, 5: 63-84.
Peters, A., and Feldman, M.L. (1977) The projection of the lateral geniculate nucleus to area 17 of the rat cerebral cortex. IV. Terminations upon spiny dendrites. *J. Neurocytol.*, 6: 669-689.
Peters, A., Feldman, M.L., and Saldanha, J. (1976) The projection of the lateral geniculate nucleus to area 17 of the rat cerebral cortex. II. Terminations upon neuronal perikarya and dendritic shafts. *J. Neurocytol.*, 5: 85-107.
Peters, A., and Jones, E.G. (1984) Classification of cortical neurons, in: *Cerebral Cortex, Cellular Components of the Cerebral Cortex*, Vol. 1, A. Peters, and E.G. Jones, eds., Plenum Press: New York and London, 107-121.
Peters, A., Proskauer, C.C., Feldman, M.L., and Kimerer, L. (1979) The projection of the lateral geniculate nucleus to area 17 of the rat cerebral cortex. V. Degenerating axon terminals synapsing with Golgi impregnated neurons. *J. Neurocytol.*, 8: 331-357.
Peters, A., and Regidor, J. (1981) A reassessment of the forms of nonpyramidal neurons in area 17 of cat visual cortex. *J. Comp. Neurol.*, 203: 685-716.
Ramón-Moliner, E. (1967) La différentiation morphologique des neurones. *Arch. Ital. Biol.*, 105: 149-188.
Rezak, M., and Benevento, L.A. (1979) A comparison of the organization of the projections of the dorsal lateral geniculate nucleus, the inferior pulvinar and adjacent lateral pulvinar to primary visual cortex (area 17) in the macaque monkey. *Brain Res.*, 169: 19-40.
Ribak, C.E., and Peters, A. (1975) An autoradiographic study of the projections from the lateral geniculate body of the rat. *Brain Res.*, 92: 341-368.
Rieck, R.W., and Carey, R.G. (1985) Organization of the rostral thalamus in the rat: Evidence for connections to layer I of visual cortex. *J. Comp. Neurol.*, 234: 137-154.
Rosenquist, A.C., Edwards, S.B., and Palmer, L.A. (1974) An autoradiographic study of the projections of the dorsal lateral geniculate nucleus and the posterior nucleus in the cat. *Brain Res.*, 80: 71-93.
Ruiz-Marcos, A., and Valverde, F. (1970) Dynamic architecture of the visual cortex. *Brain Res.*, 19: 25-39.
Saint Marie, R.L., and Peters, A. (1985) The morphology and synaptic connections of spiny stellate neurons in monkey visual cortex (area 17): A Golgi-electron microscope study. *J. Comp. Neurol.*, 233: 213-235.
Sanides, D. (1979) Commissural connections of the visual cortex of the cat, in: *Structure and Function of Cerebral Commissures*, I.S. Russell, M.W. van Hof, and G. Berlucchi, eds., MacMillan Press: London, 236-243.
Sanides, F. (1970) Functional architecture of motor and sensory cortices in primates in the light of a new concept of neocortex evolution, in: *The Primate Brain. Advances in Primatology*, C.R. Noback, and W. Montagna, eds., Appleton-Century-Crofts: New York, 137-208.
Sanides, F., and Sanides, D. (1972) The "extraverted neurons" of the mammalian cerebral cortex. *Z. anat. Entwickl. -gesch.*, 136: 272-293.
Shatz, C.J., Lindström, S., and Wiesel, T.N. (1977) The distribution of afferents representing the right and left eyes in the cat's visual cortex. *Brain Res.*, 131: 103-116.
Somogyi, P. (1978) The study of Golgi stained cells and of experimental degeneration under the electron microscope: A direct method for the identification in the visual cortex of three successive links in a neuron chain. *Neuroscience*, 3: 167-180.
Szentágothai, J. (1973) Synaptology in the visual cortex, in: *Handbook of Sensory Physiology, Central Visual Information*, Vol. 7, R. Jung, ed., Springer, Berlin, 269-324.
Szentágothai, J. (1975) The 'module concept' in cerebral cortex architecture. *Brain Res.*, 95: 475-496.
Szentágothai, J. (1978) The neuron network of the cerebral cortex: A functional interpretation. *Proc. Roy. Soc. London, B*, 201: 219-248.
Szentágothai, J. (1979) Local neuron circuits of the neocortex, in: *The Neurosciences. Fourth Study Program*, F.O. Schmitt, and F.G. Worden, eds., MIT Press, Cambridge, Massachusetts, 399-415.
Valverde, F. (1967) Apical dendritic spines of the visual cortex and light deprivation in the mouse. *Exp. Brain Res.*, 3: 337-352.
Valverde, F. (1968) Structural changes in the area striata of the mouse after enucleation. *Exp. Brain Res.*, 5: 274-292.

Valverde, F. (1971) Short axon neuronal subsystems in the visual cortex of the monkey. *Intern. J. Neurosci.*, 1: 181-197.
Valverde, F. (1976) Aspects of cortical organization related to the geometry of neurons with intra-cortical axons. *J. Neurocytol.*, 5: 509-529.
Valverde, F. (1983) A comparative approach to neocortical organization based on the study of the brain of the hedgehog (*Erinaceus europaeus*), in: *Ramón y Cajal's Contribution to the Neurosciences*, S. Grisolía, C. Guerri, F. Samson, S. Norton, and F. Reinoso-Suárez, eds., Elsevier, Amsterdam, 149-170.
Valverde, F. (1985) The organizing principles of the primary visual cortex in the monkey, in: *Cerebral Cortex, Visual Cortex*, Vol. 3, A. Peters and E.G. Jones, eds., Plenum Press: New York and London, 207-257.
Valverde, F. (1986) Intrinsic neocortical organization: Some comparative aspects. *Neurosience*, 18: 1-23.
Valverde, F. (1988) Competition for the sake of diversity. *Behav. Brain Sci.*, 11: 102-103.
Valverde, F., De Carlos, J.A., López-Mascaraque, L., and Doñate-Oliver, F. (1986) Neocortical layers I and II of the hedgehog (*Erinaceus europaeus*). II. Thalamo-cortical connections. *Anat. Embryol.*, 175: 16171-179.
Valverde, F., and Facal-Valverde, M.V. (1986) Neocortical layers I and II of the hedgehog (*Erinaceus europaeus*). I. Intrinsic organization. *Anat. Embryol.*, 173: 413-430.
Valverde, F., and Facal-Valverde, M-V. (1988) Postnatal development of interstitial (subplate) cells in the white matter of the temporal cortex of kittens: A correlated Golgi and electron microscopic study. *J. Comp. Neurol.*, 269: 168-192.
Valverde, F., López-Mascaraque, L., and De Carlos, J.A. (1989) Structure of the nucleus olfactorius anterior in the hedgehog (*Erinaceus europaeus*). *J. Comp. Neurol.*, 279: 581-600.
Valverde, F., and Ruiz-Marcos, A. (1969) Dendritic spines in the visual cortex of the mouse: Introduction to a mathematical model. *Exp. Brain Res.*, 8: 269-283.
Welker, C. (1976) Receptive fields of barrels in the somatosensory neocortex of the rat. *J. Comp. Neurol.*, 166: 173-190.
White, E.L. (1979) Thalamocortical synaptic relations: A review with emphasis on the projections of specific thalamic nuclei to the primary sensory areas of the neocortex. *Brain Res. Rev.*, 1: 275-311.
Wilson, M.E., and Cragg, B.G. (1967) Projection from the lateral geniculate nucleus in the cat and monkey. *J. Anat.*, 101: 677-692.
Winfield, D.A., and Powell, T.P.S. (1976) The termination of thalamo-cortical fibers in the visual cortex of the cat. *J. Neurocytol.*, 5: 269-281.
Winfield, D.A., and Powell, T.P.S. (1983) Laminar cell counts and geniculocortical boutons in area 17 of cat and monkey. *Brain Res.*, 277: 223-229.
Winfield, D.A., Rivera-Dominguez, M., and Powell, T.P.S. (1982) The termination of geniculocortical fibers in area 17 of the visual cortex in the macaque monkey. *Brain Res.*, 231: 19-32.
Wise, S.P., and Jones, E.G. (1976) The organization and postnatal development of the commissural projection of the rat somatic sensory cortex. *J. Comp. Neurol.*, 168: 313-344.
Wise, S.P., and Jones, E.G. (1978) Developmental studies of thalamo-cortical and commissural connections in the rat somatic sensory cortex. *J. Comp. Neurol.*, 178: 187-208.
Woolsey, T.A., Dierker, M.L., and Wann, D.F. (1975) Mouse SmI cortex: qualitative and quantitative classification of Golgi-impregnated barrel neurons. *Proc. Natl. Acad. Sci. USA*, 72: 2165-2169.
Woolsey, T.A., and Van der Loos, H. (1970) The structural organization of layer IV in the somatosensory region (SI) of mouse cerebral cortex. The description of a cortical field composed of discrete cytoarchitectonic units. *Brain Res.*, 17: 205-242.

ON THE COINCIDENCE OF LOSS OF ELECTRORECEPTION AND REORGANIZATION OF BRAIN STEM NUCLEI IN VERTEBRATES

Bernd Fritzsch

Scripps Institute of Oceanography
UCSD, Dept. of Neurosci., A-001
La Jolla, CA 92093

INTRODUCTION

Only anamniotic vertebrates, such as lampreys, sharks, bony fish, and amphibians possess electroreceptive organs. Some bony and cartilaginous fish species have developed electric organs and some bony fish communicate with electric signals. This electrocommunication parallels in many ways the auditory system. Some of these fish have developed a large brain sometimes approaching the brain-body weight ratio of mammals (Bell and Szabo, 1986). These enlarged brain areas, in part related to electrocommunication, may form as much as 50% of the brain volume but may not be recognizable at all in related species. Clearly the sometimes laminated areas of the brain devoted to the processing of electrosensory information display a very large variability in size not matched by any other sensory modality. With respect to the relative increase in size and the laminar organization these changes rival in a way those underlying the evolution of the mammalian cerebral cortex. Understanding of the mechanism(s) through which this variability is achieved in the specialized electrosensory system may help to understand more general problems of vertebrate brain evolution such as the phylogeny of the mammalian cortex.

Electroreception is not only related to large variation in the size of certain brain areas devoted to this sense but is also known to reach the forebrain in sharks (Bodznick and Northcutt, 1984; Bodznick and Boord, 1986) and in salamanders (Northcutt and Plassmann, 1989), which are together with caecilians the only tetrapods that have retained the old vertebrate sense of electroreception (Fritzsch and Münz, 1986). In contrast, amniotes have all lost this old vertebrate sense, presumably also including its telencephalic connections. If amniotes are at all able to perceive weak electric fields, like the platypus (Scheich et al., 1986), they do it with organs that are not homologous to those of anamniotic vertebrates. This pattern of distribution shows that electroreception cannot be directly related to the evolution of the mammalian cortex, the topic of this book, because this sense was lost before the telencephalon started its massive development and differentiation in ancestral mammals.

I will highlight some aspects of phylogenetic and ontogenetic repatterning underlying reorganizations in the electrosensory system to exemplify more general principles of changes related to regression, decoupling and subsequent reorganization in the octavolateral system of vertebrates. The examples given for the electrosensory and mechanosensory lateral line are meant to illustrate a more general problem of vertebrate brain evolution largely unexplored so far: what may be gained if something gets lost.

The general idea underlying this presentation is particularly well exemplified by the functional transformation of the hyomandibular bone into a middle ear ossicle (Gaupp, 1912) after the upper jaw had fused with the neurocranium (Starck, 1982). This event freed the hyomandibular bone from its ancient function to support the jaws and allowed its incorporation into the ear capsule as a middle ear bone. Although the first event, decoupling of the hyomandibular, is necessary, it has apparently not always resulted in the functional transformation of the hyomandibular bone as shown by the cases of ratfish and lungfish (Starck, 1982) where the hyomandibular bone simply forms an appendage on the cranium. In the third lineage, the tetrapods, a need arose for a sound conducting element in the middle ear to achieve impedance matching of the airborn sound to the inner ear. The availability of a functionally non-committed bone in the vicinity of the middle ear provided a fortunate opportunity for a functional transformation of this ossicle into a sound conducting middle ear ossicle.

This example outlines two aspects of how regressive evolution may lead through functional decoupling of elements to the evolution of novel function of this element, provided it can be used for this function with moderate changes. Both loss of the old function as well as requirement for a novel function together with a historical coincidence of both changes are necessary to make such a transformation a likely event.

The review provided here will stress this coincidence in loss of electroreception and rearrangement of brain stem nuclei. I will argue that complete loss of afferents may effect the brain stem nuclei and can result in reorganization and subsequent stabilization at a somewhat different, newly evolved pattern. In the last part I will speculate on possible effects of loss of electroreceptors for the evolution of thalamo-telencephalic connectivities.

Electroreceptive organs are as old as any other major vertebrate sense organ, but were lost at least three times independently

The electroreceptive organs, until ten years ago believed to be present only in some bony and cartilaginous fish species, are now known to be much more widespread (Bullock, 1986; Northcutt, 1986). This conclusion is based on a number of criteria by which the electroreceptive organs of anamniotic vertebrates other than teleosts can be defined as homologous.
- a) The ampullary organs of nonteleosts have a canal leading to an ampulla surrounded by sensory cells. The organs are excited when the outside is negatively charged. Their threshold varies between less than 1 to 100 uV/cm (Zakon, 1986).
- b) The organs are exclusively innervated from ganglion cells confined to the anterior lateral line ganglia. When organs occur on the trunk, a recurrent branch of the anterior lateral line nerve innervates them (Northcutt, 1986).
- c) Afferent nerve fibers terminate in a single nucleus of the alar plate, the dorsal or electrosensory nucleus. This nucleus has ascending lemniscal connections to the midbrain which may continue to the forebrain pallium (Bodznick and Northcutt, 1984; Bodznick and Boord, 1986) or striatum (Northcutt and Plassmann, 1989).

Based on these criteria, electroreception can be identified in at least some members of most anamniotic vertebrate taxa. Among cyclostomes, lampreys are electroreceptive but hagfishes apparently not. This can be due either to reduction, as suggested also for the mechanosensory lateral line of hagfish (Northcutt, 1989a), or due to the primitive absence of this sense in the latter. Lack of any evidence of this sense in acraniate (headless) vertebrates, the outgroup of craniate vertebrates, impedes further analysis of this issue. In any case, the available data suggest that electroreception was present in the common ancestor of lampreys and jawed vertebrates and is presumably as old as any major craniate vertebrate sense.

Among jawed vertebrates, most bony fish, all frogs and all amniotes do not possess ampullary electroreceptors. In contrast to the problem surrounding cyclostomes, outgroup comparison shows that loss of electroreception in bony fish, anuran amphibians and amniotic vertebrates is a derived feature achieved independently in each of these lineages. It is unclear how the disappearance of electroreceptors actually proceeded. Data on the regression of a comparable sense, the mechanosensory lateral line organs in amphibians (Fritzsch, 1989), indicate that both sudden loss in closely related species as well as continual diminution to a vestigial organ with subsequent loss may be possible. The following scenario assumes that the loss of electroreceptors was in each lineage a single event which caused suppression of electroreceptor formation, most likely through a change of epigenetic interactions in the developing placode.

Electroreceptors reevolved in some bony fish and are more diverse than in any other anamniotic taxon

Loss of ampullary electroreceptors in bony fish is believed to have occurred in the common ancestor of modern bony fish because the two sister groups lack electroreception (Northcutt,

1986). This implies that loss of electroreception is primitive for bony fish (i.e. neopterygian ray-finned fishes). As a consequence, we must assume that electroreception was reinvented at least twice in teleosts, once in mormyroids/ notopteroids and once in siluriforms/ gymnotiforms lineages. This suggestion is backed up by structural and functional differences, e.g. the reversed excitatory (anodal) polarity as compared to the non-teleost ampullary organs and a somewhat different central projection (Finger et al., 1986). The fact that only teleosts have evolved a second type of electroreceptor, the tuberous receptors specifically used in active electrolocation and species specific communication (Zakon, 1986), stresses the uniqueness of bony fish with respect to their electrosensory system.

Unfortunately neither loss nor reevolution of electroreceptors in teleosts is understood in terms of differential regulation of ontogenetic events (Finlay et al., 1987). This uncertainty relates to the fact that it is still unknown what the embryologic origin of any electroreceptor is. The available evidence indicates that electroreceptors, like all other organs with secondary sensory cells as a sensory element, may derive from placodal material (Northcutt, 1986, 1989a). This suggestion is probably true also for the ganglion cells (Fritzsch and Northcutt, unpublished data). It is not known how the placodal material is made to form electroreceptors, i.e. what inductive events and what genes are necessary for the formation of electroreceptors. Consequently it can not be ruled out that loss of ampullary electroreceptors may be nothing but loss of specific interactions necessary to express genes relevant for the formation of electroreceptors out of placodal material forming secondary sensory cells. If this is true, it is further conceivable that all teleosts have those genes, presumably stabilized by their pleiotrophic effects, but simply do not express them. Possibly then, reevolution of ampullary organs may represent a re-expression of genes relevant for the formation of these organs.

Alternatives such as transformation of mechanoreceptive ampullary electroreceptors were traditionally favoured both for teleosts and non-teleosts. Most likely, the evolution of electroreceptors of non-teleosts was a transformation of some kind of mechanosensory hair cell into an electrosensory hair cell. However, this assumption should be reconsidered for teleosts (Finger et al., 1986).

Two related questions add to the unsolved issue of the origin of electroreceptors within teleosts:

Where do the second type of electroreceptors, the tuberous organs, and where do the appropriate central nuclei for electroreceptive afferents come from? Tuberous organs have evolved at least three times independently in mormyrids, gymnotiformes and siluriforms (Finger et al., 1986; Andres et al., 1988), presumably out of ampullary organs. It was proposed that the genes for the teleost ampullary receptors may have been duplicated followed by a different evolution in the two alleles (Braford, 1986), i.e. one or more genes coding relevant aspects of formation of ampullary organs may, after duplication, evolve into genes specifying a tuberous receptor instead.

The issue of the formation of the central nuclei in which the afferents of these organs end is even more complex and less understood. As a part of the CNS, these cells are unlikely to be derived from placodal material, the likely source of the electroreceptors. In order to understand this issue we have to understand what happens to CNS neurons when their specific input is lost, i.e. when neither ampullary organs nor their afferents are formed. We also need to know how newly evolved electroreceptors establish a specific area of termination in the brain stem.

All available embryological data show that central neurons will be generated even in the complete absence of their peripheral input (Parks et al., 1987). Based on these data, it appears likely that this will happen also to the second order electrosensory neurons when no electroreceptors are formed in ancient bony fish. The subsequent historical and ontogenetic fate of these cells will depend on their biochemical identity. If this is such that adjacent fibers, e.g. from the mechanosensory lateral line organs or other sources can innervate these cells, they may survive and become incorporated into the mechanosensory lateral line nucleus. If this is not possible, the available evidence (Parks et al., 1987; Finlay et al., 1987) suggests that the majority of these second order electrosensory neurons will degenerate subsequent to their generation, i.e. both historically and during ontogeny these neurons disappear. Given that the loss of electroreceptors occurred presumably some 250 million years ago in the bony fish lineage, it is unfortunately impossible to study the fate of the second order electrosensory neurons in modern bony fish. This issue will be addressed again below in the discussion on electroreceptor loss in amphibians.

Explanations for the origin of second order neurons on which the afferents of the newly evolved electroreceptors of teleosts end depend critically on the interpretation of the electroreceptors themselves. The currently accepted conviction is that a group of mechanosensory second order neurons becomes selectively innervated by electroreceptive organ afferents and is transformed into an electroreceptive nucleus different from the ancient dorsal electroreceptive nucleus (McCormick, 1982). Pending the interpretation of the origin of the teleost electroreceptors, this historical event can mean the following ontogenetic changes:

a) It is in fact a part of the mechanosensory nucleus which becomes transformed into an electrosensory nucleus by invasion of the newly developed electroreceptive afferents (Finger et al., 1986), i.e. mechanosensory second order neurons are transformed by electrosensory afferents into second order electrosensory neurons.
b) Both the organs and the central neurons are newly formed by reshuffling existing genes in such a way that they result in the de novo formation of both structures.
c) The primordial electrosensory nucleus, historically incorporated into the mechanosensory nucleus when the electroreceptors were lost in ancestral teleost, receives specifically electrosensory afferents of the re-evolved ampullary organs and segregates again from the mechanosensory nucleus.

Clearly combinations of these proposals or even other, more complex alternatives are conceivable. Xenoplastic grafts between non-electroreceptive and electroreceptive teleosts such as catfish and goldfish are necessary to distinguish between these alternatives.

Again the situation is different for tuberous organs. Being uniformly considered as derived from ampullary organs, they innervate a gigantic part of the rhombencephalic medulla of teleosts which have these organs. This is particularly obvious when we compare the medulla of three related ostariophysin taxa, one without electroreceptors (goldfish), one with ampullary organs only (catfish), and one with ampullary and tuberous organs (gymnotid fish). The ampullary electroreceptors form somatotopic maps of the body surface in the electrosensory nucleus (Finger et al., 1986). Three (gymnotiformes) or two (mormyriformes) additional maps are formed by the tuberous organs, some of which are mirror symmetric to those of ampullary organs (Carr and Maler, 1986; Bell and Szabo, 1986). These maps are formed in a laminated, cortex-like area of the intermediate nucleus of the medulla. In contrast, there is no recognizable electroreceptive nucleus at all in the goldfish.

These data show that the issue of what electroreceptors are and where their central pathways come from are not at all solved. Likewise, the process of their diversification in teleosts is neither understood for the organs nor for the electrosensory nuclei or their ascending connections in terms of the ontogenetic steps taken.

Loss of electroreception may have been relevant for the formation of auditory nuclei in amphibians and amniotes

The question of the ultimate fate of second order electrosensory neurons in the absence of electrosensory afferents was already raised for teleosts (see above) and will be dealt with in amphibians in more detail. In amphibians, two out of three orders (salamanders, caecilians; Fritzsch and Münz, 1986) are electroreceptive. In contrast, frogs do not display any trace of electroreceptors or their afferents. Relevant to our context is the suggestion that a dorsal, undifferentiated cell group in the medulla of frogs is not a newly formed group of cells but may represent the transformed dorsal electrosensory nucleus of salamanders (Fritzsch, 1988). These cells apparently participate in the formation of the auditory nucleus of frogs. Irrespective of the unresolved nature of these cells, it is clear that loss of the ancient electroreceptive sense coincides with the rearrangement of the alar plate nuclei in ancestral frogs which eventually lead to the formation of the auditory nuclei. Loss of the ampullary organs is therefore at least indirectly related to the formation of auditory nuclei of frogs (Fritzsch, 1988).

The polarity of changes is in this case clear because the basilar papilla and its afferents appeared much earlier in phylogeny than auditory nuclei, namely already in the coelacanth (Fritzsch and Wake, 1988). The presence of afferents of the basilar papilla alone is therefore not sufficient to cause the immediate appearance of a specialized nucleus, an implicit corollary assumption of invasion and/or a parcellation-like processes (Ebbesson, 1984). Likewise, the suggested widespread connections of fibers during development on which selection can work

(Finlay et al., 1987) as found for thalamic afferents (Frost, this volume) is not supported in the case of lateral line and inner ear afferents of frogs which are seggregated even when one component is absent (Fritzsch, 1988; 1989b).

A comparable reorganization within the alar plate occurs in teleosts after the loss of electroreception. These parallelisms may have a single underlying cause: loss of a large sensory input will result first in instability followed by substantial rearrangement of cell groups in that area. This rearrangement may lead to formation of nuclei not seen in those groups with the "conservative" pattern of the afferents. Moreover, this reorganization of cell masses after loss of major inputs characterizes not only teleosts and frogs, but also amniotes. Therefore I suggest that loss of an ancient system results in a decoupling of cells determined to become innervated by the afferents of these organs. This can cause some rearrangement of existing nuclei, perhaps accompanied by a phylogenetic respecification of cell groups whose afferents were lost.

The case of some primitive bony fish which lack electroreception but show no apparent reorganisation of the alar plate cell masses (McCormick, 1982) indicates that decoupling of second order electrosensory neurons may not necessitate the reorganisation of neuronal material but rather offers only an opportunity to do so. A second event must coincide which can make use of the available cellular sources like the need for auditory nuclei to process the airborn sound with its novel physical features.

Experimental evidence to support the above outlined scenario for the origin of auditory nuclei in frogs may be gained by ablation experiments of one or the other organ system of the octavolateralis complex. This has been performed in amniotes, in particular the chick otocyst (Parks et al., 1987). Here, however, there is no lateral line or electroreceptive system available to take over cells depleted of their normal octaval input. Most recently otocyst ablation was performed in embryonic frogs (Fritzsch, 1990). The available data show substantial reduction in area and cell numbers of the octaval nuclei, but show no specific reinnervation of this area by the adjacent lateral line afferents. Instead, fibers running in the descending trigeminal tract apparently establish projections to this area. This shows that ocataval second order neurons proliferate and differentiate to some extent without their afferents. It also suggests that inner ear afferents and second order octaval neurons are to such an extent specified that lateral line fibers can not invade these. It remains to be shown whether this will also be true for lateral line and electroreceptive afferents in bony fish and non-bony fish like the salamanders.

Can loss of electroreception play any role for the evolution of diencephalic connections to the forebrain?

It is still unclear at which stage of vertebrate evolution thalamic afferents first reached the telencephalon and where the cells on which these afferents end come from. Jawless chordates like lampreys and hagfish apparently have already such a connection (Wicht and Northcutt, unpublished observations). Where the terminal areas of the thalamic afferents segregated from olfactory areas, i.e. was there at one stage an overlap of and a competition between olfactory and thalamic afferents won by the latter (Northcutt and Puzdrowski, 1988)? Or will newly invading thalamic afferents simply reach cells produced in the telencephalon but not specified to be second order olfactory neurons, i.e. can interneurons be transformed into neurons which process specific inputs? Or were telencephalic neurons on which the thalamic afferents impinge formed anew when these fibers first reached the telencephalon? These questions are rather similar to the problem of the origin of the electroreceptive nuclei in vertebrates at the time electroreceptors first formed during evolution. Moreover, evolution of electroreceptive nuclei in teleosts may bear directly on this question if the electroreceptors and second order neurons in this lineage turn out to be newly formed structures and not a somehow transformed ancient electroreceptive sense.

Available evidence indicates that at least in gnathostome vertebrates all sensory modalities reach the forebrain via thalamo-telencephalic projections. Data on agnathans, the sister group of gnathostome (jawed) vertebrates, are needed to show the pattern in the outgroup. Given that electroreception does reach the forebrain of sharks and salamanders (Northcutt and Plassmann, 1989), it may here be asked whether loss of this input had not similar effects on rearrangements of cell masses in the forebrain as it was outlined above for the brain stem alar plate, i.e. was there a cascade effect (Finlay et al., 1987) of the loss of electroreception throughout the brain? This question is particularly interesting for amniotes which have not only lost the electroreceptive sense but also the mechanosensory lateral line (Fritzsch, 1989). It is known that both sensory systems reach the forebrain in at least some anamniotic vertebrates (Bleckmann et al., 1988).

It appears possible that these regressive events in ancient vertebrate sensory systems do at least indirectly relate to the evolution of the amniotic forebrain, provided the thalamic and telencephalic neurons are not lost together with the entire electrosensory and mechanosensory pathway. If thalamic and telencephalic electrosensory and lateral line target neurons are formed in the absence of a specific electrosensory and mechanosensory lateral line input they may provide an opportunity for other thalamic projections, such as retinal, somatosensory or auditory afferents to expand into territories deprived of electroreceptive and lateral line afferents. Alternatively, absence of electrosensory and lateral line afferents to the thalamus may result in an instability with a subsequent reorganization of cell masses as well as thalamic afferents as proposed above for the brain stem. Data on experimental reorganization of afferents to the thalamus obtained in mammals (Frost, this volume) suggests that thalamic afferents can in fact be rearranged to drive other areas of the thalamus and through them of the forebrain. These data are compatible with the suggestion for the potential effect of loss of electroreceptors proposed above. More data on elasmobranchs, which have retained the old octavolateral senses but have an absolutely and relatively enlarged forebrain (Northcutt, 1989b), are needed to reveal whether their central forebrain organization is so different from amniotes as the above presented view of reorganization caused by loss of electroreception would require.

SUMMARY

The distribution of electroreception in vertebrates is reviewed and the implication for the evolution of the brain stem nuclei devoted to the processing of information provided by octavolateralis organs is discussed. Based on the coincidence of rearrangement of nuclei within the brain stem alar plate with the loss of the old vertebrate sense of electroreception, it is argued that the loss of electroreception may offer the opportunity for changes of an otherwise stable neuronal pattern. These changes may lead to the evolution of new nuclei such as auditory nuclei in frogs. The implications of this suggestion for the evolution of thalamic-telencephalic connections are discussed.

ACKNOWLEDGEMENTS

This paper has benefitted very much from insightful suggestions and comments of Dr. T. H. Bullock, Dr. R. G. Northcutt and G. Striedter. This work was supported by the German Science Foundation (DFG).

REFERENCES

Andres, K.H., Düring, M. von, Petrasch, E. (1988) The fine structure of ampullary and tuberous electroreceptors in the South American blind catfish Pseudocetopsis spec. *Anat. Embryol.*, 177: 523-535.

Bell, C.C. and Szabo, T. (1986) Electroreception in mormyrid fish. In: T.H. Bullock and W. Heiligenberg, (eds.). *Electroreception*. Wiley: New York, pp. 375-422.

Bleckmann, H., Bullock, T.H. and Jorgensen, J.M. (1988) The lateral linemechanoreceptive mesencephalic, diencephalic, and telencephalic regions in the thornback ray, Platyrhinoidis triseriata (Elasmobranchii). *J. Comp. Physiol.*, 161: 68-84.

Bodznick, D. and Northcutt, R.G. (1984) An electrosensory area in the telencephalon of the little skate, *Raja erinacea*. *Brain Res.*, 298: 117-124.

Bodznick, D. and Boord, R.L. (1986) Electroreception in chondrichthyes: central anatomy and physiology. In: T.H. Bullock and W. Heiligenberg, (eds.). *Electroreception*. Wiley: New York, pp. 225-256.

Braford, M.R. (1986) African knifefishes: The Xenomystines. In: T.H. Bullock and W. Heiligenberg, (eds.). *Electroreception*. Wiley: New York, pp. 453-464.

Bullock, T.H. (1986) Significance of findings on electroreception for general neurobiology. In: T.H. Bullock and W. Heiligenberg, (eds.). *Electroreception*. Wiley: New York, pp. 651-674.

Carr, C.E. and Maler, L. (1986) Electroreception in gymnotid fish: central anatomy and physiology. In: T.H. Bullock and W. Heiligenberg, (eds.). *Electroreception*. Wiley: New York, pp. 319-374.

Ebbesson, S.O.E. (1984) Evolution and ontogeny of neural circuits. *Behav. Brain Sci.*, 7: 321-366.

Finger, T.E., Bell, C.C. and Carr, C.E. (1986) Comparisons among electroreceptive teleosts: Why are electrosensory systems so similar? In: T.H. Bullock and W. Heiligenberg, (eds.). *Electroreception*. Wiley: New York, pp. 465-482.

Finlay, B.L., Wikler, K.C. and Sengelaub, D.R. (1987) Regressive events in brain development and scenarios for vertebrate brain evolution. *Brain Behav. Evol.,* 30: 102-117.

Fritzsch, B. and Münz, H. (1986) Electroreception in amphibians. In: T.H. Bullock and W. Heiligenberg, (eds.). *Electroreception.* Wiley: New York, pp. 483-492.

Fritzsch, B. (1988) The lateral-line and inner ear afferents in larval and adult urodeles, *Brain Behav. Evol.,* 31: 325-348.

Fritzsch, B. and Wake, M.H. (1988) The inner ear of gymnophione amphibians and its nerve supply: a comparative study of regressive events in a complex sensory system. *Zoomorphology.,* 108: 210-217.

Fritzsch, B. (1989) Diversity and regression in the amphibian lateral line system. In: S. Coombs, P. Görner, H. Münz (eds.) *The Mechanosensory Lateral Line.* Neurobiology and Evolution. Springer Verlag, 99-115.

Fritzsch, B. (1990) Experimental reorganization in the alar plate of the clawed toad, Xenopus laevis. I. Quantitative and qualitative effects of embryonic otocyst extirpation. *Develop. Brain Res.,* 51: 113-122.

Gaupp, E. (1912) Die Reichertsche Theorie. *Arch. Anat. Physiol., Suppl.,* 1-416.

McCormick, C.A. (1982) The organization of the octavolateralis area in actinopterygian fishes: A new interpretation. *J. Morphol.,* 171: 159-181.

Northcutt, R.G. (1986) Electroreception in nonteleost bony fishes. In: T.H. Bullock and W. Heiligenberg, (eds.). *Electroreception.* Wiley: New York, pp. 257-286.

Northcutt, R.G. and Puzdrowski, R.L. (1988) Projections of the olfactory bulb and nervus terminalis in the silver lamprey. *Brain Behav. Evol.,* 32: 96-107.

Northcutt, R.G. and Plassmann W. (1989) Electrosensory activity in the telencephalon of the axolotl. *Neurosci. Lett.,* 99: 79-84.

Northcutt, R.G. (1989a) The phylogenetic distribution and innervation of craniate mechanoreceptive lateral lines. In: S. Coombs, P. Gîrner, H. Münz (eds.) The Mechanosensory Lateral Line. *Neurobiology and Evolution.* Springer Verlag. pp. 18-78.

Northcutt, R.G. (1989b) Brain variation and phylogenetic trends in elasmobranch fishes. *J. Exp. Biol.,* 2: 83-100.

Parks, T.N., Jackson, H. and Conlee, J.W. (1987) Axon - target cell interactions in the developing auditory system, *Curr. Top. Develop. Biol.,* 21: 309-340.

Scheich, H., Langner, G., Tidemann, C., Coles, R. and Guppy, A. (1986) Electroreception and electrolocation in platypus. *Nature,* 319: 401-402.

Starck, D. (1982) *Die vergleichende Anatomie der Wirbeltiere,* pp 274, Springer,Heidelberg

Zakon, H.H. (1986) The electroreceptive periphery. In: T.H. Bullock and W. Heiligenberg, (eds.). *Electroreception.* Wiley: New York, pp. 103-156.

THE DESIGN OF STRIATE CORTEX

N.V. Swindale
Dept. of Ophthalmology
University of British Columbia
Vancouver, V5Z 3N9 Canada

INTRODUCTION

Our lack of knowledge about the functions of most regions of the brain makes it difficult to discuss how different parts of it may have evolved, and why particular brain regions are differently organised in different species. It is relatively easy to understand why the shape of a bird's beak varies from one species to the next, since the function of the beak is understood, but why is the lateral geniculate nucleus of a monkey organised differently from that of a cat or a mouse? A satisfactory explanation of the evolution of the lateral geniculate nucleus (or that of almost any other brain region) is impossible because we have very little idea of what the nucleus is for. Although discussing the evolution of the striate cortex may seem an equally impossible task, we do have some ideas about its possible function, thanks mainly to the work of Hubel and Wiesel, and my purpose in this chapter is to discuss one of their suggestions about what striate cortex does, and to show how this may have influenced some aspects of its design or evolution.

The coverage constraint

The nature of this constraint was first suggested some years ago by Hubel and Wiesel (1974b; 1977). They pointed out that the visual information present in any single small region of visual space could be analysed only by cells in a relatively small region of visual cortex, roughly 1 - 2 mm in diameter, and that this size was independent of cortical magnification factor or visual field eccentricity. Within this region, later referred to as the cortical point image (Van Essen et al., 1984) cells vary in their response properties, principally in terms of their connection strengths with either eye, a property referred to as ocular dominance, and their orientation selectivity. Any particular neuron can be characterised by some combination of ocular dominance and orientation preference, and other neurons in the same vertically disposed column of cells will share this particular combination of preferences. Cells in one column might for example have a strong input from the right eye, a relatively weak one from the left eye, and be selective for vertically oriented contours. Cells in a nearby column might be connected equally strongly to the left and right eyes and have a preference for left oblique orientations. Hubel and Wiesel suggested that columns in the striate cortex appeared to be laid out in such a way as to ensure that all combinations of orientation and eye preference would occur at least once within any region with a diameter equal to the size of the point image. This requirement will be referred to here as the coverage constraint: if the constraint is not satisfied and certain combinations of eye and orientation preference are not adequately represented at some visual field locations the animal might be expected to be perceptually blind or less sensitive to those stimuli, and this would presumably put it at a functional disadvantage.

Hubel and Wiesel (1977) proposed a layout of orientation and ocular dominance columns that would satisfy this combinatorial problem (Fig. 1): both sets of columns were hypothesized to be stripes of uniform width, with an overall repeat period of about a millimeter, intersecting most probably at 90° or a similar angle. It is easy to see that this would guarantee that all combinations would occur at least once within a region equal in size to whichever set of columns had the larger period. The hypothesis was only partly based on experimental evidence however: although it was

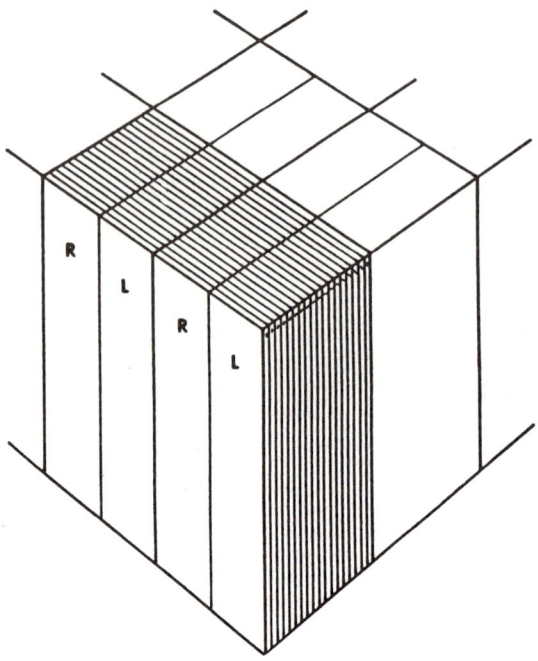

Figure 1. Diagram taken from Hubel and Wiesel (1977) illustrating a hypothetical arrangement of ocular dominance columns, labelled L and R, intersecting with a set of slab-like orientation columns at right angles. (Reproduced with permission)

known that ocular dominance columns form parallel stripes with a repeat period of about 850μm, neither the form taken by orientation columns, nor their structural relationship with ocular dominance columns was known at that time. More recent experimental evidence (Hubel et al., 1978; Blasdel and Salama, 1986) has shown that orientation columns are not parallel stripes of uniform width as shown in Fig. 1, nor do they intersect ocular dominance stripes at right angles. To further complicate matters, ocular dominance columns are not always simple stripes either, but may branch, curve, end blindly, or change width. In addition, there is evidence that they have a complex structural relationship with orientation columns, whereby regions of high spatial rates of change of orientation tend to be located in the centers of ocular dominance stripes (Blasdel and Salama, 1986). The repeat periods of orientation and ocular dominance columns are also different - about 600 μm for orientation (Hubel and Wiesel, 1974a; Hubel et al., 1978) and 850 μm for ocular dominance (Hubel et al.; LeVay et al., 1985). It should also be noted that the cortical point image is unlikely to be a cylindrical region of tissue, since the edges of receptive fields are not sharply defined, and the scatter of receptive field centers within a single cortical column probably has a Gaussian distribution.

These more complex morphological features raise the question as to whether or not the striate cortex is designed in such a way as to ensure uniform coverage, and if so, how this depends on factors such as the size of the cortical point image, the morphology and periodicity values of the ocular dominance and orientation columns, and the structural relations between the two. For example, if orientation and ocular dominance columns had the same periodicity and ran parallel to each other (which might happen by chance in some areas if the two were overlaid independently) destructive 'beats' between the two representations could occur, leading to an absence of certain combinations of eye and orientation preference (Cynader et al., 1987) and thus to poor coverage.

These questions can be answered given a quantitative definition of coverage, and the availability of accurate morphological models of orientation and ocular dominance columns. Some years ago, I showed how to generate, on a computer, patterns that closely resemble (at least on

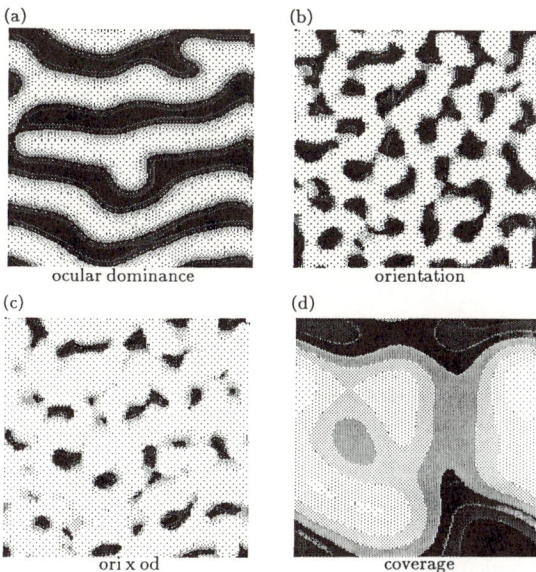

Figure 2. Illustration of the method used to calculate coverage. All four panels are modelled maps of neural activity, generated using the algorithms described in Swindale (1980, 1982).
  a) Hypothetical pattern of activity evoked in a small area of cortex by all orientations seen through one eye. Intensity of shading is proportional to amount of activity, and the pattern corresponds closely to the pattern of ocular dominance column.
  b) The activity evoked in the same area of model cortex by a single orientation seen through both eyes, derived from the pattern of orientation columns, assuming orientation tuning curves are approximately Gaussian in shape with $\sigma = 24°$.
  c) The activity evoked in the model cortex by a single orientation (present at all receptive field locations) seen through one eye. This is simply the product of the activity patterns seen in panels (a) and (b).
  d) A transformation of panel (c) into retinotopic coordinates, under the assumption that cortical magnification factor is locally uniform and isotropic. The density of shading at each point is now proportional to the total amount of cortical activity evoked by a single orientation at each retinal locus. This transformation is effected by convolving the image in (c) with the cortical point image, assuming a unit magnification factor.

the basis of a detailed visual scrutiny) the patterns of ocular dominance stripes (Swindale, 1980) and orientation columns (Swindale, 1982) found in the macaque monkey. More recently, I proposed a combined model that appears to accurately mimic the structural relations described by Blasdel and Salama (Swindale, 1990a). It is thereby possible to generate model cortices where one can vary at will the relative periodicity of the two types of column, as well as other structural features, such as the tendency of the columns (most obvious in the case of ocular dominance stripes) to run in a single predominant direction (e.g. perpendicular to the border with area 18).

Definition of coverage

Coverage can be defined in the following way (see Swindale, 1990b for the mathematical details): consider first of all a 4-dimensional stimulus space; two of the dimensions are for receptive field position (ϕ, ψ) and the third is for orientation, θ. A fourth dimension, ε, can have only two values, one for the left and the other for the right eye. A single point in this space thus represents the existence of a contour of a single orientation present in a particular receptive field location in one of the two eyes. Given a model cortex, one can calculate, for any point in this stimulus space, the total amount of neural activity likely to be evoked in the cortex (Fig. 2). Calculations are based on neural activity, since it is assumed that a cell can signal useful information about a stimulus even when it is not the one to which the cell is most sensitive. For example, a cell which is driven by both eyes, but less strongly by the left, is presumably still useful

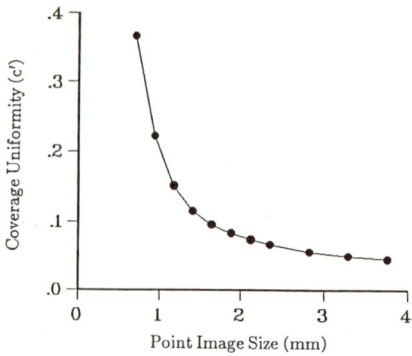

Figure 3. The effect of point image size on coverage uniformity at a constant periodicity ratio (ori/od) = .71, approximately the same as in the macaque monkey.

for vision when the right eye is closed, even though it is not maximally activated by any stimulus. To make calculations of coverage therefore, ocular dominance maps are calculated in such a way that values vary smoothly across the cortex, being binocular at the edges of the stripes, and monocular only in the centers. One also needs an estimate of the shape of the average orientation tuning curve of all the cells in a single column. Based on measurements by Schiller et al. (1976) this is probably well described by a Gaussian curve with a σ = about 24°.

Calculation of coverage proceeds as follows: for a given model cortex and its associated piece of stimulus space one calculates, for each point in the stimulus space (i.e. every combination of eye, orientation and visual field location), the total amount of neural activity, $A(\phi, \psi, \theta, \epsilon)$ evoked in the model cortex, taking into account the width of orientation tuning and the size and shape of the cortical point image. If coverage is perfectly uniform, the value of A should be the same for each point $(\phi, \psi, \theta, \epsilon)$. A simple measure of non-uniformity can be obtained by taking the ratio between the standard deviation of A and its mean, to yield a value c', which will be referred to here as 'coverage uniformity'. Coverage is good (i.e. uniform) when the value of c' is small, and poor (i.e. uneven) when c' is large.

Point image size

It is obvious that a very small point image will necessarily lead to poor coverage because each visual field location will be represented by only the center of a single ocular dominance or orientation column. This would mean that the other eye, or other orientations, would have no cortical representation at that location. On the other hand if the point image is sufficiently large, coverage is almost certain to be uniform. This is because any location in visual space will activate a large area of cortex that will contain many sets of orientation and ocular dominance columns that will probably contain all combinations of the two parameters in nearly equal numbers. While this might be desirable, a large point image would presumably have disadvantages, such as decreasing the accuracy with which objects could be localised in space. One might therefore expect point image size to be as small as was consistent with obtaining reasonably uniform coverage.

Figure 3 shows how coverage uniformity, c', varies over a range of point image sizes. For this calculation all other parameters, such as the periodicities of orientation and ocular dominance columns, were chosen to be similar to the actual values in the macaque monkey. The cortical point image was assumed to have a Gaussian profile, with a size expressed as four times the width of the underlying standard deviation. This measure of diameter is probably roughly comparable with the estimates of diameter made experimentally by Hubel and Wiesel (1974b, 1977) and Van Essen et al. (1984). The figure shows that c' increases rapidly as point image size decreases below a diameter of about 1.5 mm. This is similiar to experimental estimates of point image size, which are in the range 1 - 2 mm, and suggests that point image size may indeed be as small as allowed by the constraints on uniform coverage as defined here.

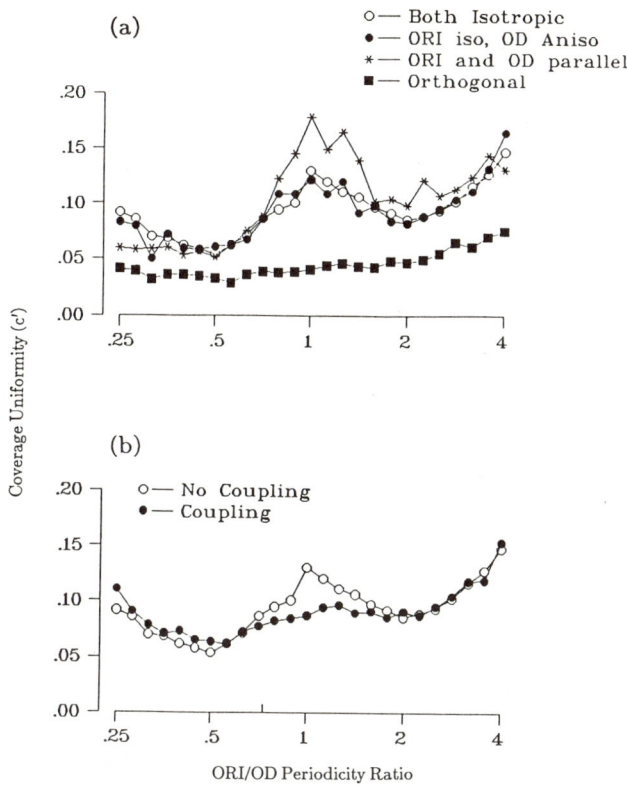

Figure 4. The effect of periodicity ratio (ori/od) on coverage uniformity for a variety of different structural conditions. See text for further description.

Column periodicity

In macaque striate cortex, orientation and ocular dominance columns have periodicities of 600 µm and 850 µm respectively, a ratio of about 0.7. In area 17 of the cat, the corresponding values are 1050 µm (Löwel et al., 1987) and 800 µm (Löwel and Singer, 1987), a ratio of about 1.3; in area 18 of the same species, the values are 1250 µm and 1860 µm, a ratio of about .67 (Cynader et al., 1987). These observations show that both the periodicity ratio, as well as the absolute periodicity values of orientation and ocular dominance columns are not fixed, but vary, within and between species. It seems intuitively likely that coverage uniformity might depend on the periodicity ratio, and, as mentioned above, may be worst when orientation and ocular dominance columns run in the same direction and have the same periodicity.

Figure 4a shows how coverage uniformity depends on the periodicity ratio between orientation and ocular dominance columns, for a variety of different structural conditions. In order to study the effect of periodicity ratio *per se*, the actual periodicity values of the two types of column were varied reciprocally, about a common geometric mean, to produce a four-fold range of ratios, keeping point image size constant at a value of about 1.5 mm. A variety of different structural conditions were simulated: in all of these the structural relations described by Blasdel and Salama (1986) were assumed to be absent, and the two sets of columns simply overlaid at random. In one condition (Fig. 4a, open circles) both sets of columns were spatially isotropic i.e. there was no tendency for either type of column to be elongated in any direction, and the two kinds of column thus intersected at random angles. In another condition, ocular dominance stripes were made anisotropic and tended to run in a single direction (filled circles), while in a third condition (asterisks) both types of column were made to run in the same direction. Coverage varied in a similar way for all three of these conditions, being most uniform at periodicity ratios (ori/od) of about 0.6 or less, and least uniform at ratios close to unity or greater than three. As was expected, coverage was worst when columns had the same periodicity and ran parallel to each other. In a fourth condition (filled squares) the two sets of columns were made to run orthogonally to each other. This condition comes closest to Hubel and Wiesel's original model of striate cortex design (Fig. 1) and produces more uniform coverage than in any other design condition, at all frequency ratios, although there is still a bias for ratios less than unity.

Figure 4b shows the effect of introducing 'coupling' between orientation and ocular dominance columns i.e. a tendency for regions within which orientation preference changes rapidly to lie in the centers of the ocular dominance stripes, as described by Blasdel and Salama (1986). This has the effect of removing the rise in the value of c' that occurs at frequency ratios close to unity, although coverage values at other ratios are unaffected.

Does the cortex optimise coverage?

The analysis presented here suggests that a number of features of striate cortex anatomy may be at least partly explained in terms of a need to optimise coverage uniformity. Overall, the calculations show that, to obtain uniform coverage, orientation columns should have a repeat period about half that of ocular dominance columns, and that the cortical point image should have a diameter equal to about twice the spacing of ocular dominance columns. Striate cortex anatomy in the macaque conforms at least roughly to this design, since orientation columns have a spacing of about 600 µm, ocular dominance columns a spacing of about 850 µm and the point image is about 1.5 mm in diameter. It is likely however, that other constraints will interact with those discussed here to determine column periodicities. For example, orientation discrimination may require the detection of some minimum change in the locus of cortical activation, and this would be related to orientation column periodicity. Reducing the average spacing of orientation columns might interfere with this task, although it would improve coverage uniformity, as would doing away with columns altogether and simply distributing orientation preferences randomly.

Coverage uniformity could also be improved by another simple change in cortical design, in which ocular dominance and orientation columns were each elongated in orthogonal directions, as illustrated in Fig. 4a (squares). The degree of elongation required to do this would not be significantly different from that found either in the orientation column system, in the tree shrew (Humphrey et al., 1980) or in the ocular dominance bands present in foveal striate cortex of the macaque. This direction of elongation however is always perpendicular to the border between areas 17 and 18, and possibly there is some developmental constraint that prevents columns from running parallel to the border.

The role of coupling

The present analysis does not provide an obvious functional explanation for the coupling between ocular dominance and orientation columns discovered by Blasdel and Salama (1986) since coupling has no effect on coverage at the periodicity ratio found in the monkey (Fig. 4b). Coupling may nevertheless play some functional role related to coverage. For example, ocular dominance columns decrease in size from fovea to periphery by a factor of about two (LeVay et al., 1985), and unless the size of the orientation columns changes similarly the periodicity ratio between the two is certain to fall into a range where coupling would significantly improve coverage uniformity. Local variations in column size or periodicity might have the same effect.

Limitations of the analysis

This paper has shown how computer generated models of columnar structure in the visual cortex can be used as a tool to explore the effects of changes in cortical structure on the efficiency with which the cortex achieves uniform representation of certain stimulus features. Although the results suggest that the visual cortex is organised in such a way as to achieve reasonably uniform coverage, some weaknesses in the analysis can be pointed out. While the computer models visually resemble the real patterns, it is possible that there are unrecognised structural differences between the real and model patterns which would have the effect of altering the coverage estimates and their dependence on periodicity ratios and point image size. One way of checking for this would be to calculate coverage values from the voltage sensitive dye patterns using the methods outlined here, and to compare the results with the values predicted by the models. Another problem is that the real stimulus space that is mapped onto the cortex is likely to include more dimensions than the four considered here. Thus most cortical neurons are sensitive not only to the orientation, retinal location and eye of origin of a stimulus, but also to properties such as length, colour, contrast, velocity and direction of movement. If selectivity along any of these, or other dimensions has an ordered columnar representation then the stimulus space to be considered in the calculation of c' should be expanded along each of these additional dimensions. This would necessarily yield additional sources of variance in A, and thus larger values of c', although the role of the factors discussed here would be expected to remain the same.

REFERENCES

Blasdel, G.G. and Salama, G. (1986) Voltage-sensitive dyes reveal a modular organisation in monkey striate cortex. *Nature*, 321: 579-585.
Cynader, M.S., Swindale, N.V. and Matsubara, J.A. (1987) Functional topography in cat area 18. *J. Neuroscience*, 7: 1401-1413.
Hubel, D.H. and Wiesel, T.N. (1974a) Sequence regularity and geometry of orientation columns in the monkey striate cortex. *J. Comp. Neurol.*, 158: 267-294.
Hubel, D.H. and Wiesel, T.N. (1974b) Uniformity of monkey striate cortex: a parallel relationship between field size, scatter, and magnification factor. *J. Comp. Neurol.*, 158: 295-306.
Hubel, D.H. and Wiesel, T.N. (1977) Functional architecture of macaque monkey striate cortex. *Proc. R. Soc. Lond. B*, 198: 1-59.
Hubel, D.H., Wiesel, T.N. and LeVay, S. (1977) Plasticity of ocular dominance columns in monkey striate cortex. *Phil. Trans. R. Soc. Lond. B,* 278: 131-163.
Hubel, D.H., Wiesel, T.N. and Stryker, M.P. (1978) Anatomical demonstration of orientation columns in macaque monkey. *J. Comp. Neurol.*, 177: 361-380.
Humphrey, A.L., Skeen, L.C. and Norton, T. T. (1980) Topographic organisation of the orientation column system in the striate cortex of the tree shrew (*Tupaia glis*). II. Deoxyglucose mapping. *J. Comp. Neurol.*, 192: 549-566.
LeVay, S., Connolly, M., Houde, J. and Van Essen, D.C. (1985) The complete pattern of ocular dominance stripes in the striate cortex and visual field of the macaque monkey. *J. Neuroscience*, 5: 486-501.
Löwel, S., Freeman, B. and Singer, W. (1987) Topographic organisation of the orientation column system in large flat-mounts of the cat visual cortex: a 2-deoxyglucose study. *J. Comp. Neurol.*, 155: 401-415.
Löwel, S. and Singer, W. (1987) The pattern of ocular dominance columns in flat mounts of the cat visual cortex. *Exp. Brain Res.*, 68: 661-666.
Schiller, P.H., Finlay, B.L. & Volman, S.F. (1976) Quantitative studies of single-cell properties in monkey striate cortex. II. Orientation specificity and ocular dominance. *J. Neurophysiol.*, 39: 1320-1333.
Swindale, N.V. (1980) A model for the formation of ocular dominance stripes. *Proc. R. Soc. Lond. B*, 208: 243-264.
Swindale, N.V. (1982) A model for the formation of orientation columns. *Proc. R. Soc. Lond. B*, 215: 211-230.
Swindale, N.V. (1990a) A model for the coordinated development of columnar systems in primate striate cortex. (submitted).
Swindale, N.V. (1990b) Coverage and the design of striate cortex. (submitted).
Van Essen, D.C., Newsome, W.T. and Maunsell, J.H.R. (1984) The visual field representation in striate cortex of the macaque monkey: asymmetries, anisotropies and individual variability. *Vision Res.*, 24: 429-448.

REPRESENTATIONAL GEOMETRIES OF TELENCEPHALIC AUDITORY MAPS IN BIRDS AND MAMMALS

H. Scheich
Zoological Institute
Technical University Darmstadt
Schnittspahnstr. 3, 6100 Darmstadt, FRG

INTRODUCTION

The evolution of the mammalian as well as of the bird brain is highlighted by the parallel acquisition of a voluminous telencephalon. Yet the routes of morphological forebrain differentiation taken by the two forms are strikingly divergent. Birds lack a true cortex with its specific cell types and a segregation of gray and white matter areas is not conspicuous. Instead the telencephalic mass is entirely composed of basal forebrain nuclei and multiple and thick dorsal layers of stellate cells, the latter being separated by several thin fibrous laminae and crossed by diffuse fiber tracts. In the layers of this dorsal roof, traditionally misnamed "striatum", local variations of cytoarchitecture are present but rarely with sharp boundaries. Only with modern connectivity studies it has been recognized over the past three decades that in spite of this low degree of overt organization the bird telencephalon follows a plan very similar to that of mammalian forms. This covers functional subsystems (including neocortical equivalents) and their intratelencephalic connections as well as connections with subtelencephalic structures (Karten, 1969). It is the aim of this contribution to characterize the functional organization of the auditory cortex analogue, Field L, of birds and to compare it to auditory cortex in the mongolian gerbil (*Meriones unguiculatus*). Field L of birds from various families has been studied in greater detail over the years in this laboratory and the gerbil has been chosen recently as a mammalian model which allows specific comparisons with bird auditory systems. These species share low frequency hearing with most space in the telencephalic auditory maps devoted to the analysis of frequencies below 10 kHz. In chicks and gerbil there seems to be no specialization for communication sounds greatly distorting the spatial organization of their auditory maps. A comparison of auditory maps seems possible since, in spite of some differences in the organization of the cochlea, birds appear to possess analogues of all mammalian nuclei of the ascending auditory pathway and very similar types of units (see Sachs et al., 1980). Therefore, I shall assume here that the steps of hierarchical processing along the auditory pathway, lead to similar functional results below the level of the forebrain. The major difference between bird and mammalian organization appears to occur in the forebrain. Birds have one multilayered tonotopic gradient to harbour mechanisms of "telencephalon-like" auditory pattern analysis, while the rule for mammals, except presumably for very basal forms, is to have multiple cortical representations of the cochlea.

Interestingly, not a single aspect of auditory pattern recognition and learning has been discovered to date, in which bird's performance, in principle, is inferior to mammals. Yet, this is achieved in probably a single telencephalic auditory map with neuronal elements and an organization different from mammalian auditory cortex.

Field L: Basic findings

In 1914 Rose described a plate-like assembly of densely packed cells in the caudal neostriatum of the bird telencephalon and called it Field L. Following some preliminary reports on evoked auditory responses in this area (Erulkar, 1955) Karten (1968) identified Field L as the telencephalic target area of the thalamic auditory nucleus ovoidalis in the pigeon using degeneration technique. Highly complex response properties of single units were subsequently revealed in Field L (Biederman-Thorson, 1970; Leppelsack 1974, Scheich et al., 1979a).

A reinvestigation of connectivities with injections of labelled amino acids in physiologically identified auditory sites in thalamus and forebrain of the Guinea fowl disclosed a more intricate organization of Field L than previously assumed (Bonke, B. et al., 1979). It turned out that the original Field L in the neostriatum was only the layer of thalamic inputs to a larger auditory area. This newly defined field L included a postsynaptic layer L_1, the intermediate input layer L_2 and a postsynaptic layer L_3, all parallel in dorsoventral sequence (Fig. 1). In addition, a strip of dorsally overlying auditory hyperstriatum ventrale (AHV), separated by the lamina hyperstriatica, was found to be an integral part on the organization receiving a projection from Field L layers and projecting back to L_2.

Systematic microelectrode penetrations provided evidence of a precise tonotopic organization. Units with similar best frequency were assembled in a thin sheet with isofrequency contours oriented radially across the lamination L_1 - L_3 including AHV and in a longitudinal dimension extending in rostrocaudal direction (Bonke, D. et al., 1979). Bands of increased metabolic activity corresponding in orientation and extent to electrophysiological isofrequency contours were visualized with the 2-deoxyglucose (2DG) mapping technique (Scheich et al., 1979b). Tonotopic 2DG-bands as obtained with tone burst stimulation turned out to be an extremely robust phenomenon in Field L and subsequently facilitated a more comprehensive exploration of functional organization of the field.

Similarities of laminar organization of Field L and mammalian auditory cortex

The laminar organization of Field L is made visible by the 2DG method and histochemical techniques with or without auditory stimulation (Fig. 1) (Scheich et al., 1979b; Scheich 1983; Heil and Scheich, 1986). The input layer L_2 spontaneously shows a higher metabolic activity than the postsynaptic layers L_1 and L_3. An overlying strip of auditory HV is also strongly labelled. The band of high 2DG activity corresponding to L_2 is an excellent landmark of the laminar organization especially in the transverse plane. L_2 is similarly pronounced as a band of high spontaneous activity in cytochrome-oxidase stain (Braun et al., 1985).

Interestingly, the metabolically active input layer L_2 is selectively spared by the Timm-stain which detects heavy metals, particularly zinc, in tissue (Fig. 1, inset) (Faber et al., 1989). The postsynaptic layers L_1 and L_3 are moderately active in this stain and clearly separble from AHV. The selective sparing of L_2 is accompanied by the same phenomenon in all primary sensory projection areas of the bird telencephalon (e.g. visual ectostriatum, somatosensory neostriatum, n. basalis, and the Wulst input layer). This and heterogeneous silver deposits distinguishing many parts of the forebrain render the Timm method superior to Nissl and other routine stains for visualizing anatomical subdivisions of the bird telencephalon (Faber et al., 1989).

A concept for differential 2DG staining of Field L layers, can be derived in part. High 2DG uptake is primarily related to high density and high electrical activity of synaptic neuropil (Heil and Scheich, 1986; Nudo and Masterton, 1986). The reason is the dominant energy consumption of the membrane sodium pump (Mata et al., 1980) which favors 2DG accumulation in cellular compartments with large surface-to-volume-ratios like synaptic neuropil and not in perikarya. Thus, local postsynaptic spike activity appears to contribute less to overall 2DG uptake than presynaptic and dendritic potentials. That thalamic input activity in L_2 must be high indeed, is indirectly shown by high spontaneous and evoked activity of L_2 neurons (Bonke, D. et al., 1979). This also means that much of the input activity from the thalamus is conveyed by a tight input-output link to local neurons. The presence of densely packed small stellate cells in L_2 (microneurons) (Saini and Leppelsack, 1981) probably being contacted by a host of terminals also suggest many input compartments all accumulating 2DG. Strong cytochrome oxidase activity in L_2 may be a reflection of high mitochondrial density in these input compartments involved in the energy metabolism of the accumulated glucose.

This same reasoning can be applied to mammalian auditory and, generally, to sensory cortex in which high spontaneous 2DG uptake in input layer IV (Hubel et al., 1977; Steffen et al., 1988) and concomitantly high cytochrome activity are found (Wong-Riley and Riley, 1983). Sparing of L_2 and of all other thalamic sensory input layers in the bird telencephalon by the Timm stain corresponds to the same phenomenon in mammalian cortex layer IV (Haug, 1973). Since Timm-

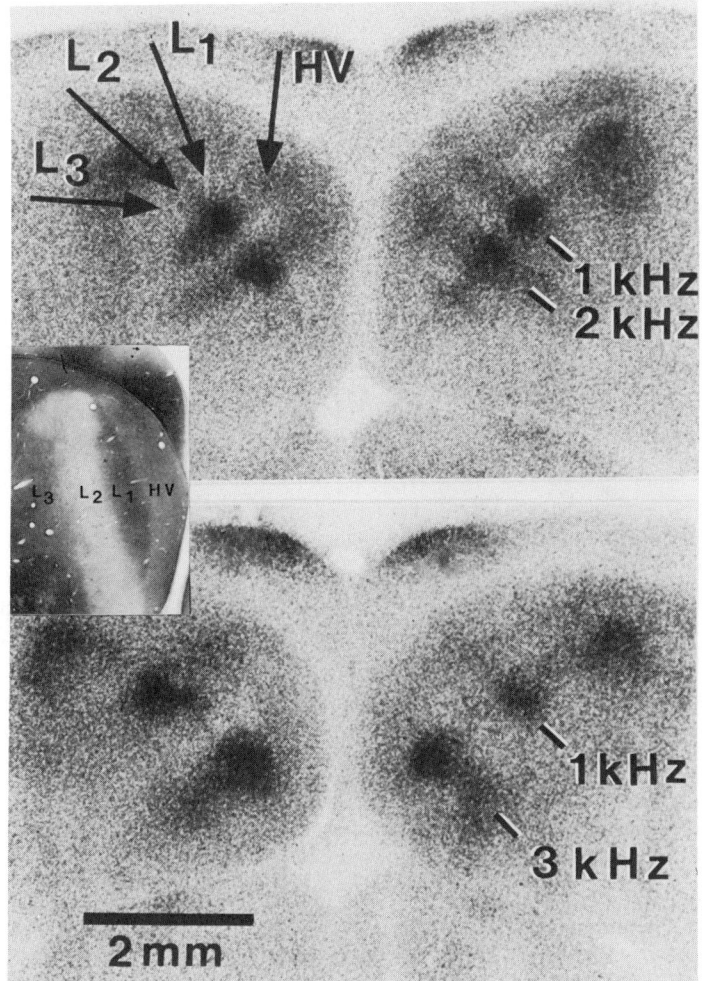

Figure 1. Laminar and tonotopic organization of Field L as seen in 2-deoxyglucose autoradiographs of Guinea fowls. The sections were cut in the transverse plane and are taken from an intermediate rostrocaudal level of Field L. Layers HV, L_1, L_2, and L_3 as well as orthogonal stripes of labelling produced by alternating tones are indicated. The bird in the top inset was stimulated with 1 and 2 kHz alternating at a rate of 3/sec. In the bottom inset stimulation was with 1 and 3 kHz. Note that the position of the 1 kHz stripe is the same along the L_2 layer as in the top case while the 3 kHz stripe has shifted towards the ventromedial end of L_2 (after Scheich, 1985). The small inset to the left illustrates the laminar organization of the chick Field L as seen in the Timm stain. Part of the left hemisphere is shown adjacent to the midline at a similar transverse plane as in the autoradiographs. Note reduced stain in L_2 and differential staining of the other layers (after Faber et al., 1989).

silver-granules are found predominantly in certain synaptic boutons this may reflect the presence of certain transmitter mechanisms working with zinc-containing enzymes. Lack of zinc both in sensory thalamic input layers of neocortex and of neocortical equivalents in bird telencephalon may be in support of a homologization of these thalamo-telencephalic systems proposed by Karten (1969).

Tonotopic organization of Field L

Field L layers with AHV on top appear to have only one common representation of the cochlea. This was shown independently with electrophysiological mapping in the Guinea fowl

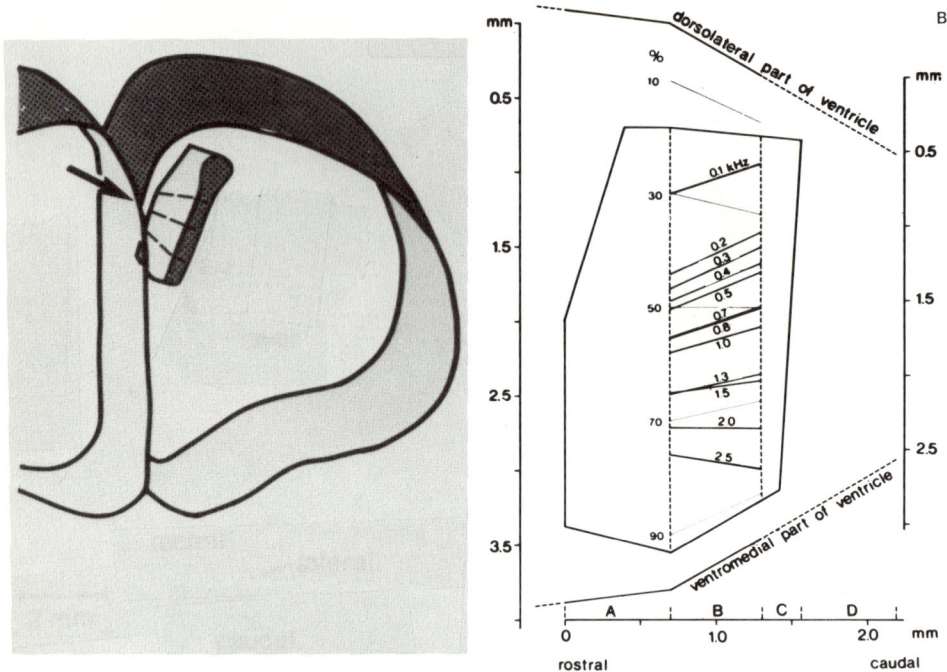

Figure 2. A schematic three-dimensional drawing of the position of the L_2 layer in the right hemisphere of the chick seen from caudal. The large arrow indicates the top view on L_2 for the two-dimensional reconstruction of its tonotopy in B. There the whole dimensions of the layer are indicated by a continuous line. Dashed lines mark the rostral and caudal borders of the range in which 2DG Frequency Band Laminae were spatially analyzed from serial sections in different experiments. The centers of these labelled laminae are drawn and the respective stimulus frequencies indicated. Note the caudorostral convergence of contours (from Heil and Scheich, 1985).

(Bonke D. et al., 1979) and with 2DG mapping in the chick (Heil and Scheich, 1985). The tonotopic gradient covers low frequencies dorso-laterally to high frequencies ventro-medially (Fig. 2). The tonotopic space to about 80% is devoted to representation of frequencies below 3 kHz in the domestic chick (Heil and Scheich, 1985). Higher frequencies do not appear to occupy vastly more space in Field L of other bird families including songbirds and parrots (Müller and Scheich, 1985, Häusler, 1989), and even in an echo-locating swiftlet (Collocalia, unpublished). This, however, does not exclude tuning of individual neurons to higher frequencies (max. 8.7 kHz in the chick, Heil et al., 1987). Frequencies below 1 kHz are represented in almost 50 % of the tonotopic space in the chick. The strongly 2DG labelled and expanded dorsolateral low frequency area of the tonotopic gradient in some birds (Fig. 1) might even respond to infrasound (Theurich et al. 1984b, Heil and Scheich, 1986). It responds to slow modulations of higher frequency carriers (FM, AM) (Heil, 1989) which could explain also high frequency input as found there in the starling (Häusler, 1989).

We could demonstrate by combined electrophysiological and 2DG experiments in the same animal that a band of 2DG labelling induced by tone stimulation indeed correspond to a band of units having a response peak at or near that tonotopic frequency (Fig. 3) (Theurich et al., 1984a). Thus, the 2DG method is a faithful correlate of electrophysiologically defined tonotopic organization. Units lying in the same tonotopic band have a strong excitatory input in common but may show quite different tuning properties depending on the layer. L_2 units usually have a clear best frequency and pronounced sidebands of lateral inhibition. L_1, L_3 and AHV units in addition to the common tonotopic frequency input may show excitatory and inhibitory convergence from various frequency channels. Broad tuning or several additional response peaks, which may be as strong as the tonotopic response, are common. Multiple tuning and inhibitory bands, in part, explain L_1 and L_3 unit selectivity to complex sounds. Around 10 % of units in a tonotopic band may not be exited by pure tones at all. These and various other properties which make units

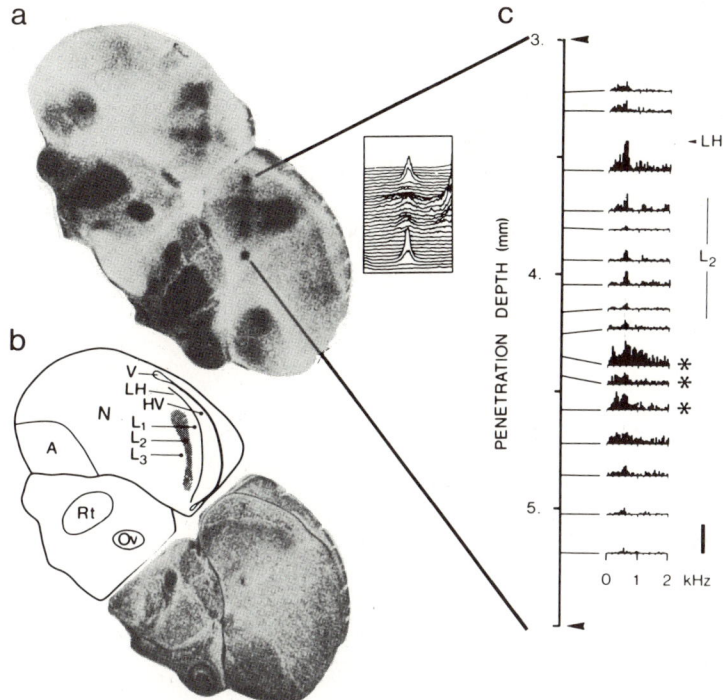

Figure 3. Correspondence of electrophysiologically determined tonotopic frequency of units and 2DG labelled Frequency Band Lamina in the chick. In c the overthreshold responses of units to different frequencies are shown along an electrode track orthogonal to the Field L layering. Common response peaks (tonotopic frequency) were around 500 Hz. In a the 2DG labelling in a transverse section of the same animal is shown which was stimulated with 500 Hz. In this autoradiograph the stripe of labelling is seen to fall exactly between two dark spots which are electrode marks produced at the beginning and at the end of the recording track in c. In b a scheme of Field L layering and a Nissl stained right hemisphere with the electrode marks are shown (from Theurich et al., 1984a).

selective for complex parameters including species-specific calls are illustrated in the following literature: Bonke D. et al., 1979; Scheich et al., 1979a; Scheich, 1977; 1985; Theurich et al., 1984a, Langner et al., 1981, Heil, 1989.

At this point it may be useful to introduce a few definitions. Following the described results tonotopic organization neither means that all units in a particulary two-dimensional array respond only to the tonotopic frequency nor that they all show the strongest overthreshold response there (best frequency). Instead they all have a "common denominator input" which is usually excitatory (besides varying other inputs). This tonotopic input may produce small or large response peaks or may be lost in a wide band response. In some units with complex stimulus requirements it may even remain undetected with pure tone stimulation.

The band of 2DG labelling across Field L lamination as produced by a given pure tone has some maximal width ($\approx$ 500 µm). This becomes a slab of labelled tissue when following the band in the rostrocaudal dimension. Units as defined electrophysiologically by a common denominator frequency are lying within this slab, which is called here "Frequency Band Lamina (FB lamina)". It is not clear whether best frequency of units differs across the width of a FB lamina e.g. in the direction of the tonotopic gradient. Since all units have some (limited) bandwidth it could be some aspect of this bandwidth which determines the width of the lamina.

Since 2DG labelling intensity has a peak at some point across the width of a FB lamina, it is assumed that electrical responsiveness to the tonotopic frequency is strongest at this maximum. Whether this maximum is mainly a spatial property of input compartments (the domains of

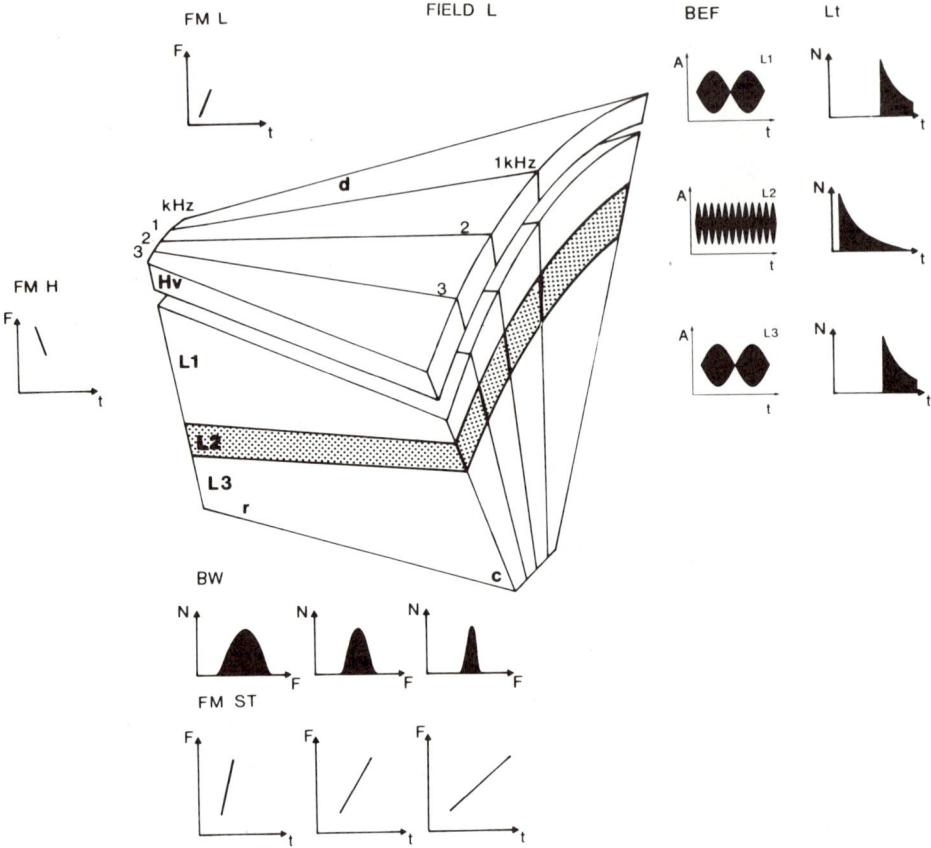

Figure 4. Scheme of multidimensional organization of right Field L (d - dorsal, r - rostral, c - caudal). The three-dimensional proportions of the structure are close to natural. Note the rostral and ventral convergence of isofrequency contours. Insets around the structure illustrate spatial gradients of functional properties of units. To the right, first column, preference for slow amplitude modulations (best envelope frequency BEF) in postsynaptic layers L_1 and HV, for fast modulations in the input layer L_2, and again for slow modulations in postsynaptic L_3 are shown. Scales: A - amplitude of signal, t - time). The second column indicates the changing response latency of units to tones across the layers. Scales: N - number of spikes, t - time. At the bottom, first row, the observed bandwidth (BW) of the overthreshold frequency response is shown as a rostrocaudal gradient. Rostrally, where isofrequency contours converge, the largest bandwidth is found. Scales: N - number of spikes, F - frequency. At the bottom, second row, the preferred steepness of frequency modulation is shown to correlate positively with bandwidth of units. Linear FM signals are given in frequency (F) versus time (t) coordinates. For simplicity only upward modulations are shown. At the top, left hand side, two insets illustrate unit preferences for direction of FM. In postsynaptic layers L_1 and L_3, low frequency tuned units tend to prefer upward modulations, middle range units have no bias, while high frequency units prefer downward modulations.

terminals and dendrites), as suggested by the spatial considerations of 2DG labelling (see above), or indicates the position of postsynaptic units with the strongest response, remains to be determined. Note, that the concept of a FB lamina is seen here as a functional ensemble of neurons or a module. It is different from that of an "isofrequency contour". The latter is an abstract line used in electrophysiological studies to connect recording sites of units with the same best frequency across some spatial dimension. It is clear that isofrequency contours are lying within FB laminae as defined by 2DG labelling.

Fig. 4 gives a schematic overview of the dimensions and tonotopic organization of Field L. Note that FB laminae converge rostrally and ventrally.

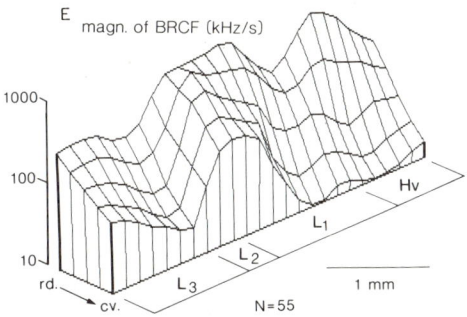

Figure 5. Two-dimensional reconstruction of unit preferences for steepness of frequency modulation (best rate of change of frequency BRCF) in chick Field L. The distribution of preferences is shown in a plane radially across layering and in rostral (rd) - caudal (cv) direction. Units from various Frequency Band Laminae are pooled and projected into this plane. Note that across all layers rostral units prefer the steepest frequency modulations. Units in postsynaptic layers HV, L_1 and L_3 prefer steeper modulations than in the input layer L_2 (from Heil, 1989)

Complex rostro-caudal organization (longitudinal tonotopic dimension)

FB laminae and isofrequency contours in chick field L are neither parallel in radial direction (across Field L layers) nor in rostro-caudal direction (Heil and Scheich, 1985). Radially they converge from dorsal towards a locus in deep layer L_3. In the longitudinal dimension they converge rostrally. Thus, a specific geometry is present which provides for more spatial representation of any given frequency band dorsally and caudally. For the caudorostral convergence two functional correlates have been found, so far. In unilaterally deafened chicks it was shown that the resulting asymmetry of 2DG labelling between left and right Field L differed in rostrocaudal direction. While in caudal sectors of the field tonotopic activation was abolished contralateral to the deafened ear, in intermediate and rostral sectors some tonotopic activity was present contralaterally in L_2 (Scheich, 1983). Thus, the results provided evidence of binaural interaction in more rostral sectors of Field L where isofrequency contours converge. Even though this was not studied explicitly it may be easily envisaged how binaural processing of sounds could benefit from a convergence of tonotopic inputs, i.e. from integration of wider frequency band by individual neurons.

The other correlate of tontopic convergence concerns selectivity of single neurons for parameters inherent in frequency modulations (FM). This was studied electrophysiologically in our laboratory using linear upward and downward FM sweeps (Heil, 1989). It was found that the steepness of FM sweeps is represented systematically along the rostrocaudal dimension of HFB laminae. The steepest modulations are preferred by units very rostral in Field L in all layers (Fig. 5). These units also have a greater bandwidth of tone responsiveness, e.g. they are more broadly tuned (Fig.4).

The gradient of preferred FM steepness along isofrequency contours may be explained to some degree by stereological considerations. Dendritic trees of local Field L neurons to a first approximation are of equal size. Tonotopic inputs to these dendritic trees, however, converge more for rostral neurons than for caudal neurons. This is shown both by the convergence of 2DG FB laminae and by the greater bandwidth of electrophysiological responsiveness of rostral neurons. Consequently, integration of postsynaptic potentials on rostral dendritic trees would cover not necessarily more frequency channels but a wider range of frequencies within the same time span. This model assumes a spatial order of frequency inputs along dendritic branches for which some indirect evidence has been obtained (Heil, 1989).

Complex organization according to depth (radial tonotopic dimension)

Responses of Field L neurons have also been analyzed systematically with sinusoidally amplitude modulated tones (AM) (Hose et al., 1987; Hose, 1988). Most units were responsive at their best frequency to some amplitude modulation of that carrier. Specializations of units for

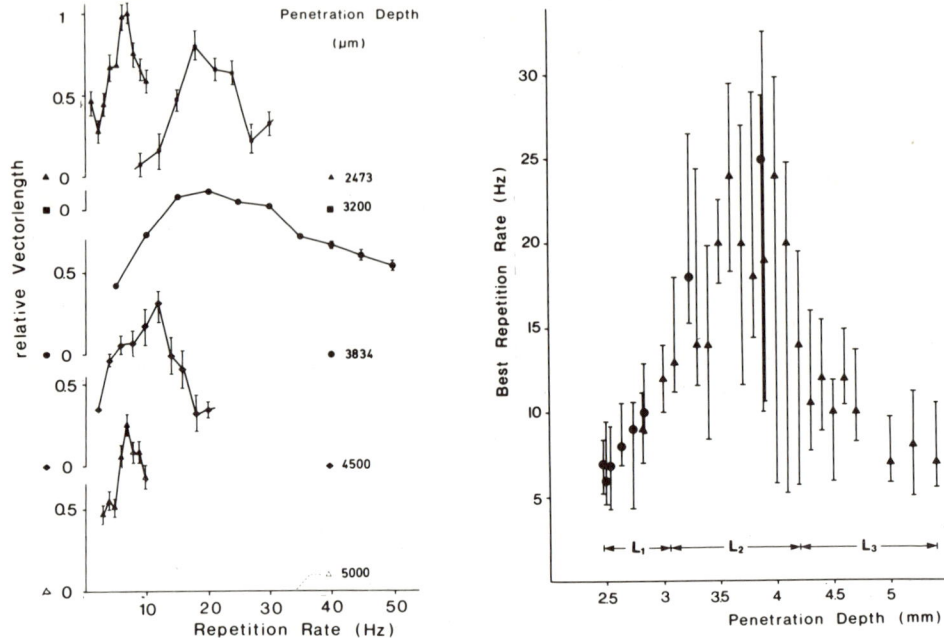

Figure 6. Radial gradient of tuning to frequencies of amplitude modulations (best envelope frequency BEF) across the layering of Field L in the mynah bird. On the left tuning of 5 units encountered at different depth of an electrode track within a given Frequency Band Lamina. Note different preferred BEF (on the x-axis indicated as Repetition Rate) and the different sharpness of tuning. Relative vector length on the y-axis is a combined measure of discharge rate and strength of synchronizaiton of units to envelopes. On the right side pooling of data from different animals in similar Frequency Band Laminae. Symbols indicate BEF of units and vertical standard deviations bandwidth of BEF tuning. Note the smooth and symmetric gradient of BEFs across the L_1, L_2, and L_3 layers (after Hose et al. 1987).

modulation frequencies covered low pass, high pass, and band pass properties. The tuning to particular modulation frequencies as seen in band pass units (Best Envelope Frequency, BEF) was found not only in terms of highest discharge rate but also in terms of the strongest phase coupling of unit discharges to modulation cycles.

BEF of units varied with respect to tonotopic organization of Field L (Fig. 6). In penetrations radially aligned with isofrequency contours BEF started in AHV and L_1 with a few Hz and gradually increased to an average of 25 Hz in L_2 (maximum 380 Hz). In L_3 the preferred modulation frequency declined again to a few Hz. Neighbouring FB laminae showed a roughly parallel distribution of preferred modulation frequencies, so that values could be pooled over radial depth of penetraton (Fig. 6, right). There was no obvious difference between L_1 and L_3 unit behavior to modulation frequency so that the distribution appeared symmetrical across the input layer L_2. Some change of BEF was also observed in the longitudinal dimension of the tonotopic map but with still undetermined geometry of respresentation.

The gradients of BEF across the lamination become more transparent by considering inputs and the basic organization of Field L. Tuning to high envelope frequencies as seen in the input layer L_2 and higher is found already at the level of the inferior colliculus analogue, Mld, of birds (Guinea fowl). There, a precise map of high modulation frequencies up to several hundred Hz was identified and related to perception mechanisms of pitch in complex sounds (Periodicity Pitch, Langner, 1983). Tuning to low envelope frequencies in terms of discharge rate was rare in the midbrain but phase coupling to slow envelope cycles was present in most cases. Consequently, it can be envisaged that by way of the thalamic n. ovoidalis this phase coupling information reaches L_2. Here envelope tuning to frequencies somewhat lower than in Mld may be established by appropriate low pass filter circuits. Similarly, by using the phase coupling to slow envelope cycles of L_2 units the preferred modulation frequency may be sequentially stepped down in the postsynaptic layers L_1 and L_3 by low pass filter mechanisms.

The distribution of different BEFs from all layers had a median at 13 Hz. Therefore, specialization for the analysis and recognition of rhythms appears to be the main feature of this BEF map in Field L. Above 20 Hz modulation frequency the resulting percept in humans is "roughness" of sound which grades into periodicity pitch around 70 Hz. The fact that these higher modulation frequencies are represented in layer L_2, suggests that field L contains combined maps of rhythm and pitch (Hose et al., 1987).

It is not unexpected that amplitude modulation analysis is integrated into tonotopic organization. According to our results different envelope frequencies of the same narrow band carrier will produce distinct areas of neuronal activation along a radial isofrequency contour. If the carrier is made wider in bandwidth, neighbouring FB laminae will be activated at the same radial depth with a given modulation frequency. This may allow various forms of lateral interaction which could be especially useful for the analysis of harmonic carriers with periodic or transient amplitude modulations. It may be speculated that the radial convergence of isofrequency contours from dorsal (L_1) to ventral (L_3) is helpful in analyzing amplitude modulations of carriers with different bandwidth. Under this assumption in L_1 narrow band signals with slow BEFs would be processed, while in L_3 wide band signals with the same slow BEFs may be preferred.

Implications of organizational geometries in Field L

Field L including AHV has been portrayed, so far, as a multilayered structure containing a tonotopic map with very specific geometry. This geometry obviously serves as a frame for systematic and spatially coherent analysis of complex auditory parameters. One of the most intriguing aspects of this single tonotopic map is its capability to harbour several different representations of complex parameters. FM, AM, and binaural mechanisms have been identified, so far.

How can different complex properties of sounds be represented unequivocally in the same map? There are at least two realizations which are compatible with the present results. For instance, FM-selective units could be clustered as small ensembles interspaced by ensembles of AM-selective and other specialized units. If rostro-caudal distances between homologous ensembles are not too large for specific connectivities a gradient of changing selectivity, for instance, for steepness of FM, could still be established. A gradient of AM selectivity could be represented by interspersed and independent clusters of units along the radial dimension. Likewise, other complex properties could be represented by neuron clusters at some angle to these orthogonal dimensions.

The other theoretical alternative would be a fine grain spatial mix of properties where neighbouring neurons are unsimiliar. Neurons could obey rules of their respective networks somewhat comparable to the independent systems of x-, y- and other cells in the retina. This fine grain mosaic could provide for a very smooth gradient of the represented parameter.

Especially the latter case in ontogeny would require precise identification of individual neurons as belonging to one or the other subsystem in order to make meaningful point to point connections. In the former case of ensembles with a more patchy distribution of specific local networks establishment of connections may be somewhat less of a problem and local mechanisms of self-organization could help to establish functionally separable ensembles of neurons. Even though in none of our studies the grain of recording was fine enough over long electrode tracks to establish systematically one or the other case it would appear from various accidental observations that neighbouring units (< 100 µm separation) had very similar properties (see for instance Scheich, 1977, Fig.2). Therefore the ensemble hypothesis appears to be more likely.

It may be that within ensembles units have various degrees of selectivity for complex parameters and even for combinations of different complex parameters. In that way ensembles could represent "crossings" of different spatial gradients of parameter representation. (For discussion of concepts of auditory feature analysis see (Scheich, 1977; Scheich et al., 1979a; Scheich, 1985).

Probably the most crucial problem which arises from an integration of several complex representations in one tonotopic map are geometrical constraints. In any given three-dimensional map, like Field L, there are only limited possibilities for shaping the geometry of tonotopic-organization in such a way that it is a good compromise for multiple representations. For instance, the caudorostral convergence of FB laminae appears to be an optimization for systematic representa-

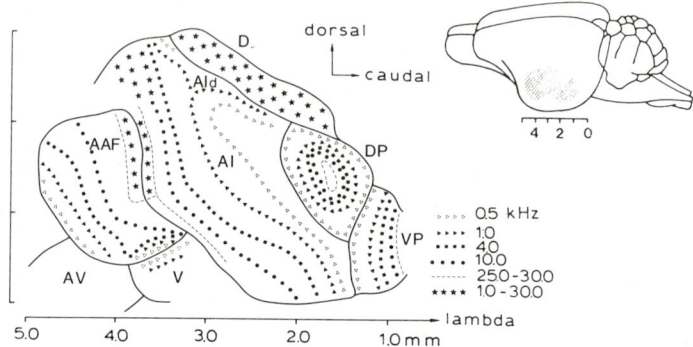

Figure 7. Summary scheme of tonotopic organization of multiple fields in gerbil auditory cortex. Isofrequency contours in the fields are marked by different symbols and were determined by connecting position of recorded units with similar best frequency (tonotopic frequency). The grid of unit recordings was 100 x 200μm. Field AV is added to this map on the basis of 2DG labelling. The small inset to the right gives the dimension of auditory cortex on the lateral surface of the gerbil telencephalon (from Thomas 1989).

tion of FM-steepness. Whether this tonotopic feature is useful at all for AM representation remains to be determined. Furthermore, the convergence of FB laminae from dorsal to ventral across the lamination is a geometrical specialization with a yet speculative functional significance. It may also be a compromise between AM parameters which are and FM parameters which are not represented along these spatial dimensions. These theoretical considerations will become more transparent with respect to mammalian auditory cortex organization where evolutionary pressures may have favored more ideal ways of representation.

Tonotopic mapping of gerbil auditory cortex with microelectrodes and with the 2DG method

The organization of auditory cortex of the mongolian gerbil (*Meriones unguiculatus*) was recently analyzed in our laboratory with fine grain microelectrode penetrations through lateral cortex as well as in 2DG experiments using tone stimuli in both cases (Steffen et al. 1988, Thomas, 1989). While responsiveness of cortical units and small clusters of units to tone bursts is usually sufficient even in anesthetized mammals to establish tonotopic maps electrophysiologically there have previously been unsuccessful attempts to obtain 2DG labeling of FB laminae in mammalian cortex (Webster et al., 1978; Ryan et al., 1982). This is astonishing especially with respect to the sharp delineation of FB laminae in bird field L with the same stimuli.

By varying temporal parameters of tone stimuli we found that time varying stimuli like narrow band frequency modulations of tones, changing rhythms of tones and alternations between two frequencies produced tonotopic labelling in several fields of gerbil auditory cortex. Presumably, auditory cortex habituates too strongly to a monotonous stimulus over the 45 min period of a 2DG experiment to accumulate enough label. The preference of auditory cortex for time varying signals appears to be also a major difference to subcortical auditory structures in mammals and to visual and somatosensory cortex in which stimulus representation is obtained with more stationary patterns (Tootell et al., 1988; McCashland and Woolsey, 1988).

Electrophysiological and 2DG mapping provided similar results in terms of location, size and number of fields (at least 8) and their tonotopic organization (Fig. 7). The largest field, termed AI, is located like a dorsoventral backbone in the center of auditory cortex. It has the longest isofrequency contours which take a roughly dorsoventral course along the lateral surface of the cortex. About half way ventral contours diverge somewhat and deviate in a caudal direction. Low frequencies are respresented caudally and high frequencies rostrally. With unit best frequencies from <100 Hz to 35 kHz the complete hearing range appears to be represented in AI. Roughly 80% of the tonotopic space of AI is devoted to representation of frequencies below 10 kHz and more than 50% to frequencies below 2 kHz. AI has the best spatial resolution for low frequencies.

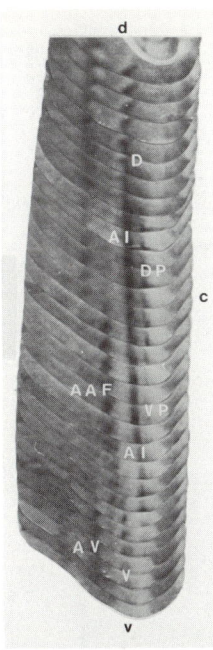

Figure 8. Reconstruction of 2DG labelled auditory cortical fields in gerbil from serial horizontal section autoradiographs of the left hemisphere (dorsal up and rostral to the left). Sections are aligned using a dorsoventral landmark in the hippocampus. The spacing between sections is such as to overlook radial labelling in each section which results in an exaggerated dorsoventral extent of auditory cortex (compare dimensions in Fig. 7). The stimulus was a tone alternating between 1 and 2 kHz. The two frequencies presented in this way are not resolved in two labelled Frequency Band Laminae but produce one broad and intense band in most fields. For the purpose of illustrating labelling in the multiple fields this type of stimulus produces the best contrast.

At the dorsal extreme of AI, low frequency contours, as determined electrophysiologically, diverge and can be interpreted to form loops. Tuning to frequencies above 4 kHz is rarely found there. Consequently this dorsal part of AI was distinguished as a subfield AId.

AI units are sharply tuned. They display short latencies and mostly onset responses to tones. Inhibitory sidebands are frequently found. Due to tangential electrode penetrations, recordings were chiefly from input layer IV. That units from all layers in a given FB laminae receive a common input is shown by the pattern of 2DG labeling (Fig. 8). The band of radial labeling reaches through all layers while forming two foci of more intense activity in layers IV and deep V. Inputs to AI are from three subdivisions of the medial geniculate body.

Rostrally adjacent to AI an anterior auditory field AAF is located where units show tuning and other basic physiological properties similar to AI. The inputs stem from at least two subdivisions of the medial geniculate body which partly overlap with AI sources. AAF has less than half the size of AI, shorter isofrequency contours and less spatial resolution for frequencies. Similar to AI the complete hearing range is represented in AAF. However, it appears that frequencies below 10 kHz are not as strongly overrepresented as in AI. Isofrequencies in AAF are roughly parallel to those in AI but form a tonotopic map mirror imaged to that of AI (Fig. 9). Low frequencies are represented rostrally and high frequencies caudally. At the ventral end low frequency contours show a slant in a caudal direction similar to contours of AI. At the common high frequency border of fields AI and AAF units are frequently found which show weak responsiveness to frequencies between 1 and 30 kHz without specific tuning. The topology of these wide band units corresponds to a dorsoventral stripe of low 2DG labeling under all studied stimulus regimes. The stripe forms a reliable landmark for the border of the two fields in 2DG autoradiographs.

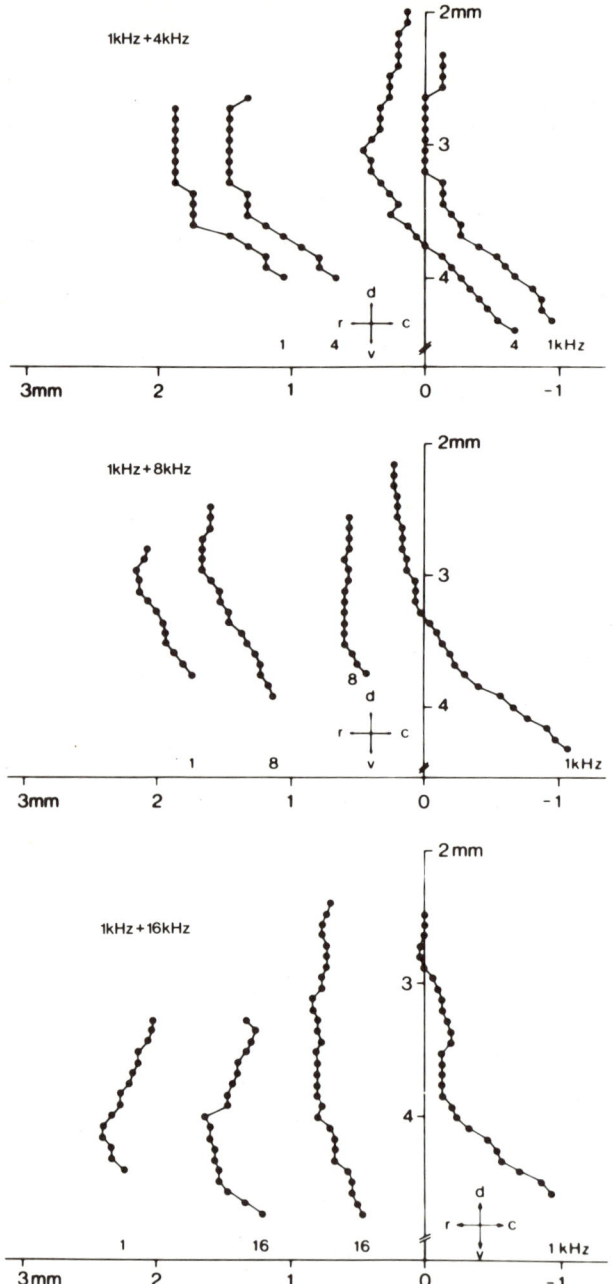

Figure 9. Plots of labelled Frequency Band Laminae in field AI and AAF as obtained in three experiments with tones alternating between 1 kHz and other frequencies (4, 8, 16 kHz). Data are from reconstructions as in Fig. 8. Due to larger frequency jumps the different Frequency Band Laminae can be resolved. Plots are in cortex surface coordinates (compare to Fig. 7) with a dorsoventral hippocampus reference line (rostrocaudal x = o). The two right hand contours are in AI and the two left hand contours in AAF. The distance between the two contours in each field increases with the magnitude of the frequency jump. The two inner contours consequently approach each other. These patterns allow the interpretation of a mirror imaged tonotopic organization with high frequencies represented around the border of the two fields. They also permit the quantitative evaluation of tonotopic organization.

Caudal to AI two small fields, a dorsoposterior DP and aventroposterior field VP, are aligned. They have similar size and are tonotopically organized but their maps have different geometries. VP has dorsoventral, roughly straight, and parallel isofrequency contours similar to AI and AAF. Low frequencies are represented rostrally so that a common low frequency boundary is formed ventral with AI.

In contrast, DP is concentrically organized with high frequencies represented in the center. This unusual type of geometry is alo reflected by the pattern of 2DG labelling. Units in DP and VP have clear best frequencies but latencies are considerably longer than in AI and AAF.

Below AAF a small ventral field V is seen both with unit recordings and in terms of a patch of increased 2DG uptake (Fig. 8). Chiefly low best frequencies are found in this field. The reconstruction of isofrequency contours from the distribution of best frequencies suggest a roughly horizontal orientation with lowest frequencies dorsal and higher linear isofrequency contours ventral. Thus, there is probably a similar geometry but a different orientation of the tonotopic gradient than in AI, AAF and VP.

Rostral to the ventral part of AAF and to V a large area is labeled with 2DG after presentation of various stimulus patterns (Fig. 8). This area is tentatively called anterior ventral field AV. Due to the curvature of the brain and vascular obstacles field AV was not yet characterized electrophysiologically.

Dorsal to AId and DP, and rostrally reaching into the boundary zone between AI and AAF a large belt region was identified. This field D harbours units which are responsive to a very large range of frequencies without obvious best frequency. The dorsocaudal aspect of field D is also labelled with 2DG under a variety of stimulus conditions (Fig. 8). It may be that D consists of various subareas with different unit selectivities.

Parcellation of mammalian auditory cortex into multiple (tonotopic) fields

The basic disposition and organization of mammalian neocortex offers opportunities of sensory representation with little spatial restraints. This appears to be facilitated not only by cortical properties as a surface structure, permitting folds as a means of expansion, but also by its composition of neuronal modules (columns) as repetitive building blocks. The potential for cortical expansion is reflected by a phylogenetic tendency for multiple field representations of various sensory modalities, notably visual, auditory, and somatosensory fields (Macko et al., 1982; Luethke et al., 1988). Furthermore as demonstrated with genetic variants of mysticial vibrissae and early ontogenetic manipulations of peripheral somatosensory input there is a large potential of compensatory representation (see Van der Loos, this volume).

For the overall phylogenetic trend towards multiple representation of sense organs in central sensory structures the Parcellation Theory by Ebbesson (1980) has provided some rationale. Selective loss of input connections could play a crucial role for determining the representation of different stimulus dimensions of a given sensory modality in multiple fields. It is not known yet whether auditory cortex as identified in basal mammals (Platypus, Tachyglossus, Didelphis, Erinaceus) has several representations of the cochlea (Bohringer and Rowe, 1977; Scheich et al., 1986; Lende, 1969; Ebner, 1969). However, already the Australian possum (Gates and Aitkin 1982) and rodents appear to have at least two fields. In mammals including those rodents which have been studied in more detail, squirrel, mouse, gerbil, and in cat and primates, a larger number of fields has been identified (see Luethke et al. 1988). A particular case is represented by auditory cortex of echolocating bats, in which a great number of specialized areas were described by Suga and collaborators (for review: Suga, 1982). It is not obvious, whether this multiplicity corresponds to the multiple representation of the cochlea of other mammals. Specialized areas could as well represent topologies beyond cochlea-related maps, e.g. categorical adaptation of the bat cortex to processing the unique echo parameters with certain spectral components only.

Funtional implications of multiple cortical fields and their tonotopic geometries

The parcellation of gerbil auditory cortex into multiple tonotopic fields appears to be similarly extensive as in cortices of cat, monkeys and bat (Merzenich et al., 1979; Merzenich and Brugge, 1973; Imig et al., 1977; Suga, 1982). The cat and bat cases are the only ones in which attempts have been made to relate tonotopic geometries to representation of more integrative parameters of acoustic stimuli. Similar to bird Field L the two basic spatial dimensions of cortical Frequency Band Laminae, the radial and longitudinal dimensions, were searched in cat fields AI, AII and AAF for representation of complex variables. For the purpose of this account it will be sufficient to shortly summarize these as yet rather limited findings.

Following the longitudinal tonotopic dimension in AI alternations between bands of units with different binaural convergence (EE unit and EI unit bands) were described (Middlebrooks et al., 1980; Schreiner and Cynader, 1984). Furthermore, gradients relating to representation of sharpness of tuning were found along isofrequency contours of AI and AII (Schreiner and Cynader, 1984). Tuning to a best modulation frequency (BEF) of amplitude modulations was seen throughout AAF of the cat (Schreiner and Urbas 1986). BEFs were positively correlated with the tonotopic frequencies (CF) of units. Thus, there was a topography across the tonotopic gradient but a topography of BEFs along isofrequency contours as in the bird was not evident (Schreiner and Urbas, 1986). So far, no gradients have been identified into depth of any auditory cortex e.g. in radial direction (Abeles and Goldstein, 1970; Goldstein and Knight, 1980).

Several spatial gradients (specialized areas) for representation of complex parameters of echo sounds have been identified in the auditory cortex of a bat (Suga 1982). For the most complicated unit properties (combination selectivities) it is not clear whether their representation relates spatially to dimensions of tonotopic geometries in fields. However, two basic parameters, amplitudes of echos and binaural properties, are represented along contours of a concentric tonotopic map. This map appears to be a blown-up sub area of primary auditory cortex with a largely conventional tonotopic gradient for non-echo frequencies. In this sense the concentric area of bat primary auditory cortex may correspond to concentric areas in the "barrel-field" of rodent somatosensoroy cortex (see Vander Loos, this volume), rather than to concentric field DP in gerbil, which is a separate representation of the cochlea.

In the functional sense of what is conveniently represented along concentric isofrequency contours field DP in the gerbil may still relate to the concentric subarea in the bat. It remains to be determined, however, whether a concentric representation of the whole peripheral receptor map as seen in DP, i.e. a concentric organization of a whole field, is functionally equivalent to concentric organization of specialized areas within a field.

From the foregoing it appears that there is not sufficient data to decide whether gradients of complex parameters are always represented along cortex isofrequency contours. Gradients of complex parameters could also follow other types of organization. Nevertheless available evidence may be sufficient to suggest the following principles. The parcellation of auditory cortex into multiple fields most frequently (especially in "higher mammals") implies a multiplication of tonotopic organizations. This is probably not trivial, since setting aside specialized areas from a primary auditory map is conceivable without tonotopic organizations as an alternative. Thus, there must be a premium on this type of organization. The simplest explanation may be that, what is useful for a primary auditory map, is also useful for additional maps e.g. maintenance of neighbourhood relationships of spectral representation (1) during ontogenesis of connectivities in the map and (2) for later coherent spectral analysis of sounds.

Particularly striking is the finding in various mammals that the multiple tonotopic maps may have different geometries with parallel linear, curvilinear divergent, and concentric isofrequency contours. Most often geometries of tonotopic gradients of neighbouring fields are mirror imaged. In the light of the results on complex representation in bird field L this suggests a higher degree of freedom for independent spatial representation of complex parameters. This aspect of optimization may not be possible with the geometric compromise of the "all purpose" map of birds.

Finally, extrinsic mechanisms beyond representation within the maps may also be relevant for the generation of multiple tonotopic gradients with different geometries. Results from the cat (Imig and Reale, 1980) and recently similar findings from the gerbil in our laboratory (Thomas, 1989) suggest that the widespread intracortical connectitivies between individual auditory fields are specifically related to tonotopic organization. Imig and Reale used a combination of electrophysiological mapping and injections of tritiated amino acids into tonotopically identified small areas. They found that cortico-cortical connections are between portions of similar best frequency representation in fields AAF, AI, P and VP in both hemispheres. In the gerbil a combination of electrophysiological recording, local injection of the retrograde tracer Fast Blue and a terminal 2DG experiment showed that (1) neurons within a FB lamina in AI or AAF are prefer-

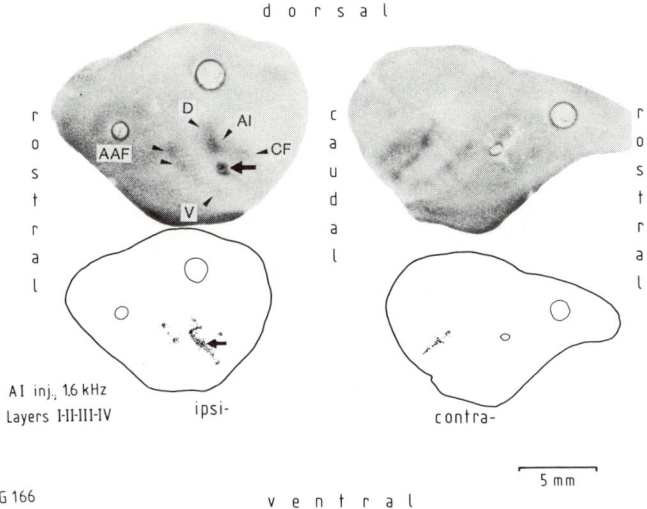

Figure 10. Demonstration of tonotopic connectivities within and between fields of gerbil auditory cortex. Top: 2DG autoradiographs of tangential sections of left and right auditory cortex of an animal stimulated with 1.6kHz tones. Labelling of Frequency Band Laminae is less even than in horizontal sections (Fig. 8) since tangential sections pass through layers with different intensity. Nevertheless, the injection site of Fast Blue on the 1.6 kHz Frequency Band Lamina in left AI can be seen as a dark spot (arrow). The best frequency of units (1.6 kHz) was determined electrophysiologically in a preceding experiment and the injection made there. The 2DG experiment with 1.6 kHz stimulation was two days later. The distribution of neurons retrogradely filled with Fast Blue is shown at the bottom in the same sections from which the autoradiographs (top) were made. The distribution of filled cells roughly matches the labelled AI and AAF Frequency Band Laminae in the injected hemisphere. There are also retrogradely filled cells in a caudal field and along the AI Lamina in the contralateral hemisphere (from Thomas, 1989).

entially interconnected and (2) that there are chiefly connectivities between corresponding FB laminae in the two fields (Fig. 10). These tonotopic connectivities suggest the importance of intra-map-geometries as well as in-between-map-geometries.

Tonotopic connectivities on the one hand demonstrate that in spite of multiplicity of maps regions therein, which are homologous in terms of spectral representation, are kept in contact. This probably serves a coherent processing of complex sound parameters which are represented in different maps. For instance, it is quite usual that natural sounds contain frequency bands which are simultaneously amplitude- and frequency-modulated and in addition contain some noise. Tonotopic connectivities could help to maintain a "unified percept" of such sound components processed in different maps. Some facets of the "Cocktail Party Effect" may be explained in this way.

Furthermore, tonotopic connectivities in conjunction with different tonotopic geometries of fields suggest additional useful principles of interaction between fields. Since AI and AAF have parallel but mirror imaged tonotopic gradients axons connecting between low frequency representations have to travel longer distances than between high frequency representations. This could be used as a mechnism in terms of delay lines for frequency-specific temporal interactions between the fields. In conclusion, tonotopic interactions between fields may be used in various ways for processing of complex sounds. They may not simply be a necessary step to overcome disadvantages of representation in multiple maps but offer additional mechanisms especially of temporal processing of sounds.

Supported by DFG, SFB 45.

REFERENCES

Abeles, M., and Goldstein, M.H. Jr. (1970) Functional architecture in cat primary auditory cortex: Columnar organization and organization according to depth. *J. Neurophysiol.*, 33: 172-187.

Biederman-Thorson, M. (1970) Auditory responses of units in the ovoid nucleus and cerebrum (field L) of the ring dove. *Brain Res.*, 24: 247-265.

Bohringer, R.C. and Rowe, M.J. (1977) The organization of the sensory and motor areas of the cerebral cortex in the platypus (*Ornitorhynchus anatinus*). *J. Comp. Neurol.*, 174: 1-14.

Bonke, B.A., Bonke, D., and Scheich, H. (1979) Connectivity of the auditory forebrain nuclei in the guinea fowl (*Numida meleagris*). *Cell Tissue Res.*, 200: 101-121.

Bonke, D., Scheich, H., and Langner, G. (1979) Responsiveness of units in the auditory neostriatum of the guinea fowl (*Numida meleagris*) to species-specific calls and synthetic stimuli. I. Tonotopy and functional zones of field L. *J. Comp. Physiol.*, 132: 242-255.

Braun, K., Scheich, H., Schachner, M., and Heizmann, C.W. (1985) Distribution of parvalbumin, cytochrome oxidase activity and 14C-2-deoxyglucose uptake in the brain of the zebra finch. I. Auditory and vocal motor systems. *Cell Tissue Res.*, 240: 101-115.

Ebbesson, S.O.E. (1980) The parcellation theory and its relation to interspecific variability in brain organization, evolutionary and ontogenetic development and neuronal plasticity. *Cell Tissue Res.*, 213: 179-212.

Ebner, F.F. (1969) A comparison of primitive forebrain organization in metatherian and eutherian mammals. *Ann. New York Acad. Sci.*, 167: 241-257.

Erulkar, S.C. (1955) Tactile and auditory areas in the brain of the pigeon. *J. Comp. Neurol.*, 103: 421-457.

Faber, H., Braun, K., Zuschratter, W., and Scheich, H. (1989) System-specific distribution of zink in the chick brain. A light- and electron-microscopic study using the Timm method. *Cell Tissue Res.*, 258: 247-257.

Gates, G.R., and Aitkin, L.H. (1982) Auditory cortex in the marsupial possum (*Trichosurus vulpecula*). *Hear. Res.*, 7: 1-11.

Goldstein, M.H. Jr., and Knight, P.L. (1980) Comparative organization of mammalian auditory cortex. In: *Comparative Studies Of Hearing In Vertebrates* (Eds: A.N. Popper and R.R. Fay) Springer-Verlag, New York, pp. 375-398.

Haug, F.M.S. (1973) Heavy metals in the brain. A light microscopic study in the rat with Timm's sulphide silver method. Methodological considerations and cytological and regional staining patterns. *Adv. Anat. Embryol. Cell. Biol.*, 47: 1-71.

Häusler, U.H.L. (1989) Die strukturelle und funktionelle Organisation der Hörbahn im caudalen Vorderhirn des Staren (Sturnus vulgaris, L.) Ph.D. Thesis, Technische Universität München.

Heil, P. (1989) Untersuchungen zur interauralen Organisation des auditorischen Systems und zur Physiologie und Topographie FM-sensitiver Neurone und ihren Beziehungen zur tonotopen Organisation des auditorischen Vorderhirns beim Haushuhn. Ph.D. Thesis, Technische Hochschule Darmstadt.

Heil, P., Langner, G., and Scheich, H. (1987) Neuronal responses in field L at the upper hearing limit of the chick. In: *New Frontiers in Brain Research*, Proc. 15th Göttinger Neurobiology Conference, Eds: N. Elsner and O. Creutzfeldt, Thieme-Verlag, Stuttgart, New York, p. 128.

Heil, P. and Scheich, H. (1985) Quantitative analysis and two-dimensional reconstruction of the tonotopic organization of the auditory field L in the chick from 2-deoxyglucose data. *Exp. Brain Res.*, 58: 532-543.

Heil, P. and Scheich, H. (1986) Effects of unilateral and bilateral cochlea removal on 2- deoxyglucose patterns in the chick auditory system. *J. Comp. Neurol.*, 252: 279-301.

Hose, B. (1987) Neuronale Verarbeitung und topographische Repräsentation sprachrelevanter zeitlicher Parameter im auditorischen Vorderhirn von Beos (*Gracula religiosa*). Ph.D. Thesis, Technische Hochschule Darmstadt.

Hose, B., Langner, G., and Scheich, H. (1987) Topographic representation of periodicities in the forebrain of the mynah bird: One map for pitch and rhythm? *Brain Res.*, 422: 367-373.

Hubel, D.H., Wiesel, T.N., and Stryker, M.P. (1977) Orientation columns in macaque monkey visual cortex demonstrated by the 2-deoxyglucose autoradiographic technique. *Nature*, 269: 328-330.

Imig, T.J., Ruggero, M.A., Kitzes, L.M., Javel, E., and Brugge, J.F. (1977) Organization of auditory cortex in the owl monkey (*Artus trivirgatus*). *J. Comp. Neurol.*, 171: 111-128.

Imig, T.J., and Reale, R.A. (1980) Patterns of cortico-cortical connections related to tonotopic maps in cat auditory cortex. *J. Comp. Neurol.*, 192: 293-332.

Karten, H.J. (1968) The ascending auditory pathway in the pigeon (*Columba livia*). II. Telencephalic projections of the nucleus ovoidalis thalami. *Brain Res.*, 11: 134-153.
Karten, H.J. (1969) The organization of the avian telencephalon and some speculations on the phylogeny of the amniote telencephalon. *Ann. N.Y. Acad. Sci.*, 167: 164-179.
Langner, G. (1983) Evidence for neuronal periodicity detection in the auditory system of the Guinea fowl: Implications for pitch analysis in the time domain. *Exp. Brain Res.*, 52: 333-355.
Langner, G., Bonke, D., and Scheich, H. (1981) Neuronal discrimination of natural and synthetic vowels in field L of trained mynah birds. *Exp. Brain Res.*, 43: 11-24.
Lende, R.A. (1969) A comparative approach to the neocortex: Localization in monotremes, marsupials and insectivores. *Ann. New York Acad. Sci.*, 167: 262-276.
Leppelsack, H.-J. (1974) Funktionelle Eigenschaften der Hörbahn in Feld L des Neostriatum caudale des Staren. *J. Comp. Physiol.*, 88: 271-320.
Luethke, L.E., Krubitzer, L.A., and Kaas, J.H. (1988) Cortical connections of electrophysiologically and architectonically defined subdivisions of auditory cortex in squirrel. *J. Comp. Neurol.*, 268: 181-203.
Macko, K.A., Jarvis, C.D., Kennedy, C., Miyaoka, M., Shinohara, M., Sokoloff, L., and Miskin, M. (1982) Mapping the primate visual system with [2-14C] Deoxyglucose. *Science*, 218: 394-397.
Mata, M., Fink, D.J., Gainer, H., Smith, C.B., Davidsen, L., Savaki, H., Schwartz, W.J., and Sokoloff, L. (1980) Activity-dependent energy metabolism in rat posterior pituitary primarily reflects sodium pump activity. *J. Neurochem.*, 34: 213-215.
McCashland, J.S., and Woolsey, T.A. (1988) High-resolution- 2-deoxyglucose mapping of functional cortical columns in mouse barrel cortex. *J. Comp. Neurol.*, 278: 555-569.
Merzenich, M.M., Anderson, R.A., and Middlebrooks, J.C. (1979) Functional and topographic organization of the auditory cortex. *Exp. Brain Res., Suppl.* (2): 61-75.
Merzenich, M.M., and Brugge, J.F. (1973) Representation of the cochlear partition on the superior temporal plane of the macaque monkey. *Brain Res.*, 50: 275-296.
Middlebrooks, J.C., Dykes, R.W., and Merzenich, M.M. (1980) Binaural response-specific bands in primary auditory cortex (AI) of the cat: topographic organization orthogonal to isofrequency contours. *Brain Res.*, 181: 31-48.
Müller, S.C. and Scheich, H. (1985) Functional organization of the avian auditory field L. A comparative 2DG study. *J. Comp. Physiol.*, A156: 1-12.
Nudo, R.J., and Masterton, B. (1986) Stimulation-induced ^{14}C-2-deoxyglucose labelling of synaptic activity in the central auditory system. *J. Comp. Neurol.*, 245: 553-565.
Rose, M. (1914) Über die cytoarchitektonische Gliederung des Vorderhirns der Vögel. *J. Psychol. Neurol.* (Lpz) 21: 278-352.
Ryan, A.F., Woolf, N.K., and Sharp, F.R. (1982) Tonotopic organization in the central auditory pathway of the mongolian gerbil: A 2-deoxyglucose study. *J. Comp. Neurol.*, 207: 369-380.
Sachs, M.S., Woolf, N.K., and Sinott, J.M. (1980) Response properties of neurons in the avian auditory system: Comparisons with mammalian homologues and consideration of the encoding of complex stimuli. In: *Comparative Studies of Hearing In Vertebrates*. Eds: A.N. Popper and R.R. Fay, Springer-Verlag, New York: pp. 323-353.
Saini, K.D. and Leppelsack, H.J. (1981) Cell types of the auditory caudomedial neostriatum of the starling, *Sturnus vulgaris. J. Comp. Neurol.*, 198: 209-229.
Scheich, H. (1977) Central processing of complex sounds and feature analysis. In: *Recognition of Complex Acoustic Signals*. Ed: T.H. Bullock, Dahlem Konferenzen, Berlin: pp. 161-182.
Scheich, H. (1983) Two columnar systems in the auditory neostriatum of the chick: Evidence from 2-deoxyglucose. *Exp. Brain Res.*, 51: 199-205.
Scheich, H. (1985) Auditory brain organization of birds and its constraints for the design of vocal repertoires. In: *Fortschritte der Zoologie*, Eds: Lindauer and Hölldobler, Bd. 31, Experimental Behavioral Ecology, G. Fischer-Verlag, Stuttgart, New York: pp. 195-209.
Scheich, H., Bonke, B.A., Bonke, D., and Langner, G. (1979b) Functional organization of some auditory nuclei in the guinea fowl demonstrated by the 2-deoxyglucose technique. *Cell Tissue Res.*, 204: 17-27.
Scheich, H., Langner, G. and Bonke D (1979a) Responsiveness of units in the auditory neostriatum of the guinea fowl (*numida meleagris*) to species-specific calls and synthetic stimuli. II. Discriminatiuon of iambus-like calls. *J. Comp. Physiol.*, 132: 257-276.
Scheich, H., Langner, G., Tidemann, C., Coles, R.B., and Guppy, A. (1986) Electroreception and electrolocalization in platypus. *Nature*, 319: 401-402.
Schreiner, C.F. and Cynader, M.S. (1984) Basic functional organization of second auditory cortical field (AII) of the cat. *J. Neurophysiol.*, 51: 1284-1305.

Schreiner, C.F. and Urbas, J.V. (1986) Representation of amplitude modulation in the auditory cortex of the cat. I. The anterior auditory field (AAF). *Hearing Res.*, 21: 227-241.

Steffen, H., Simonis, C., Thomas, H., Tillein, J., and Scheich, H. (1988) Auditory cortex: Multiple fields, their architectonics and connections in the mongolian gerbil. In: *Auditory Pathway, Structures and Functions*. Eds.: J. Syka and B. Masterton. Plenum Press, New York: pp. 223-228.

Suga, N. (1982) Functional organization of the auditory cortex: Representation beyond tonotopy in the bat. In: *Cortical Sensory Organization*. Vol. 3. Multiple Auditory Areas (Ed: C.N. Woolsey). Humana Press, Clifton, New Jersey, pp. 157-218.

Theurich, M., Langner, G., and Scheich, H. (1984b) Infrasound responses in the midbrain of the guinea fowl. *Neurosci. Lett.*, 49: 81-86.

Theurich, M., Müller, C.M., and Scheich, H. (1984a) 2-deoxyglucose accumulation parallels extracellularly recorded spike activity in the avian auditory neostriatum. *Brain Res.*, 322: 157-161.

Thomas, H. (1989) Funktionelle und anatomische Organisation des auditorischen Cortex beim Gerbil (*Meriones unguiculatus*). Ph.D. Thesis, Technische Hochschule Darmstadt.

Tootell, R.B.H., Hamilton, S.L., Silverman, M.S., and Switkes, E. (1988) Functional anatomy of macaque striate cortex. I, II, III, IV, V. *J. Neurosci.*, 8: 1500-1624.

Webster W.R., Serviáre, J., Batini, C., and Laplante, S. (1978) Autoradiographic demonstration with 2-[^{14}C]deoxyglucose of frequency selectivity in the auditory system of cats under conditions of functional activity. *Neurosci. Letters*, 10: 43-48.

Wong-Riley, M. and Riley, D.A. (1983) The effect of impulse blockage on cytochrome oxidase activity in the cat visual system. *Brain Res.*, 261: 185-193.

FLYING CATS AND FLYING PRIMATES: EVOLUTIONARY SURPRISES FROM NEUROBIOLOGY

John D. Pettigrew
Vision, Touch and Hearing Research Center
University of Queensland
St. Lucia, 4067 AUSTRALIA

INTRODUCTION

In this chapter I use two papers (Pettigrew 1979; Pettigrew 1986) to show how one can illuminate evolutionary problems with the fine spotlight provided by modern neuroscientific investigations.

In the first paper I show the evidence from which one may conclude that owls are "flying cats". Both of these avian and mammalian hunters have independently evolved an apparatus for binocular vision with great functional similarities despite the structural differences in the visual pathways which reflect their separate origins. I believe that we can learn much about the constraints which operate on the evolution of a neural system for binocular vision by noting the essential functional features shared in common between the two systems and by ignoring the many features in the "neural embroidery" which do not appear to be essential for binocular vision because they are found in one system but not in the other. One such case concerns the key feature, shared by both cat and owl, that complex visual processing is delayed until after information from both eyes converges. This also seems to be a feature of the more recent successful machine algorithms for stereopsis, which also begin comparison of the two eyes at an early stage (e.g. Frisby and Mayhew, 1980).

In the second paper, on "flying primates", I try to turn the evolutionary spotlight in the reverse direction, from the neural data onto the evolutionary process itself. The success of this venture hinges on one's interpretation of the primate features I discovered in the brain of the flying fox (Pettigrew, 1986). Are these features which have been convergently acquired by flying foxes, in the same way that owls have convergently acquired the neural apparatus for binocular vision? Or is the epithet "flying primate" to be interpreted more literally, to mean that the primate brain features are a true reflections of the flying fox's lineage and that an early branch of the primates evolved flight? As unlikely as the latter scenario sounds, this is the one which continues to gain support. With reference to "neural emboidery" with no obvious functional significance, separate investigators have now revealed many examples which are shared by flying foxes and primates, in the retino-tectal pathway, accessory optic system, cortico-spinal motor system and hippocampus, but which are not found in other mammals (Pettigrew et al., 1989). The most parsimonious interpretation of these findings is that flying foxes are closely related to primates. The implications of this interpretation are far-reaching and may explain the strong feelings aroused by this work (see for example Wible and Novacek, 1988).

The first implication, based on the complete absence of any of the primate brain features from microbats, is that flying foxes and microbats represent two independent lineages of flying mammals. The idea that mammalian flight evolved twice is not new, but it is largely discredited. That it should be revived by neural data is apparently a little inflammatory to conventional morphologists! The second implication is that the brain's wiring patterns may be inherently conservative in evolution and therefore represent a rich source of information about phylogeny. The trick, as with more conventional morphological data, is to avoid being fooled by the similarities which have been brought about by convergence. As I have tried to illustrate with the owl, a convergence of brain function in two unrelated groups of animals is generally accompanied by differences in the arrangement of the pathways which reflect their separate origins, so long as the methods are refined enough to reveal them. In view of the known superiority of the telencephalon in matters adaptive, I am more cautious about the interpretations of similarities in cortex, particularly when there appear to be good functional explanations for them (take, for example, the alterations in the somatotopic map of the forelimb of flying foxes and some microbats, discussed in Pettigrew et al., 1989). On the other hand, the striking identity of such widely separate systems as the eye-midbrain connections, hippocampus and cortico-spinal tract, along with molecular evidence, in both flying fox and primate, convinces me of their relatedness. In the same way, the many differences between the primary visual connections of the owl and cat, not to mention the obvious gross differences in all aspects of grain morphology, provide reassuring confirmation of the separate origins of mammals and birds, despite the striking functional similarity of the physiology for binocular vision in the forebrain of the cat and owl.

REFERENCES

Frisby, J.P. and Mayhew, J.E.W. (1980) Spatial frequency tuned channels: implications for structure from psychophysical and computational studies of stereopsis. *Phil. Trans. Roy. Soc. Lond. B.*, 290: 95-116.

Pettigrew, J.D. (1979) Binocular visual processing in the owl's telencephalon. *Proc. Roy. Soc. Lond. B*, 204: 435-454.

Pettigrew, J.D. (1986) Flying primates? Megabats have the advanced pathway from eye to midbrain. *Science*, 231: 1304-1306.

Pettigew, J.D., Jamieson, B.G.M., Robson, S.K., Hall, L.S., McAnally, K.I., and Cooper, H.M. (1989) Phylogenetic relations between microbats, megabats and primates (Mammalia: Chiroptera, Primates). *Phil. Trans. Roy. Soc. Lond. B.*, 334: 1-70.

BINOCULAR VISUAL PROCESSING IN THE OWL'S TELENCEPHALON

>J. D. Pettigrew
>Beckman Laboratories
>Division of Biology
>California Institute of Technology
>Pasadena, CA 91125

>Reprinted with permission from: (1979) *Proc. Roy. Soc. Lond. B*, 204, 435-454.

Single neurons recorded from the owl's visual Wulst are surprisingly similar to those found in mammalian striate cortex. The receptive fields of Wulst neurons are elaborated, in an apparently hierarchical fashion, from those of their monocular, concentrically organized inputs to produce binocular interneurons with increasingly sophisticated requirements for stimulus orientation, movement and binocular disparity. Output neurons located in the superficial laminae of the Wulst are the most sophisticated of all, with absolute requirements for a combination of stimuli, which include binocular presentation at a particular horizontal binocular disparity, and with no response unless all of the stimulus conditions are satisfied simultaneously. Such neurons have the properties required for 'global stereopsis,' including a receptive field size many times larger than their optimal stimulus, which is more closely matched to the receptive fields of the simpler, disparity-selective interneurons.

These marked similarities in functional organization between the avian and mammalian systems exist in spite of a number of structural differences which reflect their separate evolutionary origins. Discussion therefore includes the possibility that there may exist for nervous systems only a very small number of possible solutions, perhaps a unique one, to the problem of stereopsis.

INTRODUCTION

The partial decussation of optic nerve fibres, which makes binocular interaction possible in mammals, is not present in the birds (see, for example, Hirschberger 1967). Nevertheless, binocular interaction is possible in the avian brain, since thalamic efferents may cross to the

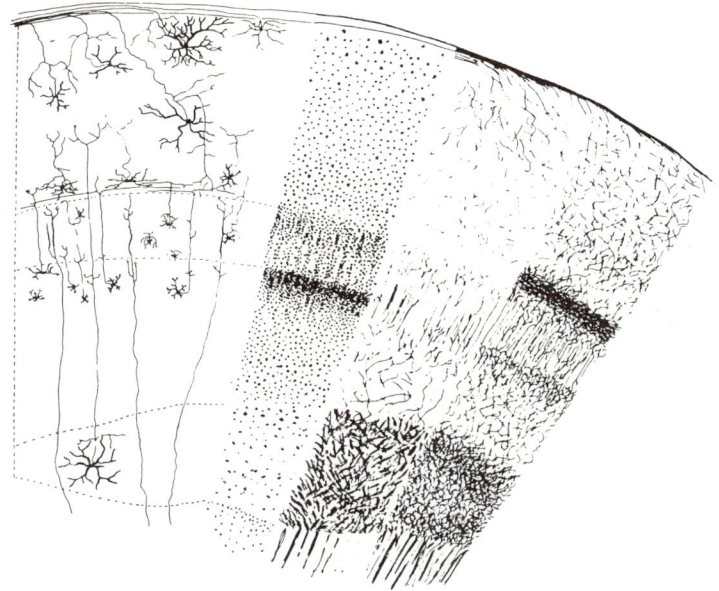

Figure 1. Laminar organization of the visual Wulst of the owl, as revealed by four different staining techniques; from left to right, Golgi, Nissl, myelin, and fibre stains. The diagram shows a representation of a coronal section through the hyperstriatum; the most inferior structure shown is the hyperstriatum dorsale (h.d.) bounded on its lower margin by the lamina suprema frontalis (dotted line at bottom). Afferent fibres enter from below and arborize in the two prominent granular layers. Efferent fibres exist on the superior surface in the tectomesencephalic tract which passes medially, becoming thicker as it picks up more efferents. The dense fibre band just above the granular layers is reminiscent of the stria of Gennari found in mammalian visual cortex.

visual Wulst of the opposite telencephalon by means of the dorsal supraoptic decussation. This was first demonstrated by Karten et al. (1973), who also speculated, on the basis of cytoarchitectonic features, that the avian visual Wulst might be an analogue of mammalian visual cortex. This speculation appears to have been borne out by neurophysiological recording from single neurons within the Wulst (Pettigrew & Konishi 1976a, b; Cooper & Pettigrew 1979). This account gives more detail of the degree to which functions within the avain Wulst has converged upon function within mammalian visual cortex.

METHODS

Data were collected from ten barn owls (*Tyto alba*), one great horned owl (*Bubo virginianus*), and two burrowing owls (*Speotyto cunicularia*), anaesthetized with ketamine (12 mg/kg). The burrowing and great horned owls were recovered after the recording session.

Details of recording technique have already been described (Pettigew & Konishi 1976a; Cooper & Pettigrew 1979).

The small eye movements were monitored by observing the superior limb of the pecten, at its intersection with the optic nerve head, as it moved with reference to a grid projected from an ophthalmoscope which was fixed in place. In both *Bubo* and *Speotyto* a dark foveal pit, surrounded by a bluish reflex, was visible in each eye and was projected to the screen to establish the visual axis. In *Tyto* the projection of the *area centralis* was inferred from the pecten position by means of data obtained from retinal whole mounts (Wathey & Pettigrew 1979).

Orientation and disparity tuning curves were obtained by pseudo-randomly interleaving stimulus trials at different orientations/disparities by the on-line use of a computer (Nova 2) which controlled stepping motors on the stimulus display and on the Risley variable biprisms.

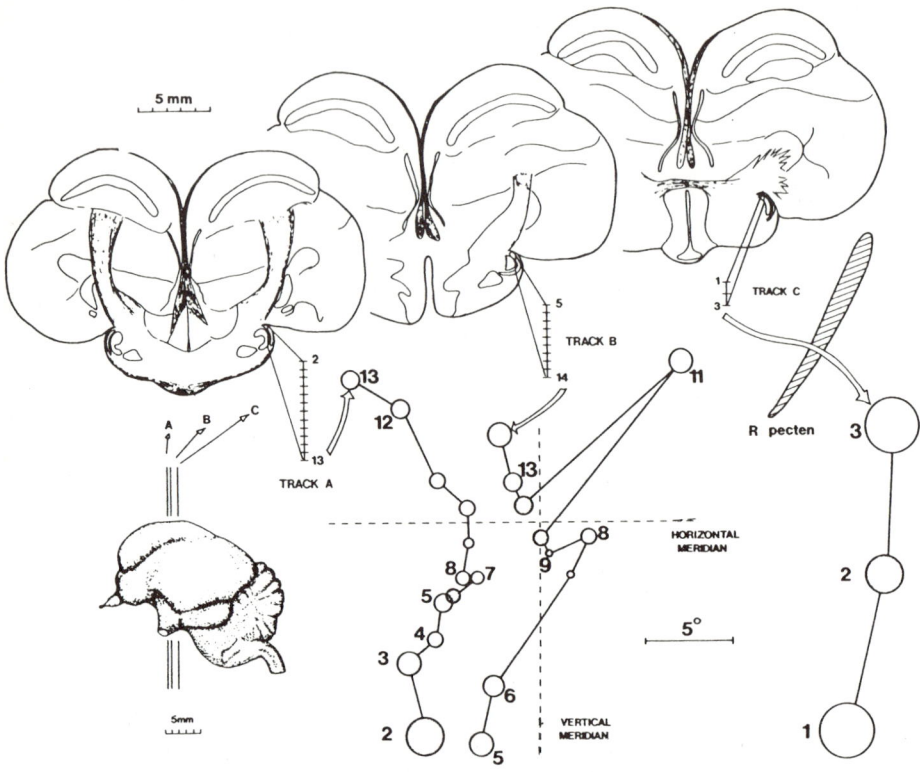

Figure 2. Topographic organization of the d.l.g.n. of the barn owl, as revealed by reconstruction of three microelectrode penetrations in coronal brain sections. Receptive field centres of single neurons are shown as circles (cross-hatched, OFF centre; open, ON centre), connected with lines to indicate the sequence in which they were encountered. The penetrations were confined to the left d.l.g.n. and all receptive fields belonged to the right eye, whose pecten base is shown as it was projected to the tangent screen. The horizontal and vertical meridians were calculated from retinal whole-mount data (Wathey & Pettigrew 1979). Note that both right and left hemi-retinae are represented within the owl's d.l.g.n., in contrast to the cat, and that the temporal retina is represented anteriorly in the nucleus. Topography along the dorsoventral axis of the d.l.g.n. is the same at it is in retina, with cells having inferior receptive fields located superiorly.

A variety of histological techniques, including Golgi, myelin, fibre and Nissl stain, were used on material available from the present series and also from donated owl brains. A cytoarchitectonic study of the visual Wulst will be the subject of a separate report (Pettigrew 1979).

RESULTS

Visual properties of the thalamic principal optic, or dorsal lateral geniculate, nucleus

As in all birds, the optic tract of the owl provides information to the first thalamic relay which is derived entirely from the contralateral eye. This conclusion was first supported by degeneration techniques after enucleation (Hirschberger 1967), is confirmed by autoradiographic studies following intraocular injection of tritiated proline (J.D. Pettigrew, unpublihed observations) and in the present series of experiments was verified by single unit recording from the thalamic relay nucleus itself. The properties of neurons were so similar to those found in the *dorsal geniculate nucleus* of mammals that is seems justified to follow Karten's original suggestion (unpublished, 1973) that the avian analogue be given the same name. Henceforward, in this paper the *nucleus opticus principus thalamicus* (Karten et al. 1973) will be called *dorsal lateral geniculate nucleus* (d.l.g.n.).

One hundred and fifty neurons recorded in 11 tracks through the d.l.g.n. of four owls showed remarkable consistency in their properties. All neurones were driven exclusively by the contralateral eye. There was no evidence of any input, even an inhibitory one, from the ipsilateral eye. All units had concentrically organized receptive fields (85 OFF centre and 65 ON centre). Units with specialized receptive field properties such as direction selectivity and local edge detection were observed on tracks passing medial and posterior to the l.g.n., but not within the body of the nucleus proper. In addition to the Kufflerian classification, l.g.n. neurons also appeared to the readily divisible into two groups if one used the battery of tests devised to separate X-like and Y-like neurons in the mammalian pathway (see Rowe & Stone 1977) and Rodieck (1979) for reviews of this rapidly expanding field). Y-like neurons were more common in the sample 96/150), perhaps because they appeared to be localized in the central magnocellular region of the nucleus where more electrode tracks passed and where cells are larger. These Y-like cells were indistinguishable on the basis of contrast-reversal of a grating target, fast flicking of a wand, resolution of a moving grating, transience of response and periphery effect, from Y-cells (or brisk transient) cells described in the d.l.g.n. of the cat. Preliminary studies on conduction velocity after intraocular electrical stimulation of the optic nerve head are so far consistent with Y-like cells having retinal inputs with shorter latency than all other visual cell types in the thalamus (around 1.5 ms compared with 10-25 ms for some of the specialized units found outside of the l.g.n.).

X-like l.g.n. cells were also closely similar to their mammalian counterparts, with smaller receptive fields, and sustained responses in comparison to the Y-like cells, as well as clearly defined null-points for contrast-reversal. No data are yet available on the latency of retinal input of these X-like cells.

The one striking difference from the mammalian d.l.g.n. was the representation within the avian d.l.g.n. of the complete contralateral retina. The anterior half of the avian d.l.g.n. contained a representation of the contralateral retina temporal to the visual pole, a part of the retina never represented in the mammalian l.g.n., except for the tiny strip near the zero meridian of normal cats (Stone 1965) and the anomalous contralateral temporal projection in albino mutants. The topography of the owl's d.l.g.n. is shown in figure 2 where these are reconstructions of three electrode tracks through the left d.l.g.n. of a barn owl. It can be seen that the first units encountered have inferior recpetive fields and that field positions have successively higher elevations as the electrode moves down. In other words, neurons are laid out along the vertical axis just as they are in the retina. Along the antero-posterior axis, one can see that field positions change in azimuth, with the most posterior neurons having receptive fields in the contralateral hemifield like the representation within mammalian d.l.g.n. However, it can be seen that more anterior tracks yield neurons with receptive fields across the vertical midline in the ipsilateral hemifield. It is this anterior half of the nucleus that departs from the mammlian pattern, since cells lying here must recross to the opposite side to bring about the unified representation of the contralateral hemifields of *both* eyes withing the Wulst, as already described (Pettigrew & Konishi 1967a) and considered in more detail below. This recrossing takes place in the dorsal supraoptic decussation, as already described by Karten et al. (1973) and can be verfied by retrograde transport studies from one Wulst, where *both* l.g.ns show labelling, the contralateral l.g.n. in the anterior half and the ipsilateral l.g.n. in the posterior half (J. Pettigrew, unpublished observations).

Visuotopic organization

In two experiments a large surface of the Wulst was exposed and a number of electrode penetrations made to determine the way in which the visual field is represented within the structure. Such experiments revealed only one visuotopic representation of the contralateral hemifield. Neurons representing the vertical meridian were found along the lateral margin of the Wulst adjacent to the vallecula, with receptive fields moving into the peripheral visual field if the electrode was placed more medially. Anterior penetrations yielded receptive fields in the inferior field and posterior penetrations yielded more superior fields. This arrangement is illustrated in figure 3 which shows the results obtained in one experiment from seven penetrations normal to the surface of the Wulst. In such penetrations normal to the surface, receptive fields occupied the same position in the visual field as the electrode advanced.

The topographic arrangement found in these mapping experiments was confirmed in subsequent experiments with long electrode tracks oblique to the surface. Such tracks yielded steady

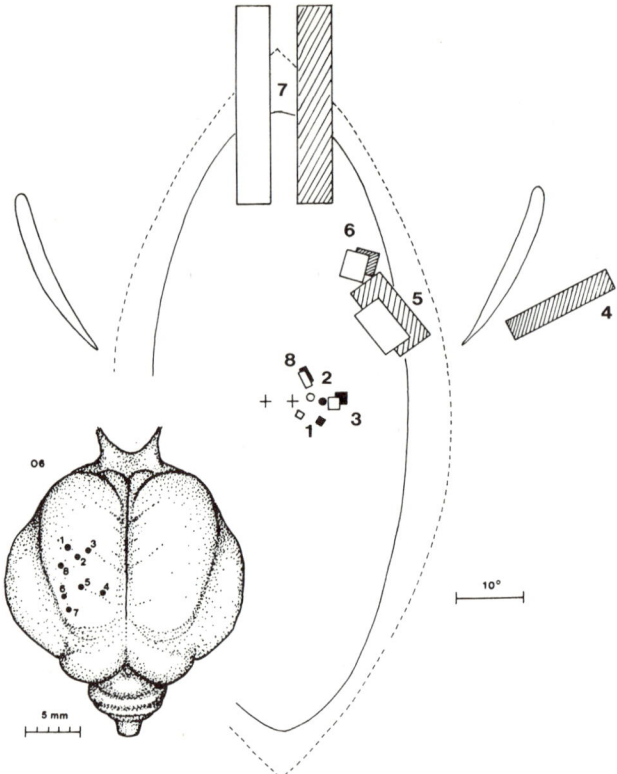

Figure 3. Topographic organization of the left visual Wuslt of barn owl 6, as revealed by seven electrode penetrations made normal to the surface in one experiment. In each penetration, receptive field position remained roughly constant; fields of single representative unit for each track are shown. Filled fields indicate contralateral eye and unfilled ipsilateral eye. The elliptical outline shows the region of binocular overlap on the tangent screen; the dotted outline is the maximal extent of this region, as revealed by reversible projection of temporal margin of the retina (visible ophthalmoscopically: the *ora terminalis*, which is smooth in the owl and not serrated as it is in some species); the continuous outline is the extent of binocular overlap obtaining during an experiment, when the barn owl's facial disk is partially collapsed because of anaesthesia. The crescent-shaped outlines show the projections of the optic nerve heads, each of which is overlain by the pecten. The crosses indicate the positions of the *area centralis* projections, as determined from retinal whole mount data. Most of the visual Wulst can be seen to be concerned with the region of binocular overlap in the contralateral hemifield. Topographic organization is the same as that found in cat area 17, with the vertical meridian represented laterally (at the border marked by the vallecular groove), the lateral field represented medially, inferior anteriorly, and superior field posteriorly.

changes in field position as the electrode was moved. The field positions tended to follow a straight line on the tangent screen whose direction depended upon the position of the electrode track and the angle made with the margin of the Wulst (figure 5). In over 20 long penetrations through the Wulst there was no instance of a reversal in the progressive shift of field positions like those commonly seen in mammalian visual cortex (e.g. at the 17/18 boundary).

By using the results obtained from all penetrations, it was possible to construct a map of the field representation with the Wulst. This map is shown in figure 4 which, in addition to providing more detail of the topography already outlined, illustrates the large overrepresentation of central visual field in the Wulst. The owl's visual field extends laterally to 90° from the vertical meridian, yet more than half of the total area of the Wulst is devoted to the central 10° adjacent to the vertical meridian. The magnification factor is 1-2 mm/deg at the area centralis, compared with 0.04 mm/deg at 30° eccentricity.

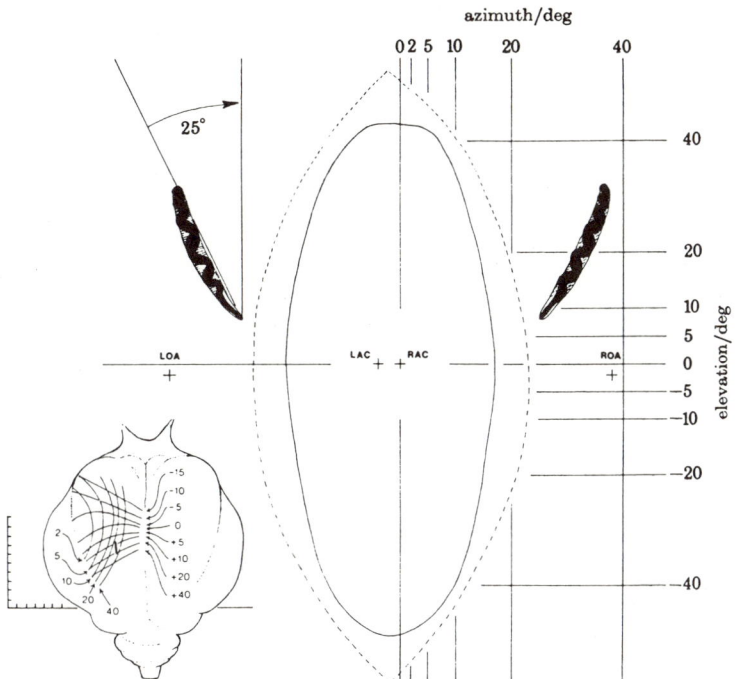

Figure 4. Summary diagram of visuotopic organization within the Wulst, obtained by pooling all the data from the present series. See figure 3 for conventions. The scale shown next to the outline of the brain is in millimetres.

Receptive fields

It was surprising to find neurons in the Wulst that had all of the receptive field properties so far described in areas 17 and 18 of cat and monkey (Hubel & Wiesel 1959, 1962, 1965, 1968, 1970), with the exception of the class of binocular neurons found in cat area 18, which have different preferred directions in each eye (Pettigrew 1973). Conversely there were no neurons whose properties were so unusual that they did not readily fit into the categories already established for mammalian areas 17 and 18. With the exception of a small number of cells recorded from the superficial layers in early experiments, all cells recorded from the Wulst could be driven visually. These early undrivable units we believe to belong to the 'obligate binocular' class, whose stringent receptive field requirements were not discovered until the third experiment (see below). The distribution of different properties among the sample of neurons studied is shown in table 1. Details of the different properties are taken up below.

(i) Non-orientated cells

The majority of these (59/63) could be excited only from one eye, had concentrically organized ON or OFF centre fields and were found in the deeper layers of the Wulst, particularly the granular layers receiving direct thalamic input. The close similarity of these units to those found by recording from within the thalamus raises the possibility that they represent recordings from the terminal arborizations of thalamic afferents. This possibility was also raised with respect to the observation of non-orientated units in layer IVc of monkey striate cortex, but was rejected on the basis of considerations such as spike waveform (Hubel & Wiesel 1968). Similar considerations in the present case make it likely that all but a small proportion of the non-orientated units are granule layer neurons, although further study with electrical stimulation of the optic radiations seems warranted to establish this with more certainty.

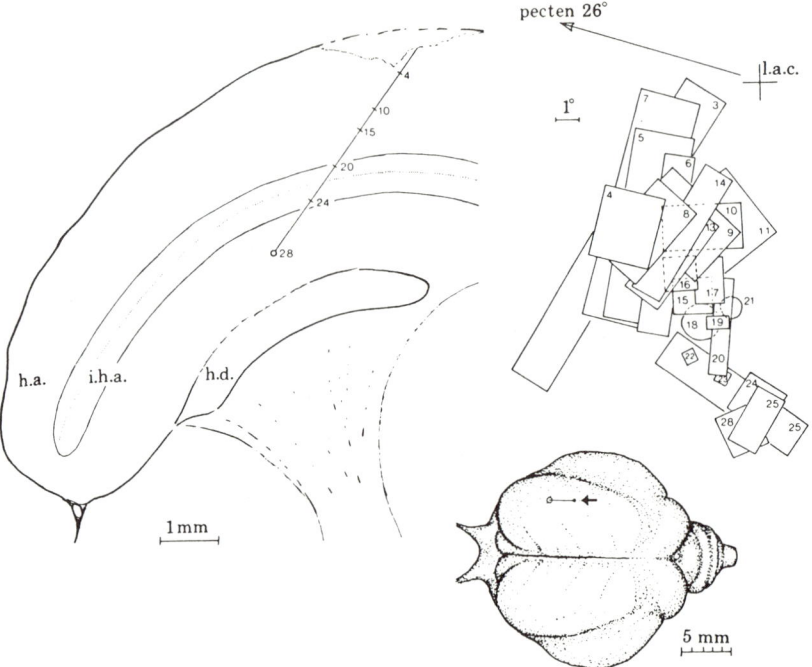

Figure 5. Reconstruction of a parasagittal electrode track in the right visual Wulst. Although most cells were binocular, only one field is shown for each unit. l.a.c., left area centralis. Note that fields are in the contralateral hemifield and move down as the electrode moves anteriorly. Note also that receptive fields of superficial neurons tend to be much larger than those of neurons located in the granular layers (i.h.a., *Hyperstriatum accessorium intercalatus*: see Karten et al. 1973), within which there are also some neurons (e.g. 18 and 21) with concentrically organized, non-orientated receptive fields. There is a tendency for neighbouring neurons to have similar orientations (for example the sequence 3-11). More evidence for this is presented in figure 6.

A tiny minority of cells (4 out of the total of 579 recorded) were both non-orientated and binocular. These cells had large, roughly circular, receptive fields (10-20° in diameter) which gave an equally brisk ON-OFF response from either eye. Despite the large size of the fields, quite good responses could be elicited by moving or flashing small targets (1/2° or smaller). Binocular interactions in this group was unremarkable, with summation of the monocular responses over a wide range of receptive field alignments.

(ii) Simple cells

A large fraction of cells recorded (126/579) has receptive field properties like those originally described for 'simple' cells by Hubel & Wiesel (1959, 1962, 1968) in the striate cortex of cat and monkey. The responses of these cells were orientation selective, according to the strict criteria now used for this designation (see Pettigrew 1974; Schiller et al. 1976). In addition, their receptive fields could be plotted with stationary flashing targets into antagonistic subregions, whose arrangements usually helped one to predict a given cell's optimal stimulus orientation but rarely led to predictions about the cell's preferred stimulus speed or direction. Linear summation could usually be demonstrated within an antagonistic subregion.

In contrast to the non-orientated cells, very few of which were binocular, the majority of simple cells (105/126) were binocular, with receptive field properties closely similar for each eye.

Simple cells were found in all layers of the Wulst, but were scarce in the superficial 2 mm.

TABLE 1. Ocular Dominance Groupings by Cell Class in Owl Visual Wulst

ocular dominance group ...	1	2-3	4	5-6	7
non-orientated	31	—	4	—	28
simple	12	60	3	42	9
simple end-stopped	—	13	16	9	—
Rosenquist and Palmer	—	2	19	—	—
'complex'	—	67	123	41	—
obligate binocular	—	—	(101)	—	—

Ocular dominance classifications follows that of Hubel & Wiesel (1962) and cell classes are described in the text. The degree of binocularity and receptive field complexity appear to be related.

(iii) Simple, end-stopped receptive fields

These cells were like the 'hypercomplex', type I cells described in cat and monkey striate cortex (Hubel & Wiesel 1968; Dreher 1972). Both single- and double-stopped varieties were seen, often in clusters about 1.5-2 mm from the surface of the Wulst. All of these cells were binocular.

(iv) Rosenquist and Palmer cells

A prominent class of cells described in cat striate cortex by Palmer & Rosenquist (1974) appears also to have its counterpart in the owl's Wulst. These cells have large, uniform receptive fields from which mixed ON and OFF responses can be elicited with small flashing targets. Responses to moving targets are direction selective, and show some degree of orientation selectivity if the target is long, but do not show a decrement if the target is shortened. The vigorous reponses to small targets that are a tiny fraction of the receptive field size, the high maintained discharge and the prolonged shower of spikes in the response, along with balanced binocularity, all make these cells easy to recognize. An interesting, but as yet unanswered, question is whether this class of cells projects to the optic tectum as their counterparts have been shown to do in the cat.

(v) 'Complex cells'

A variety of receptive field types were seen that did not fall readily into any of the categoires so far mentioned. Worthy of special mention was a group of 'black bar specialists' which could be activated only by an orientated dark bar moving against a light background. These cells did not respond very well to flashing targets and gave no response to small dark targets or to light targets of any size or shape. They appeared to be concentrated about 1 mm below the surface of the Wulst. All were binocular.

(vi) Obligate binocular cells

This class of cell had the most elaborate receptive field properties of any observed in the Wulst. They resembled the 'AND-gate' neurons described in the sheep's visual cortex (Ramachandran et al. 1977) and the 'binocular depth cells' observed in monkey visual cortex (Hubel & Wiesel, 1970), although some differences should be noted.

The most striking characteristics of these cells is their failure to respond vigorously to any form of monocular stimulation. Even binocular stimulation is ineffective unless the stimulus is appropriately orientated and moving in the appropriate depth plane in front of the owl. A variable biprism is usually required to achieve the right conditions for binocular stimulation because of the usual slight divergence of the visual axes one finds in the anaesthetized owl.

More than half of these cells preferred dark stimuli on a light background and nearly all responded best if one end of the stimulus remained within the receptive field as it was moved about. Another striking feature of these cells was their enormous receptive field size, 100-500 deg^2, in comparison to the optimal stimulus size, *ca.* 1 deg^2.

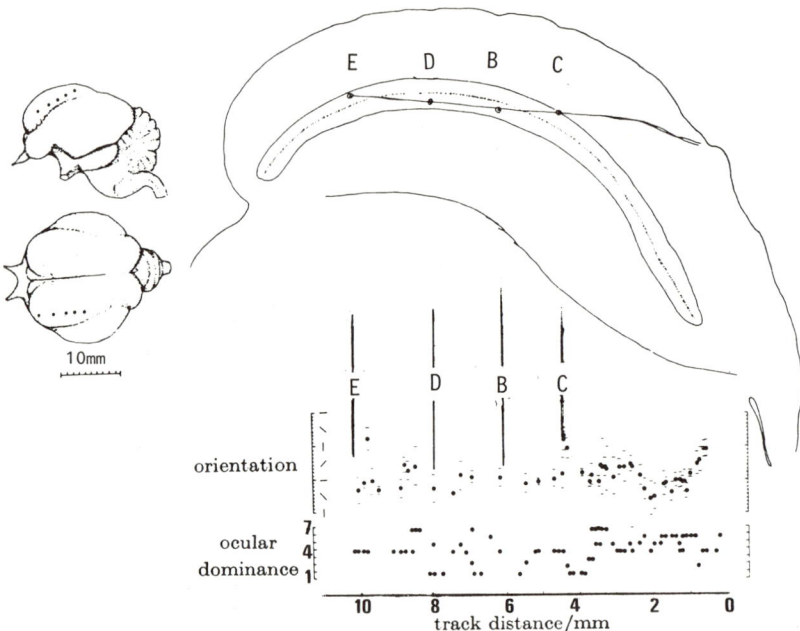

Figure 6. Reconstruction of a long parasagittal track in the left visual Wulst of barn owl 4. Dots on the diagram of the brain represent projections of lesions made by passing current through the microelectrode and shown as circles (B,C, D, and E; lesion A was not found) in the drawing of the sagittal section at left. Preferred orientation (filled circles) and the range of orientations over which a cell responded (short horizontal bars) are shown in the upper graph, and ocular dominance group (Hubel & Wiesel 1962) in the lower graph. In the upper layers, neighbouring cells have similar preferred orientations, with a gradual change as the electrode moves tangentially. In the granular layers there is a tendency toward clumps of cells with similar ocular dominance.

Perferred orientations of these cells were randomly distributed around the clock, in contrast to the situation in monkey where there appears to be a preponderance of cells with a large vertical component in their preferred orientations (Hubel & Wiesel 1970).

Binocularity

(i) Ocular dominance

Within the region of Wulst devoted to the zone of binocular overlap, it was unusual to find a neuron for which a receptive field could not be plotted in either eye. These cases were almost totally confined to the non-orientated class, of which nearly all (59/63), were monocular. Simple cells showed a tendency to be in extreme ocular dominance groups (65/126) in groups 2 and 6) and a proportion (21/126) could be excited only from one eye (table 1). None of the other classes of neuron had members which were monocular, and there was a tendency for more elaborate receptive field properties to be correlated with binocularity.

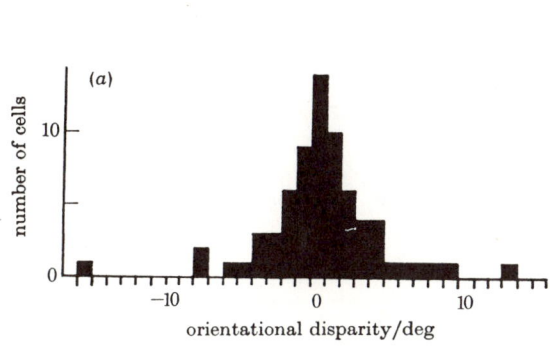
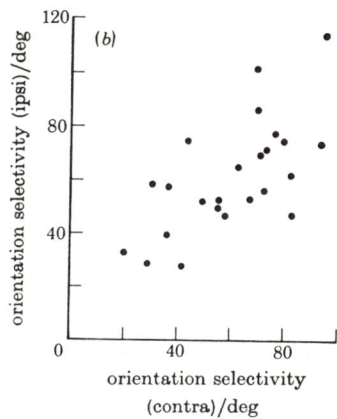

Figure 7. (a) Distribution of orientational disparities (preferred orientation of a neuron in one eye minus its preferred orientation in the other eye) for 69 neurons classified as orientation selective by strict criteria such as a response to an elongated target more selective than response to short target. The mean and mode are both zero and the standard deviation is 4.2°. (b) Match between a binocular neuron's orientation selectivity in each eye. Orientation selectivity is the total range of orientations to which a given cell will respond. (Further details of definition and determination are given by Blasdel et al. (1977).) Orientation selectivity in the contralateral eye (abscissa) is plotted against orientation selectivity in the ipsilateral eye (ordinate). The two variables are well correlated ($r = 0.66$), indicating that a given binocular neuron, in addition to having similar preferred orientations in each eye, also responds over a similar range of orientation in each eye.

There was a slight tendency for cells with similar ocular dominance to be clumped (see figure 6) and this may be correlated with the ocular dominance banding observed in autoradiographic material (Pettigrew 1979).

(ii) Matching field properties

As described for binocular neurons in cat visual cortex (Nelson et al. 1977), there was a close correlation between the receptive field properties in each eye. For example, preferred orientations were closely similar in each eye, with a mean orientational disparity of 0° and a standard deviation of ±4.2°. This finding suggests that no significant cyclotorsion takes place in the anaesthetized owl, in contrast to the cat.

Orientational specificity in one eye was also closely correlated with the same cell's orientational specificity in the other eye. This is illustrated in figure 7 where computer-generated orientation tuning curves have been used to determine the orientation selectivity of 24 neurons for each eye separately (see Blasdel et al. 1977). There is a strong correlation ($r = 0.66$), like that found in the cat striate cortex ($r = 0.77$; Nelson 1978). One has the impression that more formal analysis would reveal a similar correlation between each eye for other receptive field properties like preferred stimulus size, velocity and direction.

(iii) Receptive field disparity

After correction for residual eye movements, a significant degree to variation remains in relative retinal positions of binocular receptive field pairs. This variation has two components, a systematic decrease in receptive field separation with decreasing vertical eccentricity and a local scatter which is superimposed on the systematic change. This is shown in figure 8.

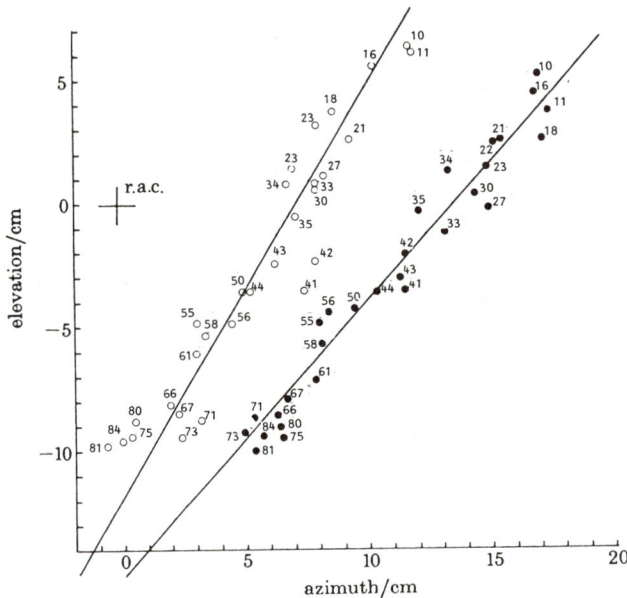

Figure 8. Receptive field centres of binocular neurons recorded in a long tangential penetration through the visual Wulst (the same as the one shown in figure 6) to show the progressive decrease in receptive field separation which occurs with decreasing elevation. Although there was some concurrent change in receptive fields azimuth as the electrode advanced, other experiments (reported in Cooper & Pettigrew 1979) enable this to be eliminated as a contributing factor in the change in binocular receptive field separation. Similarly, a systematic change in eye position was ruled out by continuous monitoring (in this experiment) and by simultaneous recording at two sites representing different elevations (see Cooper & Pettigrew 1979). Regression analysis applied to these data indicates that there is a relative rotation of 2.6° between the two sets of points which in turn implies that the vertical meridians are rotated by this amount. The geometrical consequences of this rotation for the representation in the Wulst of owl's three-dimensional space are described in detail by Cooper & Pettigrew (1979). Even after correction for the systematic change in positional disparity of the binocular receptive field pairs given by the two regression lines, a scatter of positional disparities remains (± 2.5°). r.a.c., right area centralis estimated from whole mount data and pecten projection. Both scales are in centimetres on the tangent screen 57 cm from the owl's eyes.

Since the systematic change with vertical eccentricity is linear, the geometrical consequences are easy to calculate. The resulting tilted vertical horopter is described in more detail elsewhere (Cooper & Pettigrew 1979).

A local disparity variation of ± 2.5° remains for the barn owl after one corrects for the systematic change. The significance of this variation is best considered along with disparity selectivity in the next section.

(iv) Disparity selectivity

When the specificity of binocular interaction was tested with the variable biprism technique (see Pettigrew et al. (1968) and Methods), a variety of patterns was observed. These included (*a*) a broad facilitatory curve, (*b*) a broad inhibitory curve, (*c*) a step-shaped curve with inhibition on one side and facilitation on the other, and (*d*) a sharply peaked facilitatory curve. Examples of each one of these patterns are illustrated in figure 9. The step-functions resemble those obtained by Poggio & Fischer (1977) from the 'near' and 'far' cells in areas 17 and 18 of macaque visual cortex. In this respect it may be worth noting that all such curves were obtained in the present study from 'simple' cells, although 'simple' cells did show other types of binocular interactions.

Sharply peaked disparity tuning curves were obtained from all classes, except non-orientated cells, and there was a tendency for the sharpest tuning to be associated with the property of end-stopping. Since many of these curves are 1° wide, or less, at half-height, the variation of 5° in receptive field disparity from cell to cell is likely to be of functional significance. This view is supported by the results of simultaneous recording experiments, where two cells having similar stimulus preferences with respect to orientation, movement and retinal eccentricity, can be shown to have quite different requirements for binocular disparity. In some cases it has been possible to show that the same moving bar stimulus will simultaneously excite cell A and inhibit cell B at one disparity value, yet inhibit A and excite B at another disparity value (see fig. 3, Pettigrew & Konishi 1976a).

Sharply defined binocular specificity is not associated with sharply defined receptive field boundaries. In the case of the 'obligate binocular' class a disparity tuning curve 1° wide at half-height may be seen, despite the facts that the monocular field boundaries are completely undefinable and the binocular field is over 10° across. In such cases, the range of binocular specificity seems to be related much more to the size of the optimal stimulus, which can be a very small fraction of the total field size. It should also be pointed out that different cells in this class achieve specificity for horizontal disparity without regard for optimal orientation.

In other words, obligate binocular cells with preferred orientations close to horizontal are just as common, and just as selective for horizontal disparity, as other obligate binocular cells. This contrasts with the situation described for otherwise comparable cells in monkey prestriate cortex (Hubel & Wiesel 1970) but is similar to that found in sheep VII (Clarke et al. 1976). In the present case the mechanisms is presumably related to the fact that these cells usually have a requirement that the stimulus bar be 'stopped' as well as appropriately orientated.

DISCUSSION

Similarity to mammalian visual cortex

The most remarkable aspect of the present findings is the close parallel revealed between the functional attributes of neurons in the visual Wuslt and the attributes already described in the mammalian visual cortex. Not only are the various mammalian receptive field properties all present, but they are arranged in a similar orderly way with systematic changes in preferred orientation, ocular dominance, etc.; even the visuotopic representation of space is orientated within the owl's brain just as it is in the cat's striate cortex, with the vertical meridian laterally and the inferior field anteriorly! The parallel also extends to the phenomenon of developmental plasticity (see Pettigrew & Konishi 1976b)

If we ignore, for the moment, the large difference between the ways in which the avian and mammalian visual pathways achieve binocular integration, there appear to be only two gross differences between the organiztion of mammalian visual cortex and avian visual Wulst; the superficial location of output neurons within the Wulst and the absence of a second, mirror-image field representation for the equivalent of area 18. The first point is consistent with a large number of observations which include (i) Golgi-stained sections in which the largest neurons within the Wulst are seen to lie superficially and to send their axons into the septomesencephalic tract forming the most superficial layer of the Wulst (see figure 1); the septomesencephalic tract connects the Wulst with the d.l.g.n. and the optic tectum (Karten et al. 1973); (ii) probable cortico-tectal cells, tentatively identified on the basis of the strong resemblance of their receptive field properties to those described by Palmer & Rosenquist (1974) for cortico-tectal cells in cat visual cortex, were always found superficially in the Wulst, and never deep to the granular layers receiving thalamic input; cat cortico-tectal cells are always found in layer V, deep to the granular layers; (iii) retrograde tracing techniques show labelling of superficial neurons after tectal injections (Bravo & Pettigrew 1979). The second point follows from the fact that no visual representation could be found lateral to the vallecula, the landmark which marks the most lateral boundary of the Wulst and the vertical meridian; electrode tracks passing into this region yield neurons with somaesthetic receptive fields. It seems very likely that the functional equivalent of mammalian area 18 actually lies within the Wulst, whose greater thickness could no doubt accommodate both representations stacked in register. If this view can be shown to be correct, it would support the idea already put forward that areas 17 and 18 are functional adjuncts (see Allman & Kaas 1974). It is certainly consistent with much evidence. For example, many Wulst neurons have functional attributes, like an absolute requirement for black stimuli and for simultaneous

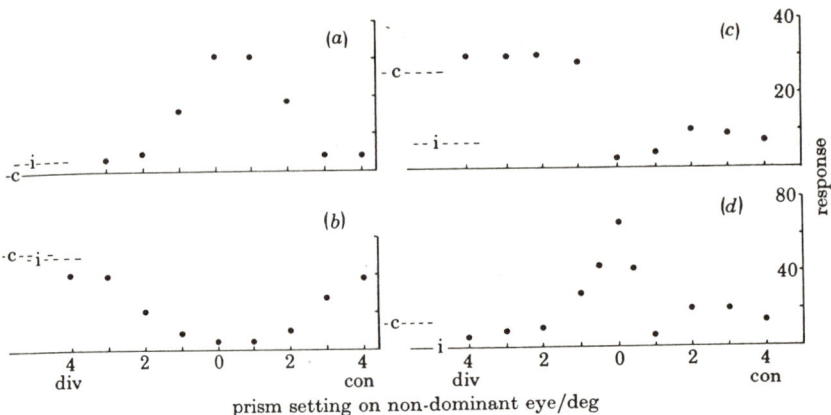

Figure 9. Disparity turning curves obtained from four binocular neurons representative of the four main kinds of binocular interaction observed in the owl's visual Wulst. Each point is the mean binocular response (spikes per second during the time the stimulus was passing over the centre of the dominant eye's receptive field, average of five trials, pseudo-randomly interleaved with trials at the other prism settings) to a moving bar stimulus at the optimal orientation and velocity. Two variable biprisms were used and the one in front of the dominant eye was always fixed during a given run while trials at different prism settings on the non-dominant eye was carried out. The control comparison for the binocular response at a given prism setting is therefore the mean monocular level (five trials) for the ipsilateral (i) or contralateral (c) eye, whichever is larger. Prism settings are given in degrees convergent (base-in) and divergent (base out). The zero is arbitrary since its absolute value depends upon the setting on the fixed prism in front of the dominant eye to bring about initial, approximate alignment. In each of these curves the changes in prism setting are all in the horizontal plane. There are four patterns of binocular interaction: (a) broad binocular facilitation; (b) broad binocular inhibition; (c) step-shaped function (both 'convergent-excited' and 'divergent-excited' varieties were seen, although an example of the latter only is illustrated); (d) sharp binocular facilitation. Each of these patterns can be shown by any class of cell, except (c), which appears to be confined to the 'simple' class (see text); (d) was more common among cells that showed some degree of 'end stopping'.

bincocular stimulation, which are more in line with those described for area 18 in mammals (Hubel & Wiesel 1965; Pettigrew 1973; Clarke et al. 1976). In addition, autoradiographic study of the thalamic input to the Wuslt reveals two sets of alternating monocular bands, one located more superficially and with a coarser pattern than the other. The coarser pattern has a repeat like that of cat area 18 while the finer pattern has a repeat like area 17 (Pettigrew 1979).

Parallel evolution

The intriguing resemblance shown between the pattern of physiological organization of the owl's visual Wuslt and the mammalian striate cortex appears to be an instance of parallel or convergent evolution. This can therefore take its place alongside other celebrated examples, like the convergence in habits, song and coloration of the African pipit (*Macronyx croceus*) with the American meadowlark (*Sturnella magna*) (Friedmann 1946). The resemblance is so strong in the latter case that the African and American birds were initially thought to be a single species (Linnaeus 1758), independent origins later being established on the basis of other more converative taxonomic criteria. In the present case it also appears possible to rule out the alternative explanation that the marked resemblance in physiological organization stemmed from a common predecessor which shared the same degree of sophisticated binocular visual processing as cat and owl. Elaboration of the striate cortex appears to be a relatively late evolutionary accomplishment with a peak of development in the primates and carnivores. Similarly, a greatly elaborated and laminated visual Wulst is a feature of some birds only, and is unlikely to have been present in such a specialized form in the earliest birds. It therefore seems very likely that birds and mammals had a common Jurassic reptilain ancestor with binocular visual processing in a neural antecedent of both Wulst and striate cortex.

A more compelling argument for parallel evolution comes from the complete contrast between the avian and mammalian patterns of optic decussation. If there were a common predecessor which achieved binocular visual processing, a reasonable expectation would be that this predecessor's strategy for binocular convergence, be it a single partial decussation (mammalian pattern) or two successive decussations (avian pattern), would now be shared by both orders. The finding that birds and mammals have such radically different strategies for achieving binocular convergence argues strongly that there were two independent lines of evolution, each leading to binocular neural processing of a remarkably similar kind.

A similar conclusion is reached by a consideration of the differences in anatomical fine structure between avian Wulst neurons and mammalian striate neurons (see, for example, figure 1). These differences, such as the total absence of the classical pyramidal cell class from the avain Wulst (Pettigrew 1979), belie the physiological similarities with respect to visual processing and support the view that there has been parallel evolution.

Parallel evolution of stereopsis?

If one accepts that cat visual cortex and owl visual Wulst represent a case of parallel evolution, one is led to the question, 'What environmental constraints were responsible for this convergence of function?' Any answer to this question that is couched in terms of a general theory of vision is unsatisfactory because there is in fact considerable divergence among the various animals with respect to their modes of visual processing. For example, it is possible for some sophisticated processing to be carried out to a large degree in the retina, where quite elaborate receptive fields can be found in a number of species, including some birds. In the present case one has to explain why complex information processing is delayed in the telencephalic pathways of both cat and owl until information from both eyes comes together, and why it is then carried out in what appears to be a number of stages, with increasing receptive field complexity going hand-in-hand with an increasing degree of binocular interaction.

A specific theory, which accounts for a satisfactorily large number of the facts, explains the parallel in terms of the selection pressures operating upon a nocturnal visually directed predator. For such a predator, the adoption of stereopsis would have the well recognized advantages of accuracy, immediacy, the absense of the need to generate monocular parallax by moving itself, the ability to 'break' camouflage, and the improved signal: noise ratio resulting from probability summation of two channels. If one accepts the view that the primate line began as a nocturnal, visually directed predator (see Allman 1977) then convergence in the interest of stereopsis can also account for the similarity of macaque visual cortex to its avian and feline analogues.

The first point in favour of this theory is the fact that areas 17 and 18 of cat and monkey cortex and the avian visual Wulst all seem to be involved in binocular disparity detection, the heart and soul of stereopsis. This point of view has been the subject of some controversy, particularly in the macaque, but the weight of evidence supports it (see Fischer & Poggio, this symposium, and Pettigrew 1978).

The second point in the theory's favour is that it provides an explanation for the arresting observation that however complex a cell's receptive field organization it is closely matched in each eye. As originally pointed out by Barlow et al. (1967), this matching could enable the relatively unambiguous selection of the two retinal image features, one in each eye, which correspond to a single feature. This selection, of the appropriate part of the images to be measured, is an essential part of the disparity measurement task required for stereopsis.

EPILOGUE

Much remains to be learned about the neural processing involved in stereopsis. In particular, while quite a lot is known about the process of local disparity detection, from studies of cat, monkey, sheep, and owl, very little is known about how global stereopsis is achieved or how horizontal disparity is extracted at the expense of vertical. It has been suggested that one solution to the last problem involves an increase in the proportion of higher-order disparity selective neurons with non-horizontal preferred orientations (Hubel & Wiesel 1970). This appears not to be the only solution to the problem, however, since the owl's 'obligate binocular' neurons achieve specificity for horizontal disparity at all preferred orientations, even horizontal, by the use of 'end stopping.'

It seems likely that further study of visual processing in the owl will further our understanding of stereopsis and of vision in general.

This work was supported by grants from the Spencer Foundation and the U.S. Public Health Service (No. EY1909 from the National Eye Institute and No. MH25852 from the National Institutes of Mental Health and Drug Abuse). The owls were provided by Masakazu Konishi, who also took part in all the early recording sessions. Invaluable technical assistance was provided by H. Adams, G. Blasdel, M. Cooper and Sarah Kennedy.

REFERENCES

Allman, J.M. (1977) Evolution of the visual system in the early primate. *Prog. Psychobiol. Physiol. Psychol.* **7**, 1-53.
Allman, J.M. and Kaas, J. (1974) The organisation of the second visual area (V II) in the owl monkey: a second order transformation of the visual hemifield. *Brain Res.* **76**, 247-265.
Barlow, H.B., Blakemore, C. and Pettigrew, J.D. (1967) The neural mechanism of binocular depth discrimination. *J. Physiol., Lond.* **193**, 327-342.
Blasdel, G.G., Mitchell, D.E., Muir, D.W. and Pettigrew, J.D. (1977) A physiological and behavioural study in cats of the effect of early visual experience with contours of a single orientation. *J. Physiol., Lond.* **265**, 615-636.
Bravo, H. and Pettigrew, J.D. (1979) A retrograde transport study of the neurones projecting to the primary visual nuclei of the owl, *Speotyto cunicularia*. In preparation.
Clarke, P.G.H., Donaldson, I.M.L. and Whitteridge, D. (1976) Binocular visual mechanisms in cortical areas I and II of the sheep. *J. Physiol., Lond.* **256**, 509-526.
Cooper, M.L. and Pettigrew, J.D. (1979) A neurophysiological determination of the vertical horopter in cat and owl. *J. Comp. Neurol.*, **184**, 1-25.
Dreher, B. (1972) Hypercomplex cells in the cat's striate cortex. *Invest. Opthal.* **11**, 355-356.
Friedmann, H. (1946) Ecological counterparts in birds. *Scient. Mon.* 43, 395-398.
Hirschberger, W. (1967) Histologische Untersuchungen an den primären viseullen Zentren des Eulengehirnes und der retinalen Repräsentation in ihen. *J. Orn., Lpz.* **198**, 187-202.
Hubel, D.H. and Wiesel, T.N. (1959) Receptive fields of single neurones in the cat's striate cortex. *J. Physiol., Lond.* **148**, 574-591.
Hubel, D.H. and Wiesel, T.N. (1962) Receptive fields, binocular interaction and functional architecture in the cat's visual cortex. *J. Physiol., Lond.* **160**, 106-154.
Hubel, D.H. and Wiesel, T.N. (1965) Receptive fields and functional architecture in two non-striate visual areas (18 and 19) of the cat. *J. Neurophysiol.* **28**, 229-289.
Hubel, D.H. and Wiesel, T.N. (1968) Receptive fields and functional architecture of monkey striate cortex. *J. Physiol., Lond.* **195**, 215-243.
Hubel, D.H. and Wiesel, T.N. (1970) Cells sensitive to binocular depth in area 18 of the macaque monkey cortex. *Nature, Lond.* **225**, 41-42.
Karten, H.J., Hodos, W., Nauta, W.J. and Revzin, A.M. (1973) Neural connections of the 'visual Wulst' of the avian telencephalon. Experimental studies in the pigeon (*Columba livia*) and owl (*Speotyto cunicularia*). *J. Comp. Neurol.* **150**, 253-278.
Linnaeus, C. (1758) *Systema Natura. Regnum Animals*, 10th edn. Leipzig: Engelmann.
Nelson, J.I. (1978) Does orientation domain inhibition play a role in visual cortex plasticity? *Expl. Brain Res.* **32**, 293-298.
Nelson, J.I., Kato, H. and Bishop, P.O. (1977) Discrimination of orientation and position disparities by binocularly activated neurons in cat striate cortex. *J. Neurophysiol.* **40**, 260-283.
Palmer, L.A. and Rosenquist, A.C. (1974) Visual receptive fields of single striate cortical units projecting to the superior colliculus in the cat. *Brain Res.* **67**, 27-42.
Pettigrew, J.D. (1973) Binocular neurones which signal change of disparity in area 18 of cat visual cortex. *Nature, Lond.* **241**, 123-124.
Pettigrew, J.D. (1974) The effect of visual experience on the development of stimulus specificity by kitten cortical neurones. *J. Physiol., Lond.* **237**, 49-74.
Pettigrew, J.D. (1978) Stereoscopic visual processing. *Nature, Lond.* **273**, 9-11.
Pettigrew, J.D. (1979) Structural organisation of the owl's visual Wulst. In preparation.
Pettigrew, J.D. and Konishi, M. (1976a) Neurons selective for orientation and binocular disparity in the visual Wulst of the barn owl (*Tyto alba*). *Science, N.Y.* 193, 675-678.
Pettigrew, J.D. and Konishi, M. (1976b) Effect of monocular deprivation on binocular neurones in the owl's visual Wulst. *Nature, Lond.* **264**, 753-754.

Pettigrew, J.D., Nikara, T.N. and Bishop, P.O. (1968) Binocular interaction on single units in cat striate cortex: simultaneous stimulation by single moving slit with receptive fields in correspondence. *Expl. Brain Res.* **6**, 391-410.

Poggio, G.F. and Fischer, B. (1977) Binocular interaction and depth sensitivity in striate and prestriate cortex of behaving rhesus monkey. *J. Neurophysiol.* **40**, 1392-1405.

Ramachandran, V.S., Clarke, P.G.H. and Whitteridge, D. (1977) Cells selective to binocular disparity in the cortex of newborn lambs. *Nature, Lond.* **268**, 333-335.

Rodieck, R.W. (1979) Visual pathways in mammals. *A. Rev. Neurobiol.* 2. (In press).

Rowe, M.H. and Stone, J. (1977) Naming of neurones: classification and naming of cat retinal ganglion cells. *Brain, Behav. Evol.* **14**, 185-216.

Schiller, P.H., Finlay, B.L. and Volman, S.F. (1976) Quantitative studies of single-cell properties in monkey striate cortex. II. Orientation specificity and ocular dominance. *J. Neurophysiol.* 39, 1320-1333.

Stone, J. (1965) The naso-temporal division of the cat's retina. *J. Comp. Neurol.* **136**, 585-600.

Wathey, J. and Pettigrew, J.D. (1979) Visual optics in the barn owl, *Tyto alba*. In preparation.

FLYING PRIMATES? MEGABATS HAVE THE ADVANCED PATHWAY FROM EYE TO MIDBRAIN

Reprinted with permission from: *Science*, 231: 1304-1306.

The pattern of connections between the retinal and midbrain has been determined with eletrophysiological and neuroanatomical methods in bats representing the two major subdivisions of the *Chiroptera*. Megachiropteran fruit bats (megabats), *Pteropus* supp., were found to have an advanced retinotectal pathway with a vertical hemidecussation of the kind previously found only in primates. In contrast, the microchiropteran bat *Macroderma gigas* has the "ancestral" or symplesiomorphous pattern of retinotectal connections so far found in all vertebrates except primates. In addition to linking primates and megachiropteran bats, these findings suggest that flight may have evolved twice among mammals.

The pattern of connections between the retina and the midbrain superior colliculus (or tectum) distinguishes primates from all other mammals so far studied (1). In strepsirhine and haplorhine primates (2), the pattern of cross-over of retinotectal fibers is like that of the retinothalamic fibers, with the result that the superior colliculus on one side of the brain subserves both eyes both but only the opposite hemifield of visual space (3). In contrast, in all other vertebrate groups so far examined, the crossover pattern of retinotectal fibers differs from that of the retinothalamic fibers, with the result that the superior colliculus subserves the whole visual field of the opposite eye (4-10). Bats have not previously been studied in this regard, although the question is of some interest because some scholars have placed bats together with primates, dermopterans, and tree shrews in the superorder Achonta (11). Moreover, threre are two distinct bat assemblages, one of which has an advanced visual organization with similarities to that of the primates. The latter are megabats, suborder *Megachiroptera*, a uniform Old World group of large, vegetarian bats reliant on their highly developed vision for foraging and obstacle avoidance (12). The microbats, suborder *Microchiroptera*, are, by contrast, diverse, world-wide, small, predominantly insectivorous bats, all of which use ultrasonic emissions for echolocation (13). I now report that fruit bars of the genus *Pteropus* have the advanced pattern of retinotectal fiber connections like that of primates. In contrast, the microbats, *Macroderma gigas*, has the plesiomorphous pattern of retinotectal projections found in most vertebrates. The findings lend support to the older classifications linking primates and bats, with the new qualifications that this applies only to the *Megachiroptera*. A corollary of this phylogenetic hypothesis is that mammalian flight has evolved independently more than once (14).

The megabats used in this study were three grey-headed flying foxes, *Pteropus poliocephalus*, two black flying foxes, *Pteropus alecto*, and one little red flying fox, *Pteropus scapulatus*, taken from the wild near Brisbane, Australia, and maintained in an outdoor aviary. The microbats were two Australian ghost bats, *Macroderma gigas*, taken from a colony of 400 breeding females at Pine Creek, Nothern Territory. *Macroderma* was chosen because, in comparison with most other microchiropterans, it has a relatively well-developed and experimentally tractable visual system, with large eyes and a temporal retinal area of increased ganglion cell density which "looks" forward like that found in *Pteropus* (Fig. 1). Two methods were used to determine the pattern of retinotectal fiber connections: electrophysiological and neuroanatomical. In

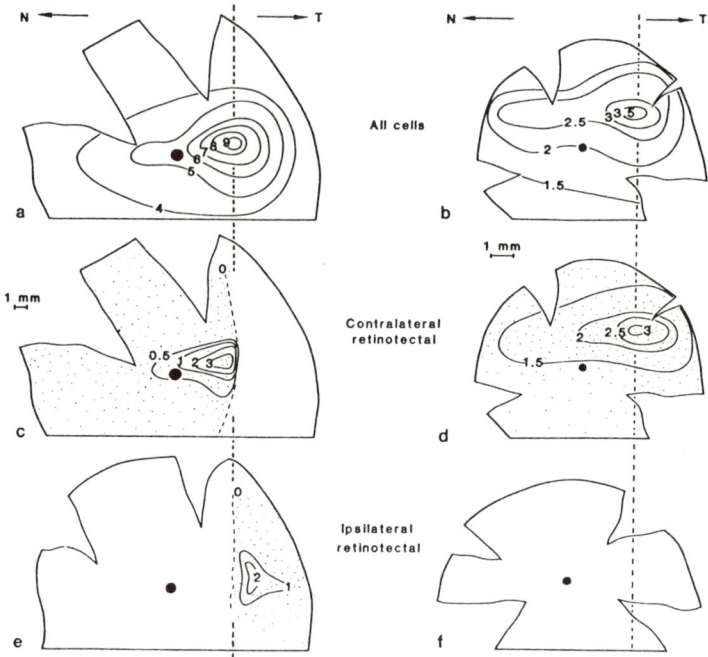

Figure 1. Differing patterns of retinotectal organization in a megachiropteran bat, *Pteropus poliocephalus* (a, c, and e) and a microchiropteran bat, *Macroderma gigas* (b, d, and f). Retinal whole mounts are shown with contours of isodensity for cells in the retinal ganglion cell layer ($\times 10^3$ mm^{-2} in a, b, d, and f; $\times 10^2$ mm^{-2} in c and e), either stained with cresyl violet (*All cells*, a and b), or reacted to show the presence of HRP in ganglion cells contralateral (c and d) and ipsilateral (e and f) to the injection site in the right superior colliculus. Abbreviations: N, nasal; T, temporal. The ipsilateral retinas (from right eye, e and f) have been mirror-reversed to facilitate comparison with the others. Stippling schematic indicates only the extent of the labeled retinal area. Both kinds of bats have a similar topography in the retinal ganglion cell layer (a and b) with a hoizontal "streak" and a clearly defined area centralis in temporal retina subserving the region of frontal visual space. Despite the similarities in overall topography in the ganglion cell layer, there are dramatic differences in the topography of the retinotectal ganglion cells. In *Pteropus* retinotectal ganglion cells are found in relatively low density compared with the total in the ganglion cell layer, but both ipsilateral and contralateral retinas have comparable densities. There is a sharp decussation line, close to the vertical meridian of both ipsi- and contralateral retinas. No labeled ganglion cells could be detected (despite a long search at high power for any weakly labeled ones) temporal to the dotted line marked (0) in the contralateral retina (c) or nasal to the dotted line marked (0) in the ipsilateral retina (e). In *Macroderma*, retinotectal ganglion cells were found only contralateral to the injection site, with no indication of a decussation at the zero vertical meridian.

the first, microelectrodes were used to record the visual responses of individual neurons in the superior colliculus of bats anestetized with intramuscular injections of ketamine and xylazine (15). The locations of visual receptive fields were plotted with respect to the zero vertical meridian for each eye, the latter having been established from the projection of the ophthalmoscopically visible optic nerve head and data from retinal whole mounts giving the relation between the area of increased ganglion cell density and the nerve head (16). A check on this estimate was provided by binocular fields recorded in overlying visual cortex (17). For the neuroanatomical studies, injections of horseradish peroxidas (HRP), in some cases conjugated to wheat germ agglutinin (WGA-HRP), were made into the superficial layers of the superior colliculus receiving retinal input to enable retrograde labeling of retinal ganglion cells (18). The previous electrophysiological determination of the location of the superior colliculus was used to guide the placement of the HRP injections in three cases. In five other cases, the overlying visual cortex was removed by suction ablation to expose the superior colliculus so that injections could be directed visually to its whole retinal projection area. After survival times of 24 to 72 hours, the animals were killed with an overdose of anesthetic, perfused with fixative, and the brain and eyes removed. Retinal whole mounts were prepared to show the pattern of labeled retinotectal ganglion cells and the brain sectioned and reacted to verify the injection site (19).

All three *Pteropus* spp. examined electrophysiologically had the primate pattern of retinotopic organization in the superior colliculus. At the caudal edge of the superior colliulus, neurons had receptive fields in the far contralateral field, whereas at the rostral edge receptive fields were located close to the zero vertical meridian. No receptive fields were found more than 5° (the accuracy inherent in the method of plotting landmarks) into the ipsilateral hemifield. A majority of neurons could be driven independently by both eyes. In *Macroderma*, the topographical arrangement in the superior colliculus was with the lateral edge representing the lower visual field, the medial edge representing the upper field and the caudal edge representing the extreme contralateral periphery as in *Pteropus*, but responses at the rostral edge could be obtained from points in visual space which were as far as 30° across the zero veritical meridian in the ipsilateral visual field. Moreover, from none of the 30 sites studied in the rostral part of the colliculus of *Macroderma* could responses be elicited from the ipsilateral eye.

These electrophysiological differences between the retintotectal organization of the two groups of bats were graphically illustrated by the results of the retrograde labeling experiments. *Pteropus* showed a vertical decussation line passing through the specialized area of both retinas, with labeled retinotectal ganglion cells on one side of the line but not the other. Densities of labeled retinotectal ganglion cells were low (2 to 4 x 10^2 mm^{-2}) compared with the densities (2 to 4 x 10^3 mm^{-2}) of the total population of neurons in the retinal ganglion cell layer (20), although both ipsilateral and contralateral densities were comparable. The same pattern was observed in all five *Pteropus* (three species) studied. *Macroderma* showed no decussation line, with labeling across the whole extent of the contralateral retina and no labeled ganglion cells at all in the ipsilateral retina. The density of labeled cells reached 2.5 x 10^3 mm^{-2} in the contralateral retina, a significant fraction of the total population of neurons in the retinal ganglion cell layer, which peaked at 3.5 x 10^3 mm^{-2} (Fig. 1).

These results show that a major, representative genus of megabats has the advanced or synapomorphous pattern of retinotectal organization found in primates. In contrast, one of the most highly visual microchiropteran bats known does not have this pattern, but shares the primitive or plesiomorphous condition with other groups of mammals (and all other vertebrates). Apart from the musculoskeletal adaptations associated with the wing itself, there are no known synapomorphous characters which unequivocally link megabats and microbats (21), so a natural question arises as to the relative weighting to be given to these two conflicting synapomorphies: the wing and the hemidecussated retinotectal pathway. Is it more likely that flight evolved in parallel in two separate lines of mammals, one of them ancestral primates, or that megabats have evolved an advanced mode of retinotectal organization independently of primates (22)?

Separate origins for the wings of megabats and microbats are supported by the presence of a number of small, but consistent skeletal differences between them (23), the presence of other synapomorphics linking megabats to primates but not to microbats (24), the appearance of sustained flight in at least three separate nonmammalian lines (25), the numerous appearances of gliding flight with three separate "inventions" in the marsupials alone (26), and a fossil record indicating an origin for microbats more than 50 million years ago (27) compared with the recent fossil megachiropterans which have been classified as primates (28). Separate origins for the advanced retinotectal organization in fruit bats and primates are not supported by the close identity between the complex details of the pathways involved in each group (29), nor by their absence in such highly visual and arboreal mammals as squirrels, cats, tree shrews, and phalangers (5-10). Serological evidence linking dermopterans and primates (30), the homologous structure and innervation of the patagium in dermopterans and megachiropterans (11), the present evidence linking primates and megachiropterans, plus the previous morphological evidence linking dermopterans, primates, and megachiropterans (23) all taken together, suggest that a fruitful line of future investigation will be the evolutionary relations of these three groups of mammals. In the meantime it seems appropriate to propose that an early branch of the primates tree may have developed the power of flight long before the hominid branch even dreamed of it.

REFERENCES AND NOTES

1. J.M. Allman, *Prog. Physiol. Psychol.* 7, 1 (1977).

2. By this test and others, members of the Menotyphla, which includes the three shrew, Tupaia, are now excluded from the primates, although they are the primates' closest "sister group" (1, 9).
3. Prosimian *Galago* [R.H. Lane, J.M. Allman, J.H. Kaas, F.M. Miezin, *Brain Res*. **60**, 335 (1973)]; new world primates, *Saimiri* [S. Kadoya, L.R. Wolin, L.C. Massopust, *J. Comp. Neurol*. 142, 495 (1972)]; and *Aotis* [R.H. Lane *et al., ibid*.]; old world monkey, *Macaca* [M. Cynader and N. Berman, *J. Neurophysiol*. **35**, 187 (1972)].
4. This is the "primitive" or plesiomorphous pattern [E.O. Wiley, *Phlogenetics* (Wiley, New York, 1981)].
5. Anurans [R.M. Gaze, Q. *J. Exp. Physiol*. **43**, 209 (1958)]; teleosts [H. Schwassman and L. Kruger, *J. Comp. Neurol*. **124**, 113 (1965)]; birds [H. Bravo and J.D. Pettigrew, *ibid*. **199**, 419 (1981)]; lizard [B.S. Stein and N.S. Gaither, *ibid*. 202, 69 (1981)].
6. Rodents: rat [K.S. Lashley, *J. Comp. Neurol*. **59**, 341 (1934)], squirrel [W.C. Hall, J.H. Kaas, H. Killackey, I.T. Diamond, *J. Neurophysiol*. **34**, 437 (1971)], and ground squirrel [C.N. Woolsey, T.G. Carlton, J.H. Kaas, F.J. Earls, *Vision Res*. **11**, 115 (1971)].
7. Rabbit [A. Hughes, *Docum. Ophthalmol. (DenHaag)* **30**, 33 (1971)].
8. Cat [M. Straschill and K.P. Hoffman, *Brain Res*. **13**, 274 (1972)].
9. Opossum [C. Rocha-Miranda, R. Mendez-Otero, A.S. Ramoa, E. Volchan, L.G. Gawryszewski, in *Development of Visual Pathways in Mammals*, J. Stone, B. Dreher, D. Rapaport, Eds. (Liss, New York, 1984), pp. 179-198].
10. Tree shrew [J.H. Kaas, J.K. Harting, R.W. Guillery, *Brain Res*. **65**, 343 (1974)].
11. W.K. Gregory, *Bull. Am. Mus. Nat. Hist*. **27**, 332 (1910).
12. W.A. Wimsatt, *Biology of Bats* (Academic Press, New York, 1970).
13. M.B. Fenton, *Rev. Biol*. **59**, 33 (1984).
14. J.D. Smith, in *Biology of Bats of the New World Family Phllostomatidae*, R.J. Baker, J.K. Jones, Jr., D.C. Carter, Eds. (Texas Tech Press, Lubbock, 1976), part 1, pp. 49-69; in *Major Patterns in Vertebrate Evolution*, M.K. Hecht, P.C. Goody, B.M. Hecht, Eds. (Plenum, New York, 1977), pp. 427-438; in *Proc. Fifth International Bat Research Conference*, D.E. Wilson and A.L. Gardner, Eds. (Texas Tech Press, Lubbock, 1980), pp. 233-244.
15. In two animals supplementation with Nembutal (pentobarbitone sodium) was used.
16. In *Pteropus* and *Macroderma* the vertical meridian was horizontally displaced from the blind spot approximately 18° and 2°, respectively. These values can be obtained from Fig. 1, given that in *Pteropus* 1 mm-**8.6°** and in *Macroderma* 1 mm =**16°** on the retina.
17. T. Nikara, P.O. Bishop, J.D. Pettigrew, *Exp. Brain Res*. **6**, 353 (1968).
18. In *Pteropus*, sex to ten separate injections were made, totaling 1 to 1.5 µl of 20 percent of HRP or 1 percent WGA-HRP solution, to involve the complete retinal projection area of the superior colliculus. In the case of WGA-HRP, which is colorless at the concentration used, fast-green dye was added to help gauge the degree of diffusion. In *Macroderma*, whose superior colliculus is only 2 mm across, two injections of 0.3 µl each were sufficient to involve the whole structure.
19. M.L. Cooper and J.D. Pettigrew, *J. Comp. Neurol*. **184**, 1 (1979).
20. Most of the remaining unlabled retinal ganglion cells are retinothalamic ganglion cells projecting to the lateral geniculate nucleus (J.D. Pettigrew, M.L. Graydon, P. Giorgi, in preparation).
21. Most of the characters usually advanced to link megabats and microbats are associated with the flight adaptation [for example, characters 51 to 60 of M.J. Novacek, in *Macromolecular Sequences in systematic and Evolutionary Biology*, M. Goodman, Ed. (Plenum, New York, 1982)], pp. 3-41. Other characters are contestable, having evolved in other unrelated mammalian (for example, fetal membrane characters 61 to 63, *ibid*.) or possibly representing plesiomorphous rather than synapomorphous characters (for example, 48 to 50, *ibid*.).
22. A third possibility is that the advanced mode of retinotectal organization arose first in meg achiropteran bats, some of which later lost their powers of flight and gave rise to the primates. The extensive and fairly continuous fossil record of primates makes this scenario highly unlikely [M. Archer, in *Vertebrate Zoogeography and Evolution in Australasia*, M. Archer and G. Clayton, Eds. (Hesperian Press, Perth, 1983), pp. 949-993; F. Szalay and E. Delson, *Evolution and History of the Primates* (Academic Press, New York, 1979)].
23. J.D. Smith and A. Starrett, in *Biology of Bats of the New World Family* Phyllostomatidae, part 3, R.J. Baker, J.K. Jones, Jr., D.C. Carter, Eds. (Texas Tech Press, Lubbock, 1979), pp. 229-316; J.D. Pettigrew, K.S. Robson, K.I. McAnally, in preparation.
24. J.D. Smith and G. Madkour, in *Proceedings of the Fifth International Bat Research Conference*, D.E. Wilson and A.L. Gardner, Eds. (Texas Tech Press, Lubbock, 1980) pp. 347-365; J.E. Hill and J.D. Smith, *Bats: An Natural History* (British Museum, London, 1984).
25. K. Padian, *Paleobiology* **9**, 218 (1982); J.M.V.

26. M. Archer, in Archer and Clayton [in (22), pp. 633-807].
27. M. Novacek, *Nature (London)* **315,** 140 (1985).
28. S. Hand, in Archer and Clayton [in (22), pp. 851-904]; A. Walker, Nature (London) 223, 647 (1969).
29. There may be corollary developments in the telencephalic visual processing of movements by primates accompanying the advances in retinotectal organization, such as the MT (middle-temporal) visual cortical area found in primates, but in no other mammals so far studied (1). By the diagnostic criteria for MT that it be located in the temporal lobe and recieve a major direct input from the primary visual cortex, this cortical area seems to be present in *Pteropus* [M.B. Calford et al., Nature (London) 313, 477 (1985)] but not in *Macroderma* (unpublished observations). The similarity of so many different and complex details of visual organization in primates and pteropids strengths the argument against parallel appearance of the two systems.
30. J.E. Cronin and V.M. Sarich, in *Recent Advances in Primatology*, vol. 3, *Evolution*, D.J. Chivers and K.A. Joysey, Eds. (Academic Press, New York, 1978), pp. 287-288.
31. Supported by grants from the Australian Research Grants Scheme and the National Health and Medical Council of Australia. L. Wise and M. Calford helped with some of the electro-physiological recording experiments and provided critical comments on the manuscript. R. Collins provided expert technical assistance. Staff of the Conservation Commission of Northern Territory gave invaluable assistance in the collection of *Macroderma*, which were obtained under permit D85-5633.

8 July 1985; accepted 15 December 1985

EMERGENCE OF RADIAL AND MODULAR UNITS IN NEOCORTEX

Mathew E. Diamond and Ford F. Ebner

Neurobiology Section and Center for Neural Science
Brown University
Providence, R.I. 02912

INTRODUCTION

Our strategy here is to compare the cortical organization of phylogenetically distant species. We can view the shared features as themes conserved during the course of evolution and the features of cortical organization that differ among species as variations on a common theme. We will compare the most primitive dorsal cortex found among living vertebrates, the visual cortex of turtles, with the somatic sensory field of mammalian neocortex. The comparison is based on classical cell morphology combined with modern intracellular and extracellular electrophysiology, tract tracing, and receptive field mapping.

The most fundamental difference between the reptilian and mammalian cortex is the organization of thalamocortical projections. In the turtle, individual thalamic fibers form synapses across the entire horizontal extent of the dorsal cortex. In contrast, thalamocortical axons in mammals originate in distinct cell clusters and terminate within strips of discrete columnar modules. Each cortical module, roughly one half millimeter in diameter, is in turn assembled from a set of narrow radial units. Receptive field studies of the somatic sensory cortex of the cat provide insights into the functional significance of radial and modular units.

The evolution of the cluster-to-module projection is accompanied by a new cell type in mammalian neocortex. Description of the cellular morphology of reptilian and mammalian cortex has a distinguished history; our approach here is to forgo traditional classifications in favor of a scheme consisting of three broad classes of neuron. Turtle dorsal cortex is made up of just two of these classes, the excitatory output cell and the inhibitory local cell. Mammalian neocortex includes both of the above and a third class as well, the excitatory local cell. The excitatory local cell of neocortex replaces the tangential thalamic fibers of turtle dorsal cortex as the chief substrate for the horizontal propagation of activity; while this cell is specialized for spreading activity vertically among the cells in the various layers of a radial unit, its unique integrative function is to interconnect cells located in separate radial units and modules.

We suggest that mammalian somatic sensory cortex is a mosaic of modular and radial units, both types of unit demonstrable by anatomical and receptive field methods. When viewed in this way, the rodent cortical whisker representation — rather than a unique pattern of organization — is best considered a modest specialization of the general mammalian plan of neocortex.

CELL TYPES COMMON TO PRIMITIVE CORTEX AND NEOCORTEX

Excitatory output cell

Turtle dorsal cortex is a three-layered cortex; a single layer of tightly packed cell bodies with a dendritic zone above and below (Johnston, 1915). The vast majority of the cells in the main cell layer are *excitatory output cells*: pyramidal cells whose axons enter the white matter and form asymmetric synapses with distant (more than a millimeter away) targets. The apical den-

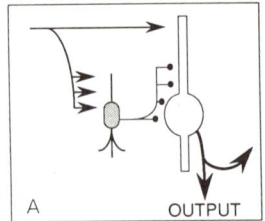 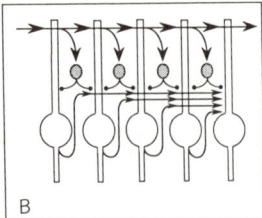

Figure 1. Synaptic and topographic organization of turtle dorsal cortex. (A) Thalamic afferents (originating in top left corner of frame) terminate densely upon the inhibitory local cell (shaded) and less densely upon the apical dendrite of the excitatory output cell. The inhibitory cell forms synapses on the dendrites and soma of the output cell. (B) Each thalamic afferent courses through the molecular layer in the anterolateral to posteromedial direction and forms synaptic contacts across a wide extent of dorsal cortex. The excitatory output cells give rise to intracortical collaterals.

drites of the excitatory output cells form a molecular layer as they extend toward the pial surface, and their basal dendrites form a subcellular neuropil between the lamina of cell bodies and the ependymal cells lining the ventricle.

The dominant input to the excitatory output cells is the thalamic projection. Thalamic fibers enter the molecular layer just beneath the pial surface in the anterolateral region of cortex and form synaptic contacts on successively more posterior and medial apical dendrites as they travel tangentially (Figure 1). Thus, one key function of the thalamocortical projection in turtles is the *horizontal* dissemination of sensory information; circumscribed clusters of cortical cells never constitute the target of lateral geniculate nucleus fibers in the turtle.

The output cell axons course in the medial direction, giving rise to collateral branches to other cortical excitatory output cells. The axons depart the cortex through the subcellular zone. They descend from cortex through the medial wall cortex, and course through the hypothalamus as far posteriorly as the midbrain tegmentum (Hall et al., 1977).

In mammals, as in turtles, there exists a cell with at least part of its axon projecting through the white matter, and whose axon terminals contain clear round vesicles and form asymmetrical (presumably excitatory) synaptic contacts (Winfield et al., 1981). This *excitatory output cell* is well suited to transfer information to distant sites; that is, to cortical targets more than a millimeter away or to subcortical structures. Previous studies using a variety of techniques have consistently shown that an excitatory output cell is nearly always a pyramidal neuron. Pyramidal cells are the most numerous elements in mammalian cortex (Winfield et al., 1980) and are found in all cellular layers.

Before the excitatory output cell axon reaches the white matter it gives rise to numerous collateral branches that distribute within the grey matter. The extent of spread varies widely from neuron to neuron (Gilbert and Wiesel, 1981; Lorente de No, 1949). After the axon enters the white matter it projects to an ipsilateral cortical, a contralateral cortical, or a subcortical target.

Inhibitory local cell

Scattered throughout the three layers of turtle cortex is a population of aspiny stellate cells which contains glutamic acid decarboxylase (GAD; the synthetic enzyme for GABA) and forms symmetrical (presumably inhibitory) synaptic contacts (unpublished observations). These neurons receive a high density of thalamic fiber terminals and, in turn, project to nearby cells in the cortex. The GABAergic neurons in turtle cortex are *inhibitory local cells* in as much as their axons have never been found to project out of cortex.

The neocortex of mammals also contains a population of GAD-positive cells which forms symmetric synapses and whose axons do not travel in the white matter (Ribak, 1978). The *inhibitory local cells* make up about 20-25% of all neurons in neocortex (Hendry et al., 1987) and exist as several subclasses, each with a distinct morphology. For the present purpose it suffices to say that most inhibitory local cells have the morphology of a nonpyramidal neuron with aspiny or sparsely spiny dendrites (Somogyi et al., 1981).

SYNAPTIC ORGANIZATION OF THALAMIC INPUTS

Certain fundamental features of the connectivity between thalamic inputs and cortical excitatory and inhibitory neurons have been faithfully conserved during the evolution of neocortex. In both turtle dorsal cortex (Ebner and Colonnier, 1978) and mammalian neocortex (Colonnier, 1968) thalamic fibers terminate as round vesicle containing profiles that form asymmetrical contacts and are presumed excitatory synapses. In the turtle, thalamic fibers synapse upon both inhibitory and excitatory neurons; terminations are particularly dense upon the dendrites of aspiny stellate cells (Figure 1A). The total number of thalamic synapses upon excitatory output cells is six times the number of thalamic synapses upon inhibitory local cells. However, because excitatory output cells greatly outnumber inhibitory local cells, there is, on average, a six fold greater number of thalamic synapses per inhibitory local cell (Smith et al., 1980). Inhibitory local cells in turn synapse densely upon nearby excitatory output cells.

In neocortex, as in turtle dorsal cortex, fibers from thalamic primary sensory relay nuclei terminate densely upon inhibitory local cells (White, 1979). For example, following destruction of the ventrobasal (VB) thalamic nucleus in mouse, degenerating thalamic axon terminals made the most synapses per unit length of dendrite with aspiny stellate cells, followed in decreasing order by layer IV spiny stellate cells, layer III pyramidal cells, and layer V pyramidal cells (White, 1978). On the soma of one aspiny bipolar cell, 67% of the synapses were made with thalamic fibers (Keller and White, 1987). Inhibitory local cells synapse upon excitatory output neurons, and typically the position of these synapses (e.g., the axon hillock or the initial segment) allows the inhibitory cell to exert a particularly strong influence on the excitatory output cell (Jones and Powell, 1969).

EXCITATORY LOCAL CELLS OF NEOCORTEX

Our argument began with the principle that in turtle cortex all excitatory cells are output neurons (the only local cells in turtle cortex are inhibitory) while in mammalian cortex an additional excitatory cell type is present whose axon remains within the grey matter. The axon of the *excitatory local cell* terminates in its entirety within a few hundred μm (horizontally and vertically) of the cell body. Nearly every excitatory local cell is a spiny stellate cell (Lund, 1984). Typically, the axon descends toward the white matter for a short distance before turning upward and arborizing. The functional significance of a cell type specialized for excitatory communication with nearby neurons will become evident as we discuss the modular and radial units of neocortex.

MODULAR UNITS OF NEOCORTEX

Thalamocortical projection as the anatomical basis of the modular unit

It is worth reiterating the contrast between the topographic organization of the mammalian thalamocortical projections and that of the turtle. In the turtle, fibers from the lateral geniculate nucleus course horizontally across several millimeters of cortex. Individual axons from the geniculate contact the dendrites of a series of cortical neurons, each axon influencing a wide domain of cortex (Figure 1B). As a consequence, receptive fields of turtle dorsal cortex neurons typically include nearly the entire representation of the retina (Mazurskaya, 1972; Orrego and Lisenby, 1962). A thalamocortical axon which terminates in the molecular layer is conserved in mammalian neocortex; it arises mainly in "nonprimary" thalamic relay nuclei, such as the posterior nucleus or the pulvinar. However, the projection from primary thalamic relay nuclei takes a form not found in reptiles. A case in point is the somatic sensory system. Individual fibers from VB of cat (Landry and Deschenes, 1981) and monkey (Garraghty et al., 1989) ascend orthogonal to the cortical surface and terminate mainly in layers IIIB and IV, often with collateral branches to other layers. In the horizontal dimension, *the terminal field of each fiber is restricted to one or a few circumscribed terminal arbors* ranging from 200 to 800 μm in diameter (Figure 2A, B).

What is the relationship between the terminal fields of neighboring thalamic cells? Is there evidence that multiple thalamocortical arbors are grouped together into a larger unit? The findings of Jones and colleagues (Friedman and Jones, 1980; Jones et al., 1982) in monkeys, based on anterograde and retrograde transport methods, do indeed provide support for an organizing principle larger than the single thalamocortical cell. Narrow cortical domains, or *modules*, are the the targets of rodlike clusters of thalamic cells. The same conclusion was reached based

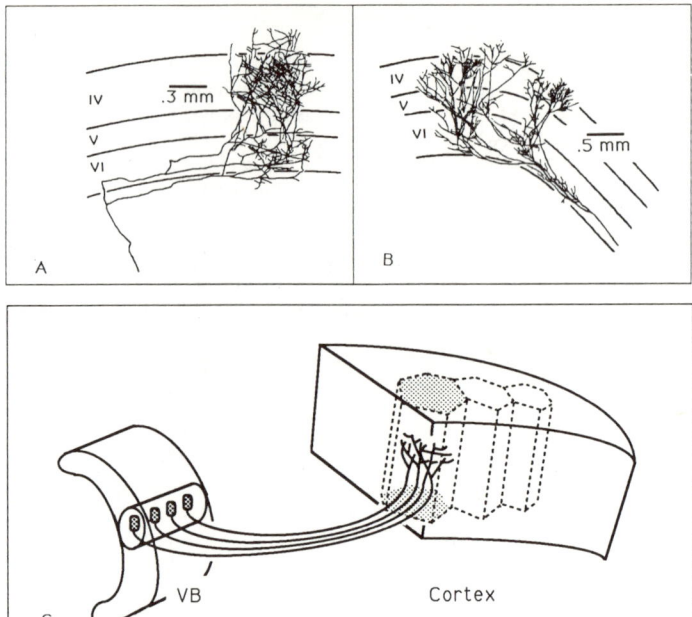

Figure 2. Mammalian plan of the projection from VB thalamus to somatic sensory cortex. In (A) the fiber arborizes in a single discrete module. In (B) the fiber arborizes in two separate modules. The thalamocortical topography is illustrated in (C). A rodlike cluster of cells in VB projects to a cortical module (collaterals to other modules not illustrated). The cell cluster in VB is one of several which form a lamellae and project, as a group, to a strip of modules. Parts (A) and (B) adapted from Figures 5 and 7 of Landry and Deschenes (1981).

on cortical projections from cat VB (see Figure 1 of Kosar and Hand, 1981). This is not to say that the projection of each cell cluster in the thalamus is restricted to a single cortical module; rather, the set of clusters contained within a lamella of VB projects to a strip of modules, each cluster projecting most densely to one module in particular (Figure 2C). The diameter of a module, as defined by bulk labeling of thalamic fibers, is between 200 and 800 μm — the same size range as the individual VB terminal arbors. It should be emphasized that the patterns of retrograde thalamic label and anterograde cortical label do not support the idea that the cortical terminal fields of neighboring thalamic cells are partially shifted and overlapping relative to one another. If that were the case, as the size of the cortical horseradish peroxidase (HRP) injection increased, the label would be transported to an enlarged but continuous thalamic region. Instead, multiple separate, distinct thalamic cell clusters are labeled. Similarly, as the size of the thalamic injection is increased, the resulting transport to cortex labels multiple separate patches (modules) rather than a single field of ever larger diameter.

Physiological detection of the modular unit

The question we take up here is whether the module identified by anatomical methods also can be defined by physiological methods. We continue to concentrate on the somatic sensory field of neocortex, and we review in some depth recent results from studies on the topography of the cortical body representation (Diamond, 1989; Favorov and Diamond, 1990; Favorov et al., 1987). The starting point in our search for a module is the assumption that neurons located within the same module will have relatively similar receptive fields, and neurons located on opposite sides of a module boundary will have less similar receptive fields. Using two different receptive field methods we demonstrate that the above assumption is valid, and that the module is therefore a physiological as well as an anatomical unit.

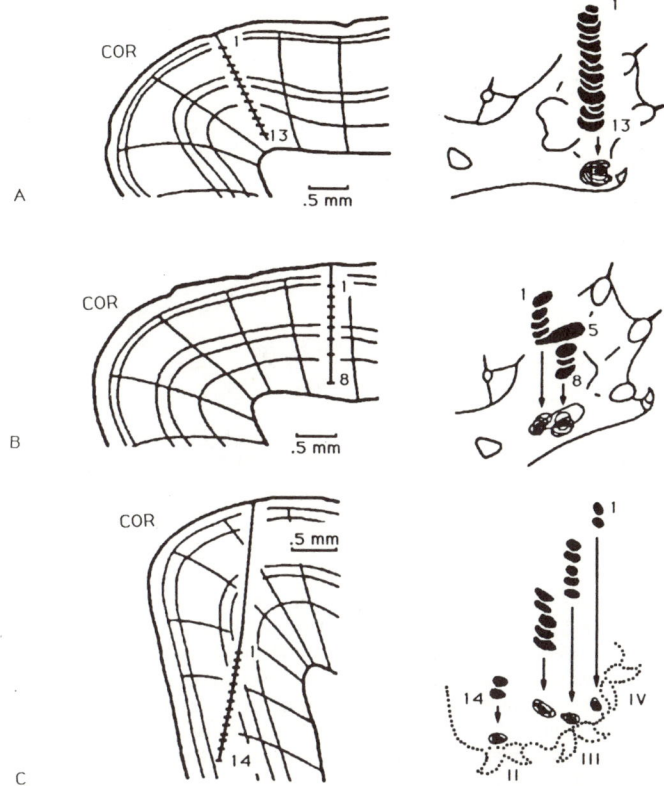

Figure 3. Sequences of minRFs reveal boundaries of modular units. MinRFs were mapped every 100-200 μm along the microelectrode track. Left column: drawing of cortical sections at low magnification to show the location of recording sites along each penetration (dashed lines). Thin lines parallel to the cortical surface indicate cortical layers and lines orthogonal to the surface indicate orientation of vertical cell cords. Right column: minRF sequences; the numbers from top to bottom identify the location of the recording site in cortex. The minRFs are also shown projected onto the skin of the distal forelimb. (A) Penetration remains within a single module. (B) Penetration crosses a modular boundary. At the depth of the fifth recording site, the electrode tip lies on the boundary between two modules. (C) A nearly tangential penetration crosses three boundaries. The minRFs shift from digit IV to III to II. COR- coronal sulcus. Adapted from Favorov and Diamond (1990).

The first method, the minimal receptive field (*minRF*) method, uses ketamine general anesthesia coupled with gentle (barely suprathreshold) tactile stimulation to render the receptive field of cortical cells as small as possible. The method is designed to identify the main input to the small cluster of cortical neurons within roughly 50 μm of the recording site. In penetrations of the primary somatic sensory cortical field (cytoarchitectonic areas 3b and 1) minRFs were mapped at intervals of about 150 μm along the electrode track. When the penetrations were very close to radial, all minRFs occupied nearly the same position on the skin (Figure 3A). In penetrations with an oblique trajectory, successive minRFs fell into groups that resemble separate stacks of pancakes: within a stack minRFs are nearly completely overlapping while adjacent stacks are separated by an abrupt change in the location of the minRF to a different skin site (Figure 3B). Penetrations tangential to the cortical surface resulted in multiple distinct stacks of minRFs (Figure 3C).

In no penetration (out of 21), either radial off-radial or tangential, were minRFs distributed across the skin in a continuous, gradually shifting manner. Instead, each time a shift of the minRF occurred— from one stack of pancakes to the next — the move happened abruptly, as the electrode tip travelled only about 40-50 μm in the plane of the cortical surface. Our interpreta-

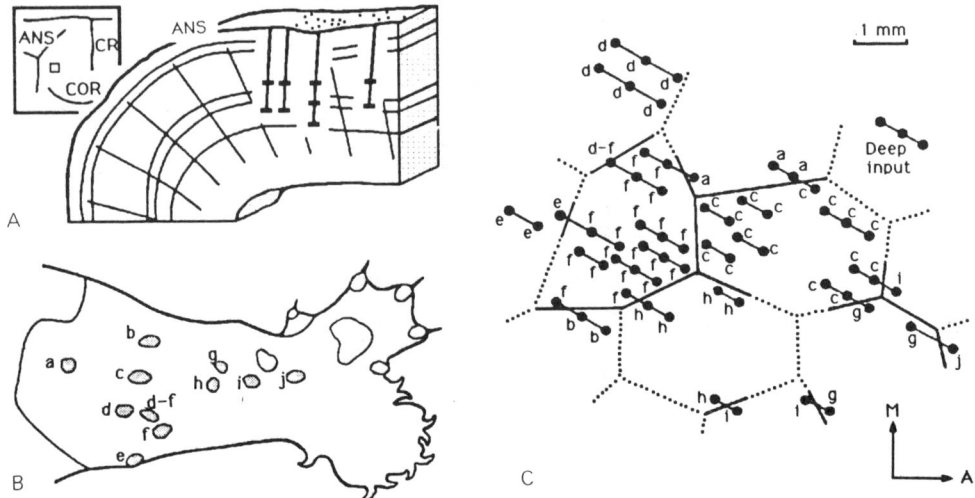

Figure 4. Closely spaced grid of penetrations reveals the size and shape of modules. (A) Diagram to show the location of penetrations between the ansate (ANS), cruciate (CR) and coronal (COR) sulci (small square in inset). (B) Each of 61 minRFs was located within one of 10 discrete spots, labeled a-j (one minRF was located midway between d and f). (C) Surface view of cortex to show recording sites (filled circles). Connected recordings sites are from the same penetration. Each recording site is labeled according to the location of its minRF, a-j. Boundaries crossed by a penetration are shown as solid lines and estimated boundaries are shown as dotted lines. Adapted from Favorov and Diamond (1990).

tion is that the point in cortex where the minRF shifts from one location to another marks a topographic boundary — presumably the edge of a modular unit. When the electrode tip lies exactly on the boundary between two modules (as in the fifth recording site of Figure 3B) neural activity originating in both modules is detected by the recording electrode. This results in a "transitional" minRF encompassing the minRFs of two modules.

Size and shape of the module

To determine the geometry of the modular unit in cat somatic sensory cortex, minRFs were mapped in a grid of closely spaced penetrations. The main result is that all the minRFs were grouped within 10 discrete spots on the skin, labeled **a-j** (Figure 4B). When each cortical recording site is labeled according to its minRF, it becomes clear that the recording sites yielding the same minRF are grouped near one another. Based on those penetrations which crossed topographic boundaries, it is possible to draw the outlines around two distinct cortical regions, labelled **c** and **f**. The geometry of regions **c** and **f** gives us the size and shape of the module: a discrete column with a roughly hexagonal outline. Within a single module, the minRF is the same throughout — this shared minRF is the *receptive field center* of the module. Shifts in the minRF occur only at the borders of a module. Additional experiments indicated that the size of a module depends on the location of its receptive field center: when the receptive field center is on the distal forelimb the module's diameter is about 450 μm, and when the receptive field center is on the forearm (e.g. near the elbow) the module's diameter is about 250 μm.

To this point we have reviewed both anatomical and physiological evidence for modular units in somatic sensory cortex. The units are similar in size whether defined by i) the topography of the skin representation (Diamond et al., 1987; Favorov and Diamond, 1990; Favorov et al., 1987; Favorov and Whitsel, 1988), ii) the projection of clusters of VB cells (Friedman and Jones, 1980; Jones et al., 1982; Kosar and Hand, 1981), or iii) the terminal fields of individual VB fibers (Landry and Deschenes, 1981; Garraghty et al., 1989).

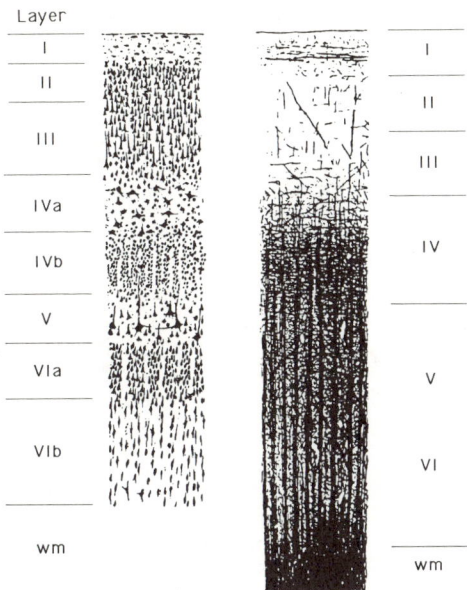

Figure 5. Anatomical correlates of the radial unit. Left: Cell bodies aligned in radial units in the visual cortex of man. Nissl stain. Adapted from Figure 1 of Cajal (1899). Right: Myelinated fibers aligned in vertical arrays in the sphenoidal gyrus of man showing the same radial orientation. Weigert-Pal method. Adapted from Figure 2 of Cajal (1900).

RADIAL UNITS OF NEOCORTEX

Anatomical basis of the radial unit

Apart from three obvious horizontal layers, the distribution of neurons in turtle dorsal cortex seems unpatterned; there is no reason to suspect any grouping of cells into multicellular units smaller than a cortical field or subdivision. In the mammalian neocortex, the opposite is true. An unmistakable grouping of cell bodies into vertical cords one to four cells wide, extending from layer II to VI, is frequently evident, as illustrated in the drawings of the human cerebral cortex by Cajal (Figure 5, Left). It was also apparent to Cajal that axons and dendrites in cortex are grouped into distinct vertical cords (Figure 5, Right). Typically, the centers of adjacent cellular cords are separated by about 30-50 µm. Bundles of apical dendrites of pyramidal cells are one component of the vertical arrays in cortex (Peters and Walsh, 1972). The apical dendrites of layer V cells form the core, and as they ascend toward the cortical surface they are joined by the apical dendrites of more superficial cells. Other nonpyramidal cell types, for example double bouquet or bipolar cells, also have narrow dendritic trees oriented toward the cortical surface and probably contribute to the organization of such vertical cords (Jones, 1975).

Functional correlate of the radial unit

It is reasonable to anticipate that the anatomically defined narrow vertical cord of cell bodies and processes — *the radial unit* — has a physiological counterpart. In this section we raise the question, How is the radial unit manifest in neuronal response properties? We will provide evidence that the neurons within a radial unit in somatic sensory cortex discharge in a cooperative manner during tactile stimulation.

Because the minRF is based on the responses of a cluster of neurons located within an approximately 50 µm diameter of the electrode tip (Favorov and Diamond, 1990), the method does not have adequate spatial resolution to identify an entity as narrow as the radial unit. Detecting a

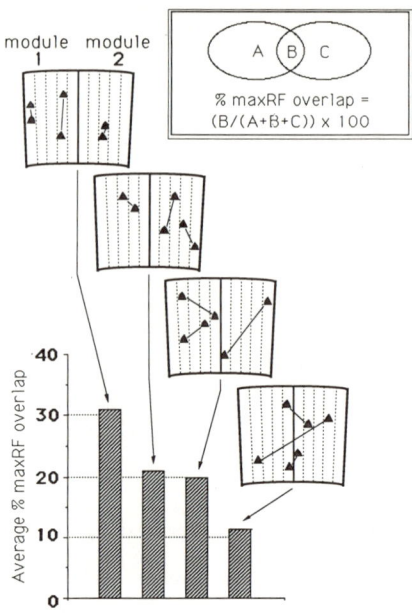

Figure 6. Radial and modular units are reflected in receptive field measurements. The inset in the upper right gives the formula for computing precent maxRF overlap. The columns represent two separate cortical modules, labeled module 1 and module 2; dashed lines delineate radial units. Four possible relative locations of pairs of neurons are illustrated from top left to bottom right: two cells in the same module and same radial unit; two cells in the same module and adjacent radial units; two cells in the same module and nonadjacent radial units; two cells in different modules (the two radial units may be adjacent or remote).

The histogram shows the average maxRF overlap for pairs of neurons with the indicated relative locations. Note that neuron pairs from the same radial unit have a realtively high degree to maxRF similarity (31% overlap). Also note that for neuron pairs from the same module but different radial units, the degree of maxRF similarity is independent of the distance between the two neurons.

functional correlate of the radial unit requires a receptive field method that selects and analyzes neural activity with higher precision. The maximal receptive field method (*maxRF*) is one such technique: unlike the minRF method, the maxRF method maps the entire skin field projecting, even if weakly, to a single cortical neuron; in practice it is necessary to record single cell responses in the absence of anesthesia to reveal the weak inputs. If the set of neurons making up a radial unit constitutes a functional entity — such that they tend to fire synchronously during tactile experience — then their coactivity will be reflected by a large shared area of peripheral receptor input. We can predict therefore that, on average, two neurons located in the same radial unit will have more similar maxRFs than two neurons located in different radial units. MaxRFs are usually much larger and more variable than minRFs (Favorov and Diamond, 1990). The amount of input common to pairs of neurons is measured here by the *percent of maxRF overlap*, as illustrated in Figure 6, inset.

Analysis of the maxRFs of 1111 pairs of neurons (where both members of the pair were located within the same module) indeed identifies a functional correlate of the radial unit. Pairs of neurons separated by less than about 10 μm in the horizontal dimension exhibit an average of 31% maxRF overlap, whereas pairs of neurons separated by 40-100 μm in the horizontal dimension exhibit an average of only about 21% maxRF overlap. Our interpretation is shown in Figure 6: two neurons located in the same radial unit share, on average, a greater area of maxRF overlap than two neurons located in separate radial units, *even when the two radial units are contiguous.* This finding implies that neurons located in the sameradial unit are more likely to be coactive during tactile experience than neurons in different radial units. The estimate of maxRF overlap for members of the same radial unit is likely an underestimate, since some pairs separate by just 10 μm may reside in separate radial units.

Based on Figures 3 and 4 (which showed that the same minRF is mapped throughout a module) we should anticipate some kinship between the maxRFs of two neurons located within the same module, even if the two neurons are located on opposite sides of the module. The specific question is, What is the nature of the movement across the skin of successive maxRFs as the recording electrode advances at an oblique angle across a modular unit? The answer is that, for any two neurons located within the same module, the average maxRF overlap is about 20%, *independent of the horizontal distance separating the two neurons*: the pair of neurons may be located in two neighboring radial units (i.e., separated by just 40-100 μm in the plane of the cortical surface), or in two radial units at opposite edges of the module (i.e., separated by any distance between 100 and 500 μm). This relationship among the radial units of a module means that a recording microelectrode traveling horizontally across a sampling single neurons from successive radial units, will fail to detect any systematic and continuous shift in maxRF position. The jitter in maxRF position can be pictured as a "random walk".

At the boundary of the module the electrode tip passes from a radial unit belonging to one module to a radial unit belonging to a second module. With this last, seemingly equivalent, advance of the electrode the maxRF position shifts markedly: pairs of neurons located on opposite sides of a module boundary exhibit an average of only 12% maxRF overlap, even if the two neurons are located in adjacent radial units (Figure 6). Comparable results have been obtained in the somatic sensory cortex of monkeys (Favorov and Whitsel, 1988).

DISTRIBUTION OF ACTIVITY WITHIN AND AMONG RADIAL UNITS: THE EXCITATORY LOCAL CELL REVISITED

What is the role of the excitatory local cell, the cell type posited as novel to mammalian neocortex? On the basis of their morphology and connections, it is reasonable to suspect that many types of excitatory local cells are specialized for the vertical distribution of information within a radial unit. For example, spiny stellate cells in layer IV receive a dense thalamic input and some of these cells possess narrow cartridge-like axonal arbors which terminate in asymmetric synapses on the dendritic spines of pyramidal and perhaps stellate neurons above and below layer IV (Jones, 1975). Thus, it is easy to imagine that a thalamic volley arriving in layer IV of a radial unit would rapidly and powerfully influence the neurons located above and below layer IV. This idea is supported by comparing in reptiles (Kriegstein and Connors, 1986) and mammals (Connors et al., 1982) the sequence of membrane potentials recorded from cortical neurons after stimulation of thalamic afferents. Output cells in an in vitro slice of turtle cortex respond to low intensity orthodromic stimulation with a pure inhibitory postsynaptic potential (IPSP). As the intensity of the stimulus is increased toward threshold, an excitatory postsynaptic potential (EPSP) is evoked, but it is still preceded by an IPSP (Figure 7A). In an *in vitro* slice of mammalian neocortex *the earliest response to a weak afferent volley is an EPSP rather than an IPSP* (Figure 7B). Since the amplitude of the EPSP is increased in the presence of bicuculline (Chagnac-Amitai and Connors, 1989), our interpretation is that in neocortex the same short latency inhibitory influences seen in turtle cortex are at work, but they are overridden by a concomitant burst of excitatory input, arising in part from the excitatory local cells. The role of the excitatory local cell in the vertical propagation of activity can also be seen in the analysis of maxRF overlap; recall the comparatively high measure of maxRF overlap (31%) observed within a narrow radial unit (Figure 6).

The morphology of the excitatory local cell also is appropriate for spreading activity across horizontal distances of roughly less than a millimeter. Strong interconnection of the excitatory local cells residing within the same module could be a contributing factor in the receptive field similarity observed within a module (20% average maxRF overlap). In other words, one function of excitatory local cells could be to link the cells within a module so that they become activated by the same tactile event. On the other hand, sets of neurons located on opposite sides of a module border (12% average maxRF overlap) would be less likely to be activated by the same stimulus: these neurons receive significantly different thalamic inputs and their excitatory local cells may be weakly interconnected. The effectiveness of the synapses between neurons of the same or different modules may be regulated through postulated rules of synaptic modification (Bear et al., 1987).

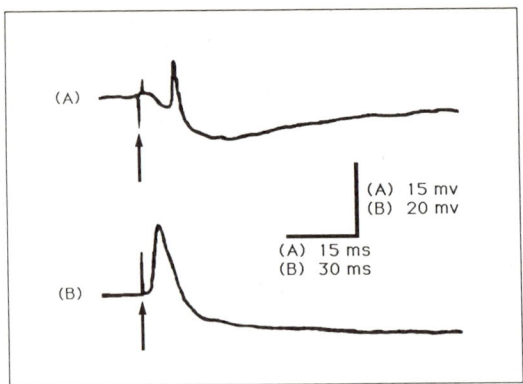

Figure 7. Influence of the excitatory local cell on excitatory output cell responses is seen in the PSP sequence evoked by an afferent volley. Intracellular recordings of the membrane potential of excitatory output cells in response to electrical stimulation of input fibers in vitro. Arrows indicate stimulus artifact. (A) In the turtle, even near the spike threshold, the initial response is an IPSP, which is followed by an EPSP. Adapted from Figure 2 of Kriegstein and Connors (1986). (B) In guinea pig, the initial event is an EPSP generated, we would argue, by the influence of excitatory local cells. Adapted from Figure 12 of Connors et al. (1982).

SPECIALIZATION OF THE MAMMALIAN PLAN OF CORTEX

The modular and radial organization of cat somatic sensory cortex is the general mammalian plan which allows us to look for variations on the theme. In some rodents the extraction of information from the environment is dominated by the whisker sensory system (Vincent, 1912). The somatic cortical field of many rodent species exhibits anatomical and physiological specializations which provide marvelous examples of the adaptation of the nervous system to their environmental niche. While these specializations are commonly viewed as setting the somatic cortex of rodents apart from that of other mammals, we interpret them as modifications of the same plan of organization that characterizes the cortical body representation in other mammals, such as cat and monkey.

The mystacial vibrissae of mice and rats are distributed across five rows on the muzzle, designated row A to row E. Each densely innervated whisker follicle — e.g. C2 — is associated with a corresponding cluster of neurons along the central somatosensory pathway. For example, in the C2 cluster of the principal trigeminal nucleus all neurons are best activated by movement of whisker C2. In turn, each cell cluster of the principal trigeminal nucleus projects to a distinct cluster of cells (called a "barreloid") in VB. A discrete bundle of fibers arises from each barreloid and arborizes within a circumscribed cortical locus of about 300-400 μm diameter (Jensen and Killackey, 1987; Killackey, 1973). The terminal field of barreloid cells is correlated with dense clusters of stellate cells in layer IV; the "barrels" of barrel field cortex (Woolsey and Van der Loos, 1970).

From this outline, the barrel, together with the column of cells above and below it, can be seen as analogous to the modular unit of the cat. As in the cat, the modular unit of rodent cortex has been identified by two receptive field methods — the minRF method (Welker, 1971) and the maxRF method (Simons, 1978). The receptive field center of a modular unit in rodent barrel field cortex is a single whisker, the "principal" whisker. The major difference between the cat and the rodent, according to our interpretation, is that in the rodent fiber bundles from VB are strictly segregated one from another. Indeed, the segregation is so absolute that the axon terminals from the medial division of the posterior nucleus (PO), a second thalamic nucleus carrying information from whiskers to cortex, are specifically excluded from the VB termination zone; instead, PO fibers project to the "septa" surrounding each barrel (Lin et al., 1987). Van der Loos has suggested that in the embryo the density of sensory receptors in the developing whisker follicle induces these discrete, segregated cellular groupings at subsequent stations along the sensory pathway (see Van der Loos in this volume).

Armstrong-James and colleagues (Armstrong-James and Callahan, 1990; Armstrong-James et al., 1990a; Armstrong-James et al., 1990b; Armstrong-James and Fox, 1987) have formulated a model of the vertical and horizontal integrative function of the modules in rodent cortex. Their conclusion is that input from the receptive field center - the principal whisker - arrives in a module as a direct thalamic volley. Input from the surrounding receptive field - the non-principal whiskers - arrives in a cortical module only at longer latencies after relays in surrounding modules. While this scheme is fundamentally the same as that proposed for cat and monkey cortex (Diamond, 1989), it stands in sharp contrast to the scheme of integrative function of turtle dorsal cortex shown in Figure 1.

Figure 1 returns us full circle to the purpose of this chapter, the comparison of neocortex with primitive cortex. In mammals the type of afferent pathway found in turtles — a superficial trajectory coursing parallel to the cortical surface — remains only in the olfactory and hippocampal regions. In neocortex the multiple waves of cell migration lead to a vastly increased number of cells in comparison to turtles. Yet these additional cells are not merely a multiplication of the same cell types found in reptiles. Rather, an entirely new cell type, a short axon cell with excitatory local connections, emerges. We have argued here that the function of the excitatory local cell is intricately related with the cluster-to-module thalamocortical projection. Thus, the main idea of this chapter is that a novel type of thalamocortical projection has evolved hand in hand with a novel type of intrinsic cortical organization. One implication of this idea is that the evolution of a specialized form of thalamic projection (one related to a species' unique behavior) can stimulate the evolution of a new cortical area. Examples of specialized forms of thalamocortical projection include those pathways segregated by function (e.g. the X, Y, and W pathways through the cat lateral geniculate nucleus and the tone-specific pathways through the bat medial geniculate nucleus) or topography (e.g. the barreloids-to-barrels pathways of rodent). The specialized thalamic projection induces a new cortical area: new modules with specialized form and function.

REFERENCES

Armstrong-James, M., and Callahan, C.A. (1990) Spatiotemporal convergence in the thalamic venteroposterior medial nucleus (VPm) of the rat. *J. Comp. Neurol.*, in press.

Armstrong-James, M., Callahan, C.A. and Friedman, M. (1990a) The role of intracortical mechanisms in the construction of centre and surround receptive fields of layer IV neurons in the rat barrel field cortex. *J. Comp. Neurol.*, in press.

Armstrong-James, M., Das-Gupta, A. and Fox, K. (1990b) Laminar latency analysis of center and surround receptive fields in the rat SI barrel field cortex. *J. Comp. Neurol.*, in press.

Armstrong-James, M. and Fox, K. (1987) Spatiotemporal convergence and divergence in the rat SI "barrel" cortex. *J. Comp. Neurol.*, 263: 265-281.

Cajal, S. Ramon y (1899) Estudios sobre la corteza cerebral humana I: Corteza visual. *Rev. trim. Microcraf., Madrid.*, 4: 1-63.

Cajal, S. Ramon y (1900) Estudios sobre la corteza cerebral humana III: Corteza acustica. *Rev. trim. Microcraf., Madrid.*, 5: 185-198.

Chagnac-Amitai, Y. and Connors, B. W. (1989) Horizontal spread of synchronized activity in neocortex, and its control by $GABA_A$-mediated inhibition. *J. Neurophysiol.*, 61: 747-758.

Collonier, M. (1968) Synaptic patterns on different cell types in the laminae of the cat visual cortex: An electron microscope study. *Brain Res.*, 9: 268-287.

Connors, B.W., Gutnick, M.J. and Prince, D.A. (1982) Electrophysiological properties of neocortical neurons *in vitro*. *J. Neurophysiol.*, 48: 1302-1320.

Diamond, M.E. (1989) Organization of somatic sensory cortex: The detection of discrete topographic units and evidence for their integrative function. Ph. D. Dissertation, University of North Carolina.

Diamond, M., Favorov, O. and Whitsel, B. (1987) The body surface is represented in SI by a mosaic of segregates. *Soc. Neurosci. Abst.*, 13: 471.

Ebner, F.F. and Colonnier, M. (1978) Quantitative studies of synapses in turtle visual cortex. *J. Comp. Neurol.*, 179: 263-276.

Favorov, O.V. and Diamond, M.E. (1990) Demonstration of discrete place defined columns - segregates - in cat SI. *J. Comp. Neurol.*, in press.

Favorov, O.V., Diamond, M.E. and Whitsel, B.L. (1987) Evidence for a mosaic representation of the body surface in area 3b of the somatosensory cortex of cat. *Proc. Natl. Acad. Sci. USA*, 84: 6606-6610.

Favorov, O. and Whitsel, B.L. (1988) Spatial organization of the peripheral input to area 1 cell columns. I. The detection of "segregates." *Brain Res. Rev.*, 13: 25-42.

Friedman, D.P. and Jones, E.G. (1980) Focal projection of electrophysiologically defined groupings of thalamic cells on the monkey somatic sensory cortex. *Brain Res.*, 191: 249-252.

Garraghty, P.E., Pons, T.P., Sur, M. and Kaas, J.H. (1989) The arbors of axons terminating in middle cortical layers of somatosensory area 3b in owl monkeys. *Somatosens. Mot. Res.*, 6: 401-411.

Gilbert, C.D. and Wiesel, T.N. (1981) Laminar specialization and intracortical connections in cat primary visual cortex. in: *The organization of the cerebral cortex* (Schmitt, F.O., Worden, F.G., Adelman, G. and Dennis, S.G., eds.) The MIT Press, Cambridge, MA, pp. 163-191.

Hall, J.A., Foster, R.E., Ebner, F.F. and Hall, W.C. (1977) Visual cortex in a reptile, the turtle (*Pseudemys scripta and Chrysemys picta*). *Brain Res.*, 130: 196-216.

Hendry, S.H.C., Schwark, H.D., Jones, E.G. and Yan, J. (1987) Numbers and proportions of GABA-immunoreactive neurons in different areas of monkey cerebral cortex. *J. Neurosci.*, 7: 1503-1519.

Jensen, K.F. and Killackey, H.P. (1987) Terminal arbors of axons projecting to the somatosensory cortex of the adult rat. I. The neuronal morphology of specific thalamo-cortical afferents. *J. Neurosci.*, 7: 3529-3543.

Johnston, J.B. (1915) The cell masses in the forebrain of the turtle *Cistudo carolina*. *J. Comp. Neurol.*, 25: 393-468.

Jones, E.G. (1975) Varieties and distribution of non-pyramidal cells in the somatic sensory cortex of the squirrel monkey. *J. Comp. Neurol.*, 160: 206-267.

Jones, E.G., Friedman, D.P. and Hendry, S.H.C. (1982) Thalamic basis of place- and modality-specific columns in monkey somatosensory cortex: a correlative anatomical and physiological study. *J. Neurophysiol.*, 48: 545-568.

Jones, E.G. and Powell, T.P.S. (1969) Synapses on the axon hillocks and initial segments of pyramidal cells in the cerebral cortex. *J. Cell Sci.*, 5: 495-507.

Keller, A. and White, E.L (1987) Synaptic organization of GABAergic neurons in the mouse SmI cortex. *J. Comp. Neurol.*, 262: 1-12.

Killackey, H.P. (1973) Anatomical evidence for cortical subdivisions based on vertically discrete thalamic projections from the ventral posterior nucleus to cortical barrels in the rat. *Brain Res.*, 51: 326-331.

Kosar, E. and Hand, P.J. (1981) First somatosensory cortical columns and associated neuronal clusters of nucleus ventralis posterolateralis of the cat: An anatomical demonstration. *J. Comp. Neurol.*, 198: 515-539.

Kriegstein, A.R. and Connors, B.W. (1986) Cellular physiology of the turtle visual cortex: Synaptic properties and intrinsic circuitry. *J. Neurosci.*, 6: 178-191.

Landry, P. and Deschenes, M. (1981) Intracortical arborizations and receptive fields of identified ventrobasal thalamocortical afferents to the primary somatic sensory cortex in the cat. *J. Comp. Neurol.*, 199: 345-371.

Lin, C.-S., Lu, S.M. and Yamawaki, R.M. (1987) Laminar and synaptic organization of terminals from the ventrobasal and posterior thalamic nuclei in barrel cortex. *Soc. Neurosci. Abst.*, 17: 248.

Lorente de No (1949) Cerebral cortex: Architecture, intracortical connections, motor projections. in: *Physiology of the Nervous System.* (Fulton, J.F., ed.) MIT Press, Cambridge, MA, pp. 288-313.

Lund, J.S. (1984) Spiny stellate neurons in: Cerebral Cortex, Volume 1, *Cellular components of the cerebral cortex.* (Peters, A. and Jones, E.G., eds.) Plenum Press, New York, NY, pp. 255-308.

Mazurskaya, P.Z. (1972) Organization of receptive fields in the forebrain of *Emys orbicularis*. *Zh. Evol. Biokhim. Fiziol.*, 8: 617-624. (Translated in *Neurosci. Behav. Physiol.* [1973] 6: 311-318).

Orrego, F. and Lisenby, D. (1962) The reptilian forebrain. IV. Electrical activity in the turtle cortex. *Arch. Ital. Biol.*, 99: 425-445.

Peters, A. and Walsh, M. (1972) Study of the organization of apical dendrites in the somatic sensory cortex of the rat. *J. Comp. Neurol.*, 144: 253-268.

Ribak, C.E. (1978) Aspinous and sparsely-spinous stellate neurons in the visual cortex of rats contain glutamic acid decarboxylase. *J. Neurocytol.*, 2: 361-368.

Simons, D.J. (1978) Response properties of vibrissa units in rat SI somatosensory neocortex. *J. Neurophys.*, 41: 798-820.

Smith, L.M., Ebner, F.F. and Colonnier, M. (1980) The thalamocortical projection in Pseudemys turtles: A quantitative electron microscopic study. *J. Comp. Neurol.*, 190: 445-461.

Somogyi, P., Freund, T.F., Halasz, N. and Freund, T.F. (1981) Selectivity of neuronal [^{3}H]-GABA accumulation in the visual cortex as revealed by Golgi staining of the labelled neurons. *Brain Res.*, 225: 431-436.

Vincent, S.B. (1912) The function of the vibrissae in the behavior of the adult white rat. *Behavior Monographs*, 1: 1-85.

Welker, C. (1971) Microelectrode delineation of fine grain somatopic organization of SmI cerebral neocortex in rat. *Brain Res.*, 26:259-275.

White, E.L. (1978) Identified neurons in mouse SmI cortex which are postsynaptic to thalamocortical axon terminals: A combined Golgi-electron microscopic and degeneration study. *J. Comp. Neurol.*, 181: 627-662.

White, E.L. (1979) Thalamocortical synaptic relations: A review with emphasis on the projection of specific thalamic nuclei to the primary sensory areas of the neocortex. *Brain Res. Rev.*, 1: 275-311.

Winfield, D.A., Brooke, R.N.L., Sloper, J.J. and Powell, T.P.S. (1981) A combined Golgielectron microscopic study of synapses made by the proximal axon and recurrent collaterals of a pyramidal neuron in the somatic sensory cortex of the monkey. *Neurosci.*, 6: 1217-1230.

Winfield, D.A., Gatter, K.C. and Powell, T.P.S. (1980) An electron microscopic study of the types and proportions of neurons in the cortex of the motor and visual areas of the cat and rat. *Brain*, 103: 245-258.

Woolsey, T.A. and Van der Loos, H. (1970) The structural organization of layer IV in the somatosensory region (SI) of mouse cerebral cortex. The description of a cortical field composed of discrete cytoarchitectonic units. *Brain Res.*, 17: 205-242.

3. NEUROEMBRYOLOGY OF THE NEOCORTEX

ONTOGENY AND STRUCTURE OF THE RADIAL GLIAL FIBER SYSTEM OF THE
DEVELOPING MURINE CEREBRUM

Verne S. Caviness, Jr. *
Jean-Paul Mission *,**
Takao Takahashi *
Jean-Francois Gadisseux *,***

*Department of Neurology
Developmental Neurobiology
Massachusetts General Hospital
Harvard Medical School
Boston, MA 02114

**Department of Developmental Neurobiology
University of Liege
Belgium

*** Developmental Neurology Unit
University of Louvain Medical School
Brussels, B-1200 Belgium

Neurons of cerebral cortical structures undergo their terminal divisions in a periventricular generative epithelium, the ventricular zone (VZ: Boulder Committee, 1970; Sidman and Rakic, 1973). Its final mitosis completed, the young cell must migrate centrifugally to achieve its position in the developing cortex. It is thought that the cell is guided in its migration by contact with the surfaces of specialized cells of astroglial lineage, the radial glial cells (Rakic 1972, 1988; Nowakowski and Rakic, 1979; Misson et al., 1989a). A monopolar radial glial form, the Bergmann glial cell, is thought to serve an analogous role as guide to neuronal migration in the developing cerebellar cortex (Rakic, 1971; Edmondson and Hatten, 1987). These bipolar and monopolar radial astroglial forms are thus essential to cellular interactions of fundamental significance for histogenesis of cortical structures of the central nervous system.

From the earliest phases of neurogenesis, radial glial cells may be identified throughout the CNS (Schmechel and Rakic, 1979; Levitt and Rakic, 1980; Houle and Fedoroff, 1983; Cochard and Paulin, 1984; Choi, 1988; Misson et al., 1988a; Edwards et al., 1989). From the outset these cells form an array of fibers which spans the matricial ventricular epithelium and the subpial limiting glial membrane at the surface of the nervous system. With the exception of a few specialized glial forms which persist into adult life, eg., Bergmann glial cells and tanycytes (Rakic, 1971; del Cerro and Swarz, 1976; Levitt and Rakic, 1980; reviewed in Edwards et al. 1989) the bipolar radial glial cells persist no longer than this early period of CNS development when neurons are actively migrating outward from the VZ (Misson et al., 1989b; Edwards et al., 1989).

The interrelationship of migrating neurons and bipolar radial glial cells has been considered closely in the developing neocortex and hippocampal formation in species as diverse as primates and rodents (Rakic, 1971; Gadisseux and Evrard, 1985; Gadisseux et al., 1989a-c). Neurons migrating to these cerebral cortical structures are in close contact throughout their length to surfaces of radial glial fibers. The constancy of this cell to cell relationship has, in fact, been the principal observation which supports the hypothesis that the fibers guide neuronal migration (Rakic, 1972; Misson et al., 1989a).

Radial glial fibers are aligned in parallel throughout their transcerebral span (Levitt and Rakic, 1980; Misson et al., 1988a). Despite the substantial distortions of cerebral structures occurring with rapid growth of the developing brain, the neighborhood relations of fibers at their ventricular point of insertion must, therefore, be maintained (at least approximately) at the surface of the cerebrum. Thus, the radial fiber system is organized in such a way that it would assure an orderly topological translation of the cell population of the VZ upon the developing cortex. To the extent that a proto-organization of the neocortex is anticipated in the organization of the VZ, its transfer in the course of neuronal migration depends upon this parallel arrangement of the radial glial fibers (Rakic, 1988). We review here, findings of current investigations relating to the ontogeny and structural organization of the system of fibers formed by the radial cells of the murine cerebral wall. Our observations indicate that the system of fibers formed by radial glial cells of the cerebral wall is dynamic in its development and structurally complex. These properties act in concert to implement the histogenetic role served by the system of fibers.

HISTOLOGICAL MARKERS ESSENTIAL TO THE ANALYSIS

Our analysis of the radial glial cell system has been critically served by the development and application of three cytological markers. These include two monoclonal antibodies, RC1 (Edwards et al., 1989) and RC2 (Misson et al., 1988a; Gadisseux et al., 1989a), which are selective for cells of astroglial lineage when applied in the prenatal murine brain. RC2 binds an epitope carried on antigens which are robust in the face of tissue fixation with aldehydes and is, therefore, suitable for high resolution microscopy (Misson et al., 1988a). These ligands are substantially more sensitive as markers for radial glial cells than antibodies against GFAP. The latter are not known to stain radial cells before the onset of cortical histogenesis in primates (Levitt and Rakic, 1980) and not reliably until cortical histogenesis is nearly completed in rodents (Valentino et al., 1984; Choi, 1988). RC1 and RC2 are more selective than antibodies against vimentin (Dahl et al., 1981; Cochard and Paulin, 1984; Pixley et al., 1984) and Rat 401 (Hockfield and McKay, 1985) which stain progenitors of neuronal and non-neuronal elements during the earliest stages of cytogenesis of the central nervous system.

The cytologic features of the bipolar radial glial cells presented by staining with RC1 and RC2 are identical to those revealed in classic golgi impregnations (Ramon y Cajal, 1911; Schmechel and Rakic, 1979). Throughout much of the early period of neocortical histogenesis, the oval soma is located in the ventricular or subventricular zones (Boulder Committee, 1970; Misson et al., 1988a). A descending process extends to the ventricular margin. An ascending process ascends radially through the overlying cerebral wall.

The third marker upon which our observations are based is glycogen. This substance is highly concentrated in radial glial fibers and may be visualized confidently in electron micrographs of tissue perfused by dialdehydes followed by treatment with acid-thiocarbohydrazide-silver proteinate (Gadisseux and Evrard, 1985). Though less sensitive as a marker for radial glia than the monoclonal antibodies RC1 and RC2, glycogen preserved in this way has proved versatile in combination with other histological methods and highly selective for radial glial fibers in ultrastructural studies of mid and later stages of active neuron migration in the cerebral hemisphere.

ONTOGENY OF THE CEREBRAL RADIAL FIBER SYSTEM

The bipolar radial glial cell, stained with either RC1 or RC2 and defined in terms of its configuration, may be delineated explicitly in the spinal level of the neural tube of the mouse at E9 (Misson et al., 1988a; Edwards et al., 1989). This date is 24 hours after neurulation has occurred. To our knowledge, currently available ligands identify no other cell class-specific antigens within the mammalian central nervous system at this relatively early developmental stage. The radial glial cell may, therefore, be the earliest neuroglial cell of the developing mammalian central nervous system to diverge from a pluripotential progenitor population. By E10 - E11, these ligands mark typical radial glial cells at all levels of the developing CNS. These cells are abundantly present in the cerebral vesicles of the mouse when they evaginate at E11.

The radial glial cell of the cerebral wall forms a structurally complex and dynamically evolving multicellular system. From the time of evagination of the cerebral vesicle, the cerebral wall is densely saturated with the processes of bipolar radial glial cells. From this earliest period of cerebral development through E14 when the cortical plate (CP, Caviness, 1982) is formed throughout the neocortical anlagen, the descending and ascending processes of the bipolar radial

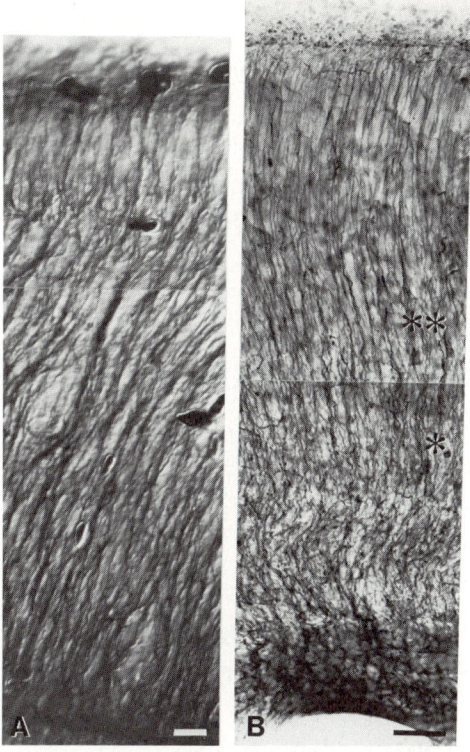

Figure 1. RC2 stained fascicles of radial glial fibers span the cerebral wall. 1A. At E14 the fascicles are uniform throughout the transcerebral span. 1B. At E17 there is a step drop in fiber density at the ESS (*) with a further decline in fiber density at the SP-CP transition (**). Coronal, 50µm sections; Bar in A = 5µm and in B = 50µm.

glial cells are grouped in fascicles (Fig. 1A) (Gadisseux and Evrard, 1985; Misson et al., 1988a). The fasciculated descending processes partition cells of the ventricular zone into radially aligned columns. The fasciculated ascending fibers are clustered in "loges" (Gadisseux and Evrard, 1985; Gadisseux et al., 1989a) between the large blocks of axonal bundles of the intermediate zone. At this earliest stage of development of the cortical plate, the ascending glial fiber fascicles appear to be uniform in their composition and patterns of spacing, an observation which suggests that at E14 the typical radial glial fiber has a full transcerebral span (Gadisseux et al., 1989a,b). On average each ascending fascicle includes 4-5 fibers with a uniform interfascicular spacing of 8 - 10 microns (Gadisseux et al., 1989a).

In the interval E14 - E15, neurons destined to form the infragranular layers complete their migrations into the CP (Caviness, 1982). During this 48 hr interval a change begins to be evident in the fascicular organization of the radial glial fibers and this subsequently progresses to reach its maximum expression at E17 (Fig. 1B) (Gadisseux et al., 1989a,b). Specifically, there appears by E15 the suggestion of a drop in the density of radial glial fibers in the outer zone of the cerebral wall. By E17 this progresses to become a dramatic "step" drop (Fig. 1B). The step drop is positioned just at the external sagittal stratum (ESS), a conspicuous fiber plane which defines the interface between intermediate zone (IZ) and the inferior border of future neocortical layer VI. This fiber plane includes, among other components, axons of the thalamocortical projection (Crandall and Caviness, 1984). Even after correction for fiber dilutions by differential centrifugal growth, the differential in fiber density of the upper cortical plate with respect to the upper IZ is as low as 60% at E17 (Fig. 2; Gadisseux et al., 1989b). This implies that at E17, in contrast to the condition at E14, no more than 40% of the ascending fibers of the radial glial system have a full transcerebral span.

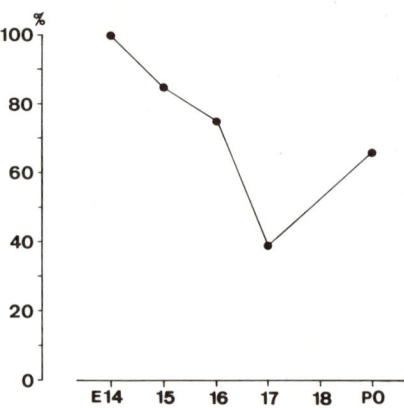

Figure 2. Graphic representation of the density of radial glial fibers at the level of the upper CP as a percentage of the fiber density at the IZ-ESS transition. Values are corrected for centrifugal dilution due to tissue growth (adapted from Gadisseux et al., 1989b).

The steep gradient in fiber density in ascent through the ESS and subplate (SP) is reflected in a decrease in the number of fibers per fascicle. In the trans-subplate span, fascicles become reduced in fiber composition in the E15 - E17 interval. By E17 only single fibers, and not fiber fascicles, span the cortical plate itself (Gadisseux and Evrard, 1985; Gadisseux et al., 1989a). Beyond E17, however, the steepness of the fiber density gradient lessens across the ESS-SP zone. By the day of birth (E19 = P0 in the mouse), there is a reversal of the process, actually an augmentation in the number of glial fibers spanning the IZ and the cortical strata. Thus, the differential in fiber density between IZ and CP, which is 60% at E17, becomes reduced to approximately 30% by the time of birth (Fig. 2; Gadisseux et al., 1989b). The interval E17 - P0, when this gradient reversal occurs, corresponds to the time of assembly of the granular and supragranular layers (Caviness and Sidman, 1973; Caviness, 1982).

Postnatally there is a rapid decline in the overall radial glial fiber number within the cerebral wall. By P8 these cells are no longer identifiable by staining with RC1 or RC2 (Misson et al., 1989b; Edwards et al., 1989). They have not been seen in the rodent cortex by staining with GFAP or impregnation by the Golgi method.

DYNAMIC PROPERTIES OF THE RADIAL GLIAL CELLS

The changes occurring in the structure of the radial glial fiber system during the final week of gestation in the mouse, are the expression of dynamic properties of the radial glial cell class. These dynamic properties are expressed at two levels. First, with respect to the individual radial glial cell, even fully differentiated forms with somata within the VZ of the murine cerebral wall are actively proliferative (Misson et al., 1988b). Secondly, at the level of multicellular organization, the overall fiber system is substantially and continuously restructured in the course of neocortical histogenesis (Gadisseux et al., 1989a,b; Takahashi et al., 1989).

Proliferative activity of radial glial cells

The radial glial cells are not a stable population composed of a specific and fixed set of cells with their radial fibers. Fully differentiated radial glial cells, unambiguously delineated as such by staining with RC1 or RC2, have been demonstrated to pass through the DNA phase of the mitotic cycle (Misson et al., 1988b). For this demonstration, the RC1 or RC2 labeled radial glial cells were "double labeled" in S-phase by *in utero* exposure to H 3-thymidine 2-4 hours before sacrifice and preparation for histological study (Fig. 3). This interval between exposure to the radiotracer and sacrifice is substantially less than the S-phase through M-phase interval which is at least 8 - 10 hr in the VZ of the rodent cerebrum (Waechter and Jaensch, 1974).

It had been previously demonstrated by staining with either GFAP or Rat 401 that the somata of radial glial cells in M phase could be identified in the VZ (Levitt et al., 1981, 1983; Hockfield

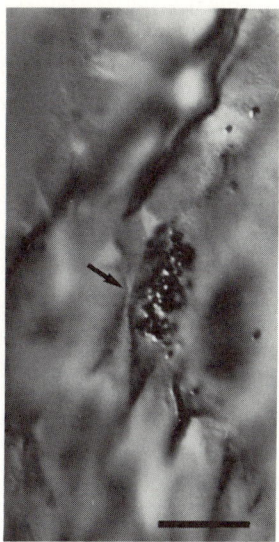

Figure 3. The soma of an RC2 stained, radial glial cell is marked by arrow. The base of the descending process is below. Intranuclear DNA was labeled by exposue to H3Thymidine 2 hours before sacrifice. Coronal, 2µm Epon-Araldite embedded section processed for autoradiography. Bar = 10µm.

and McKay, 1985). Such single labeling methods have established that cells of the VZ become committed to the astroglial lineage, presumably in the form of radial glial cells, at least as early as a terminal division. These single label methods had not established that such cells were mitotically cycling, that is, competent in their fully differentiated form to cycle through S and G2 phases.

Structural change of the radial fiber system

The radial glial cells of the murine cerebral wall are continuously and abundantly generating new growth cone tipped processes throughout the terminal week of gestation (Takahashi et al., 1989). The majority of these processes are fasciculated (Fig. 4A) and, presumably, are guided in their ascent by contact with fascicles very much in the way that migrating neurons are guided by this contact. Through E17 the density of such growth cones is greatest in the intermediate zone. A dramatic drop in their concentration occurs with ascent through the ESS and SP, an observation concordant with the drop in concentration of fibers themselves occurring in this same zone at E17 (Gadisseux et al., 1989b). Subsequent to E17, there is a centrifugal shift in the maximum density of growth cones into the SP and lower CP. This surge occurs concurrently with the surge of radial fibers into these cortical levels (Figs. 2, 5).

DECLINE IN THE RADIAL GLIAL CELL POPULATION

In the E17-P0 interval, a series of transformations and regressive changes involve the bipolar radial glial cells (Misson et al., 1989b). The somata of these cells, previously confined to the VZ and SVZ, become richly distributed throughout the width of the IZ and even appear as far superficially as the SP. This shift in the position of the cell soma occurs apparently through translocation within the ascending fiber (see also, Schmechel and Rakic, 1979).

It is apparent, further, that the growth cone tipped fibers of the radial glial cells include two separate populations (Takahashi et al., 1989). One set of growth cone tipped fibers remains fasciculated and contributes to the surge of fibers into the cortical plate during the terminal period of neuronal migration as discussed above. The other group of growth cone tipped fibers, a minority, appear to break free of the fiber fascicles in their trans-ESS and SP segments (Fig. 4B). These fascicle "free" fibers are observed to become deflected virtually orthogonally from the alignment of the parent fascicle. From the point of deflection, the growth cone "leads" the fiber tangentially across the SP - to a blood vessel or, possibly, to some other target.

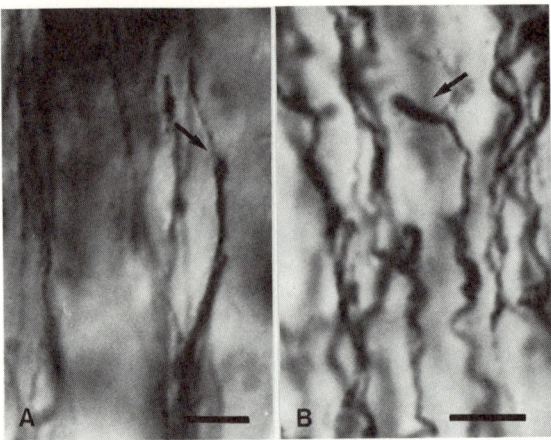

Figure 4. RC2 stained growth cones at the level of the SP. In 4A the glial fiber with its growth cone tip is closely held in a glial fiber fascicle. The growth cone tipped fiber in 4B is not in a fiber fascicle. Coronal 50μm section; Bar = 5μm.

Concurrently with the surge of additional fibers through the ESS into the overlying cortical strata during the final 3 days of intra-uterine life, the radial processes also begin to arborize (Takahashi et al., 1989). Arborization occurs both terminally and intersegmentally along the fibers. By the end of the first postnatal week, beaded and fragmented, presumably degenerating segments of radial fibers may be stained with RC2 in the cortical plate (Misson et al., 1989b).

At the same time that the bipolar radial cells have begun to arborize monopolar radial forms appear in increasing numbers in the superficial zone of the cerebral wall, spanning upper IZ and SP (Misson et al., 1989b). Increasingly through P2 and P4 the monopolar radial cells are succeeded by multipolar non-radial astroglial cells. These later forms appear first in the infragranular layers but spread centrifugally throughout the full cortical strata by P4. Terminal and intersegmental branching occurs explosively from the growing processes of the multipolar astroglial cells. By the end of the first postnatal week the multipolar cells are essentially the only RC2 staining cells in the murine cerebral cortex. They have established at this time a rich network of non-fasciculated glial processes which spreads through all neocortical layers.

HISTOGENETIC IMPLICATIONS

The enlarged perspective of the radial glial system provided by our observations invites certain amplifications of the role assigned to radial glial fiber in neocortical histogenesis. These amplifications relate both to the pattern of assembly of neurons within neocortical layers and to the specific features of the path that cells follow in their migrations.

Amplification of the transcortical fiber system after E17

The most direct implications relate to the dynamic evolution of the glial fiber system. The present observations in the mouse suggest that there is an amplification of the transcortical glial fiber during the several days prior to birth. In this species, the amplification occurs during assembly of the supragranular layers of the neocortex.

We have previously drawn attention to the abundance within supragranular layers of the mature murine cortex of large numbers of radially clustered neurons which are not part of the larger fully transcortical radially aligned clusters (Escobar et al., 1986). Conceivably, these intercalated clusters within the supragranular layers form around the radial fibers that enter the cortex in the interval E17 - P0. Should this be the case the effect would be to increase the fineness of grain of the cortical map but without affecting its underlying topology (Takahashi et al., 1989). That is, large, fully transcortical neuronal clusters might be assembled around radial fibers which are in place from the outset of neuronal migrations. The intercalated columns within and confined to supragranular layers might be assembled in relation to radial fibers formed subsequently.

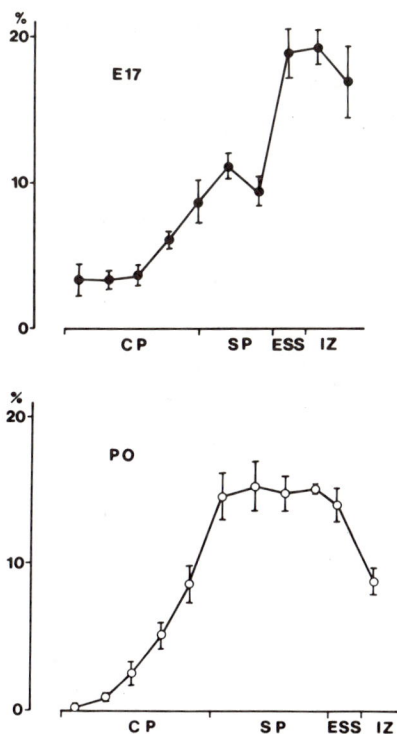

Figure 5. Graphic representation of the density of RC2 stained growth cones tipping fasciculated radial glial fibers at E17 (above) and P0 (below). Counts were made at 50μm intervals through the full depth of a column of 100μm width and extending radially inward from the outer margin of the cortical plate to the upper IZ. The depth profiles are related to the respective strata along the abscissa. Each point is the averaged value for counts from a total of 5 histological sections from 3 animals (2 sections from each of 2 animals and 1 section from the third animal). Vertical bars = S.E. of the mean. (Permission pending.)

The migratory path

The relationship of the migrating cell to radial fibers changes as it makes its ascent through the succession of strata of the developing cerebral wall (Gadisseux and Evrard, 1985). This is particularly dramatic toward E17 when the fiber system undergoes its most conspicuous changes. Within the IZ, where fibers remain largely grouped in dense fascicles, the migrating cell has an *inter*-fascicular relation to the fiber system. Throughout its circumference, the cell migrating across the IZ will be in contact with multiple glial fiber fascicles. In fact the space between fascicles is essentially that of the transverse diameter of migrating cells (Gadisseux et al., 1989c).

As the cell moves outward through the ESS and SP, into the zone of attenuation of the fiber system, it penetrates among the fibers of a single fascicle. That is, it adopts an *intra*-fascicular relation to the fibers. Beyond E17, in the final leg of its migration across the cortical plate, the cell will be in contact with the surface of only a single radial fiber.

It is to be expected, therefore, that cells arising from the same point in the VZ would ultimately enter the cortex along different radial glial fibers. That is, the radial fiber system is probably not a rigid corridor of ascent for cells arising at a common point in the VZ. Thus, as a consequence of the structural organization of the fiber system itself, a given proliferative cell line might provide neurons for different though closely adjacent radial cell groupings within the cortex. This appears to be the case in mice and rats where the cortical destinations of clones of neurons have been traced after retroviral vector insertion of a gene coding for beta-galactose (Walsh and Cepko, 1988; Luskin et al., 1988; Price and Thurlowe, 1988; Misson et al., 1989a).

ACKNOWLEDGMENTS

The authors wish to thank Margaretha Jacobson for technical and photographical assistance, and Ginny Tosney for editing the manuscript. This work was supported by NIH Grant NS 12005 and the Foundation Princesse Marie-Christine and NATO Grant 27B85BE.

REFERENCES

Boulder Committee. (1970) Embryonic vertebrate central nervous system: revised terminology. *Anat. Rec.*, 166: 257-261.
Caviness, V.S., Jr., and Sidman, R. L. (1973) Time of origin of corresponding cell classes in the cerebral cortex of normal and reeler mutant mice: an autoradiographic analysis. *J. Comp. Neurol.*, 148: 141-151.
Caviness, V.S., Jr. (1982) Neocortical histogenesis in normal and reeler mice: a developmental study based upon [3H] thymidine autoradiography. *Dev. Brain Res.*, 4: 293-302.
Choi, B.H. (1988) Prenatal gliogenesis in the developing cerebrum of the mouse. *Glia*, 1: 308-316.
Crandall, J.E., and Caviness, V.S., Jr. (1984) Axonal strata of the cerebral wall in embryonic mice. *Dev. Brain Res.*, 14: 185-195.
Cochard, P., and Paulin, D. (1984) Initial expression of neurofilaments and vimentin in the central and peripheral nervous system of the mouse embryo in vivo. *J. Neurosci.*, 4: 2080-2094.
Dahl, D., Rueger, D. and Bignami, A. (1981) Vimentin, the 57,000 molecular weight protein of fibroblast filaments, is the major cytoskeletal component in immature glia. *Eur. J. Cell Biol.*, 24: 191-196.
del Cerro, M., and Swarz, J.R. (1976) Prenatal development of Bergmann glial fibers in rodent cerebellum. *J. Neurocytol.*, 5: 669-676.
Edmondson, J.C., and Hatten, M.E. (1987) Glial-guided granule neuron migration in vitro: A high-resolution time-lapse video microscope study. *J. Neurosci.*, 7: 1928-1934.
Edwards, M.A., Yamamoto, M. and Caviness, V.S., Jr. (1989) Organization of radial glial and related cells in the developing murine CNS: An analysis based upon a new monoclonal antibody marker. *J. Neurosci.* in press.
Escobar, M.I., Pimienta, H., Caviness, V.S., Jr., Jacobson, M., Crandall, J.E. and Kosik, K. S. (1986) Architecture of apical dendrites in the murine neocortex: Dual apical dendritic systems. *Neuroscience*, 17: 975-989.
Gadisseux, J-F., and P. Evrard (1985) Glial-neuronal relationship in the developing central nervous system. *Dev. Neurosci.*, 7: 12-32.
Gadisseux, J-F., Evrard, P., Misson, J-P. and Caviness, V.S., Jr. (1989a) Dynamic structure of the radial glial fiber system of the developing murine cerebral wall: An immunocytochemical analysis. *Dev. Brain Res.* in press.
Gadisseux, J-F., Evrard, P., Misson, J-P., Caviness, V.S., Jr. (1989b) Dynamic changes in the density of radial glial fibers of the developing murine cerebral wall: A quantitative immunohistological analysis. *J. Comp. Neurol.*, in press.
Gadisseux, J.-F., Kadhim, H.J., Van de Bosch de Aguilar, P. Caviness, V.S., Jr., and Evrard, P. (1989c) Neuron migration within the radial glial fiber system of the developing murine cerebrum: An electron microscopic autoradiographic analysis. *Dev. Brain Res.*, in press.
Hockfield, S., and McKay, R.D.G. (1985) Identification of major cell classes in the developing mammalian nervous system. *J. Neurosci.*, 5: 3310-3328.
Houle, J., and Fedoroff, S. (1983) Temporal relationships between the appearance of vimentin and neural tube development. *Dev. Brain Res.*, 9: 189-196.
Levitt, P., and Rakic, P. (1980) Immunoperoxidase localization of glial fibrillary acidic protein in radial glial cells and astrocytes of the developing rhesus monkey brain. *J. Comp. Neurol.*, 193: 815-840.
Levitt, P., Cooper, M.L. and Rakic, P. (1981) Coexistence of neuronal glial precursor cells in the cerebral ventricular zone of the fetal monkey: An ultrastructural immunoperoxidase analysis. *J. Neurosci.*, 1: 27-39.
Levitt, P., Cooper, M.L. and Rakic, P. (1983) Early divergence and changing proportions of neuronal and glial precursor cells in the primate cerebral ventricular zone. *Dev. Biol.*, 96; 472-484.
Luskin, M.B., Pearlman, A.L. and Sanes, J.R. (1988) Cell lineage in the cerebral cortex of the mouse studied in vivo and in vitro with a recombinant retrovirus. *Neuron*, 1: 635-647.

Misson, J-P., Edwards, M.A., Yamamoto, M. and Caviness, V.S., Jr. (1988a) Identification of radial glial cells within the developing murine central nervous system: Studies based upon a new immunohistochemical marker. *Dev. Brain Res.*, 44: 95-108.

Misson, J-P., Edwards, M.A., Yamamoto, M. and Caviness, V.S., Jr. (1988b) Mitotic cycling of radial glial cells of the fetal murine cerebral wall: A combined autoradiographic and immunohistochemical study. *Dev. Brain Res.*, 38: 183-190.

Misson, J-P., Austin, C., Takahashi, T., Cepko, C. and Caviness, V.S., Jr. (1989a) Migrating neurons of the murine cerebrum ascend in parallel to radial fiber: Analysis based upon double-labeling of migrating neurons and radial fibers. *Soc. Neurosci. Abstr.* 15: 599.

Misson, J-P., Takahashi, T. and Caviness, V. S., Jr. (1989b) Early ontogeny of radial and other astroglial cells in murine cerebral cortex. *J. Comp. Neurol.* submitted.

Nowakowski, R.S., and Rakic, P. (1979) The mode of migration of neurons to the hippocampus: A Golgi and electron microscopic analysis in foetal rhesus monkey. *J. Neurocytol.* 8: 694-718.

Pixley, S.R., Kobayashi, Y. and Vellis, J.D. (1984) A monoclonal antibody against vimentin: Characterization. *Dev. Brain Res.*, 15: 185-199.

Price, J., and Thurlowe, L. (1988) Cell lineage in the rat cerebral cortex: A study using retroviral-mediated gene transfer. *Development*, 104: 473-482.

Rakic, P. (1971) Neuron-glia relationship during granule cell migration in developing cerebellar cortex. A Golgi and electron microscopic study in Macacus rhesus. *J. Comp. Neurol.* 141: 283-312.

Rakic, P. (1972) Mode of cell migration to the superficial layers of fetal monkey neocortex. *J. Comp. Neurol.*, 145: 61-84.

Rakic, P. (1988) Specification of cerebral cortical areas. *Science*, 241: 170-176.

Ramon y Cajal, S. (1911) *Histologie du Systeme Nerveux de l'Homme et des Vertebres*. Vol II, Maloine, Paris. Reprinted by Consejo Superior de Investigaciones Cientificas, Instituto Ramon y Cajal, Madrid.

Schmechel, D.E., and Rakic, P. (1979) A Golgi study of radial glial cells in developing monkey telencephalon: Morphogenesis and transformation into astrocytes. *Anat. Embryol.*, 156: 115-152.

Sidman, R.L., and Rakic, P. (1973) Neuronal migration, with special reference to developing human brain: A review. *Brain Res.*, 62: 1-35.

Takahashi, T., Misson, J-P., Caviness, V.S., Jr. (1989) Glial process elongation and branching in the developing murine neocortex: A qualitative and quantitative immunohistochemical analysis. *J. Comp. Neurol.* submitted.

Valentino, K.L., Jones, E.G. and Kane, S.A. (1984) Expression of GFAP immunoreactivity during development of long fiber tracts in the rat CNS. *Dev. Brain Res.*, 9: 317-336.

Walsh, C., and Cepko, C.L. (1988) Clonally related cortical cells show several migration patterns. *Science,* 241: 1342-1345.

Waechter, R.V. and Jaensch, B. (1972) Generation time of the matrix cells during embryonic brain development: An autoradiographic study in rats. *Brain Res.*, 46: 235-250.

NONPYRAMIDAL NEURONS IN THE MAMMALIAN HIPPOCAMPUS: PRINCIPLES OF ORGANIZATION AND DEVELOPMENT

Eduardo Soriano*, José A. Del Río* and Isidro Ferrer**

* Unidad de Biología Celular
  Facultad de Biología
  Universidad de Barcelona
  Diagonal 645, Barcelona 08028

** Unidad de Neuropatología
  Departamento de Anatomía Patológica
  Hospital Príncipes de España
  Hospitalet de LLobregat
  Barcelona 08097, Spain

INTRODUCTION

The hippocampus has traditionally been considered a simple cortex, with some primitive structural characteristics. For example, its particular cytoarchitectonics, with the main cell types densely packed in single layers and the segregated distribution of many hippocampal afferents, resembles the organization of phylogenetically more primitive cortices, like those of reptiles (see Lohman, this volume). In this paper we show that, in spite of this relatively simple organization, the hippocampal cortex contains essentially the same types of nonpyramidal cells or local circuit neurons (as we will call them here) as the neocortex. However, in contrast to the neocortex, there is a particular laminar distribution of nonpyramidal cells in the hippocampus and area dentata. We then focus on the neurogenesis of hippocampal nonpyramidal cells. We will provide evidence that the simpler organization of the hippocampal cortex may help us to understand some general developmental mechanisms related to the genesis and maturation of cortical interneurons.

Intrinsic organization of the hippocampus

The hippocampus can be divided into three main anatomical areas: regio inferior, regio superior and area dentata (Blackstad, 1956). The first two subdivisions together are referred to as the hippocampus proper and correspond to fields CA1, CA2 and CA3 as described by Lorente de Nó (1934). The principal neurons of the hippocampus proper are typical pyramidal-shaped neurons, which in rodents are densely packed in the pyramidal layer, but are somewhat more loosely distributed in primates. The dominating neuronal type in the area dentata is the granule cell. The perikarya of granule cells are densely packed to form the granular layer. Granule cells extend numerous apical dendrites into the molecular layer. In primates or under certain experimental conditions they may also develop basal dendrites (c.f., Frotscher et al., 1988a).

A large number of studies with the Golgi method have shown that all the subdivisions of the hippocampus are populated by a rich variety of local circuit neurons (Schaffer, 1892; Cajal, 1911; Lorente de Nó, 1934; Amaral, 1978; Tömböl et al., 1978; Ribak and Seress, 1983; Seress and Ribak, 1985; Schlander and Frotscher, 1986; Lübbers and Frotscher, 1987). Following the various classifications used at present, the hippocampal local circuit neurons may be grouped under three main headings according to the pattern of their efferent connectivity: basket cells, chandelier cells, and neurons with a less specialized pattern of axonal arborization (Fig. 1A, 1B).

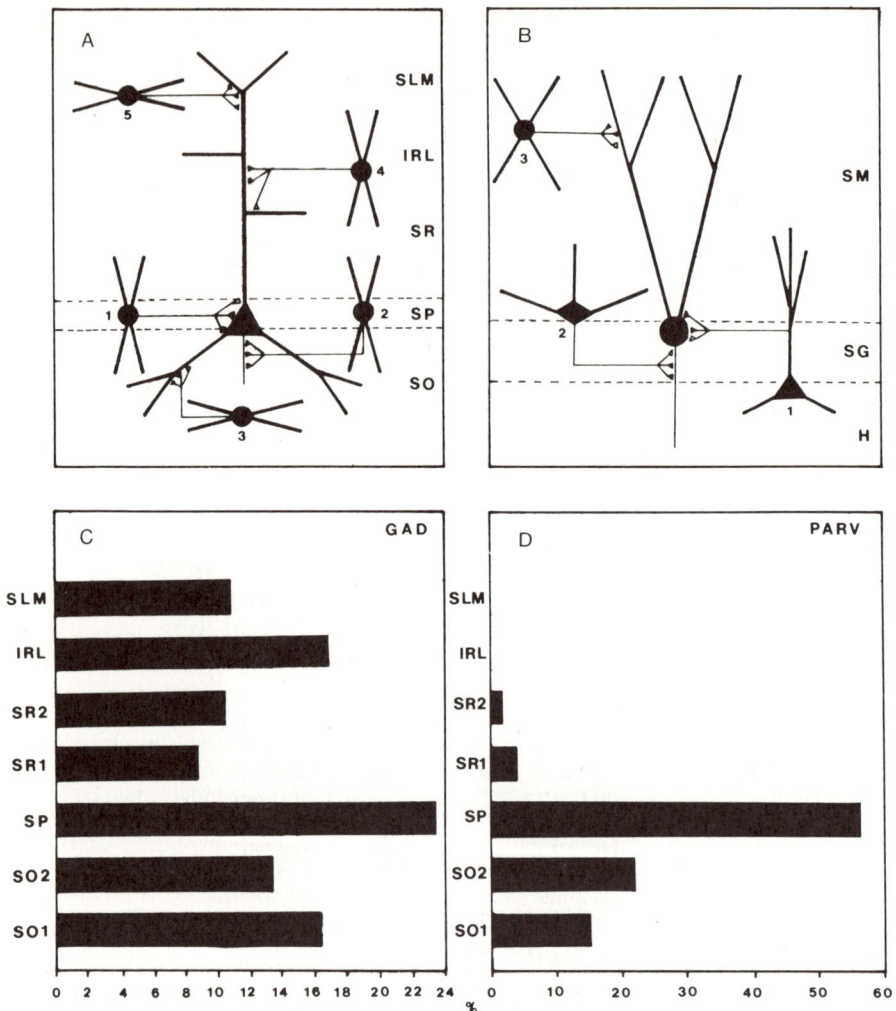

Figure 1. (a, b) Schematic drawings illustrating the efferent connectivity and location of basket cells (1), chandelier cells (2), and neurons with less specific patterns of axonal arborization (3-5) in the hippocampus proper (a) and area dentata (b). (c, d) Radial distribution of glutamic acid decarboxylase- (c) and parvalbumin-immunoreactive (d) neurons in the regio superior of the mouse hippocampus. so, stratum oriens; sp, stratum pyramidale; sr, stratum radiatum; irl, interphase radiatum-moleculare; slm, stratum lacunosum-moleculare; h, hilus; sg, stratum granulare; sm, stratum moleculare. The stratum oriens and radiatum are represented halved in (c) and (d).

Perhaps the various types of basket cells are the most characteristic group of hippocampal nonpyramidal neurons. By definition, basket cells have an axonal plexus that surrounds the cell bodies of pyramidal neurons and granule cells, forming typical pericellular arrangements. Recent Golgi-electron microscopy and immunohistochemical studies have in fact shown that the axon of these nonpyramidal neurons terminates on the perikarya and proximal dendrites of the principal neurons (Ribak et al., 1978; Ribak and Seress, 1983; Frotscher et al., 1984; Kosaka et al., 1984; Seress and Ribak, 1985; Lübbers and Frotscher, 1987).

The second group of nonpyramidal cells is formed by the axo-axonic or chandelier cells. This type of neuron was not recognized by researchers until 1974 when they were described for the first time in the cingular cortex (Szentagothai and Arbib, 1974). They can be identified by the terminal pattern of their axons which are distributed in such a way as to form characteristic vertical aggregations of boutons. Later on, chandelier cells were found in numerous neocortical areas, always establishing synapses specifically with the axon initial segment of pyramidal neurons (Peters, 1984). The existence of a similar type of cell in the hippocampus proper of cats and monkeys was demonstrated by Somogyi and co-workers (Somogyi et al., 1983a, 1983b, 1985). It is now known that this type of inhibitiory neuron is present in all subdivisions of the rodent hippocampus (Soriano and Fariñas, 1987) including the area dentata, where specialized chandelier cells connect with the axon initial segment of granule cells (Soriano and Frotscher, 1989; Soriano et al., 1989c).

Finally, all subdivisions of the hippocampus contain a heterogeneous population of local circuit neurons whose axons do not form any characteristic morphological arrangement but rather arborize exclusively in the plexiform layers (Cajal, 1911; Lorente de Nó, 1934). This pattern of arborization suggests that their postsynaptic targets most probably include the medial and distal dendrites of granule cells and pyramidal neurons. In fact, for a subgroup of these interneurons (those that are found in the stratum lacunosum-moleculare), this suggestion has been confirmed by electron-microscopic analysis of intracellularly labeled cells (Lacaille and Schwartzkroin, 1988). Thus, most likely the terminals of this heterogeneous group of nonpyramidal cells establish synapses with the dendrites of principal neurons which are the most numerous components of the neuropil in the plexiform layers.

All the groups of nonpyramidal cells mentioned above are represented in the neocortex where they show similar axonal arborizations and target selectivities. Minor morphological differences do exist which probably reflect the different arrangements and distribution of the target cells in the two cortical structures (c.f., Fairén et al., 1984). Furthermore, the same types of local circuit neurons can be identified in the neocortex of mammals phylogenetically distant (Valverde, 1983). Thus, it seems reasonable to propose that nonpyramidal cells with a similar pattern of efferent connectivity are essential, constant and evolutionary-old components in the structural and functional design of the mammalian cerebral cortex.

Finally, as in the case of the neocortex, immunohistochemical studies have shown that the vast majority of nonpyramidal cells are GABAergic (Ribak et al., 1978; Seress and Ribak, 1983; Kosaka et al., 1985; Misgeld and Frotscher, 1986; Lübbers and Frotscher, 1987; Sloviter and Nilaver, 1987). Further, the same neuroactive peptides are detected almost invariably in subsets of nonpyramidal cells in both cortices (Swanson et al., 1987). These chemical similarities, together with the coincidence of target connectivity, suggest that comparable nonpyramidal cells exert similar influences on postsynaptic neurons in both cortical structures. In this context it would be interesting to explore whether both populations of neurons are driven by comparable afferent circuits. Whereas a considerable amount of data is available for the hippocampus (c.f., Frotscher, 1988), studies on afferent connections of nonpyramidal cells in the neocortex are still fragmentary which makes comparison difficult. However, Freund and co-workers have recently demonstrated that GABAergic subcortical afferents make similar connections on nonpyramidal cells in both cortical regions (Freund and Antal, 1988; Freund and Gulyás, 1989).

In spite of these similarities, local circuit neurons in the hippocampus and neocortex are organized in a different way. For example, the radial distribution of GABAergic cells in the regio superior of the hippocampus indicates that about 75% of these neurons are located in the plexiform strata, outside the pyramidal layer (Fig. 1C). In contrast, the vast majority of GABAergic nonpyramidal cells in the neocortex were observed in layers II to VI, together with the principal neurons (c.f., Cobas et al., 1987). In addition, the main types of local circuit neurons in the hippocampal cortex are more concentrated in certain layers. This could have been anticipated by Golgi studies which showed that many basket and chandelier cells are located in or close to the cell layers. In contrast, nonpyramidal cells with a less-specialized pattern of arborization predominate in the plexiform layers (see the above references). Further evidence for this radial segregation comes from immunocytochemical studies for the Ca^{++}-binding protein parvalbumin. This protein is present in a subset of cortical GABAergic neurons which correspond to the basket and chandelier cells (De Felipe et al., 1989; Soriano et al., 1989c). In the hippocampus, parvalbumin antibodies only stain the GABAergic boutons that form pericellular baskets and multiple synaptic contacts on axon initial segments (Kosaka et al., 1987; Katsamaru et al., 1988; Sloviter 1989; Nitsch et al., 1989; Soriano et al., 1989c). In agreement with Golgi data, PARV-positive cell bodies are located within or close to the cell layers, especially in the regio superior and area dentata (Fig. 1D). The situation is dissimilar to that in the neocortex where the various groups of nonpyramidal cells have a more widespread distribution throughout the cortical layers.

Development of hippocampal nonpyramidal cells

Provided that the main groups of cortical local circuit neurons are represented in the hippocampus and show a relatively precise distribution, this region seems to be a good model for the analysis of the development of these cells. Moreover, studies in the hippocampus permit to analysis of this process in the area dentata where the majority of the principal neurons cells originate postnatally (Angevine, 1965). First, we were interested in determining the period of neurogenesis of nonpyramidal neurons and how their laminar distribution appears during development. The combination of 3H-thymidine autoradiography with immunohistochemical markers for GABAergic cells appears to be a suitable tool for such an analysis. This approach was first used in the neocortex by Miller (1985) and Fairén et al. (1986) demonstrating that GABAergic neurons and pyramidal cells are generated concurrently following a similar "inside-out" gradient of positioning. In a series of experiments in collaboration with A. Fairén (Institute Cajal, Madrid), we have analyzed the birthdates and the gradients of neurogenesis for GABAergic cells in the mouse hippocampus (Soriano et al., 1986, 1989a, 1989b). The main results are summarized in the following paragraphs.

GABAergic neurons in all three hippocampal subdivisions are formed prenatally. The peaks of neurogenesis are statistically earlier (Fig. 2A) than those of principal cells which corroborated previous data for the rat (Lübbers et al., 1985; Amaral and Kurz, 1985). This is particularly evident in the area dentata where granule cell generation takes place mostly during the first few weeks after birth. The laminar analysis of neurogenesis shows that GABAergic neurons are positioned along a radial "sandwich" gradient in the hippocampus proper, especially in the regio superior where neurogenesis lasts for a longer period (Fig. 2B). According to this spatio-temporal order, the first neurons to be generated are destinated to the plexiform layers, predominantly those furthest from the stratum pyramidale; successive generations have a more widespread distribution throughout the plexiform strata and pyramidal layer; at last, nonpyramidal cells mainly appear in the pyramidal layer and adjacent strata. In addition, both GABAergic cells and pyramidal neurons are positioned following an "inside-out" sequence within the pyramidal layer (Angevine, 1965; Soriano et al., 1989a). Thus, in the hippocampus proper GABAergic nonpyramidal cells are positioned according to two different and simultaneous spatio-temporal gradients along the radial axis: the entire population of nonpyramidal cells follows a "sandwich" gradient, those located in the stratum pyramidale follow the well-known "inside-out" sequence.

In contrast, no conspicuous gradients are observed for GABAergic cells in the area dentata Lübbers et al., 1985; Soriano et al., 1989b). This is surprising because it does not appear to follow the general rule in the cortex and it is well known that granule cells are positioned according to a precise "outside-in" gradient (Angevine, 1965). If we consider that the neurogenesis of GABAergic cells in this region takes place when only a few granule cells are generated, one may speculate that a substantial number of principal neurons are needed to create a clear spatio-temporal gradient in the genesis of nonpyramidal cells.

Relation between birthdates and classes of nonpryamidal cells

To address this aim we first examined the area dentata as it is well-known that the main groups of nonpyramidal cells in this region can be recognized using immunocytochemical techniques for GABAergic cells (c.f., Seress and Ribak, 1983; Lübbers and Frotscher, 1987). The results showed that most GABAergic neurons in the mouse were labeled with 3H-thymidine within a period of two days, irrespective of their morphology and location. This indicates that basket cells, chandelier cells and other types of local circuit neurons are formed simultaneously in this region (Soriano et al., 1989b). However, because of the peculiar neurogenesis of the area dentata we cannot extrapolate this result to other subdivisions of the hippocampus. To explore this relationship further, by using a procedure similar to that described by Cobas et al., (1987) and Cobas and Fairén (1988), we carried out a morphometrical analysis of the GABAergic cells in the regio superior which were labeled with 3H-thymidine at different embryonic days (Soriano et al., in preparation). This study showed that neurons with a horizontal orientation are generally formed before those with a vertical orientation and before spheroidal cells (Fig. 2C). Also, larger neurons tend to become postmitotic first as illustrated in Fig. 2D. According to the "sandwich" gradient of neurogenesis, most basket cells and chandelier cells which are present in the pyramidal layer have a later period of neurognesis than the nonpyramidal cells with a less specialized axonal pattern located in the plexiform layers. However, the various morphological classes of GABAergic neurons in individual layers are labeled with 3H-thymidine during the same period

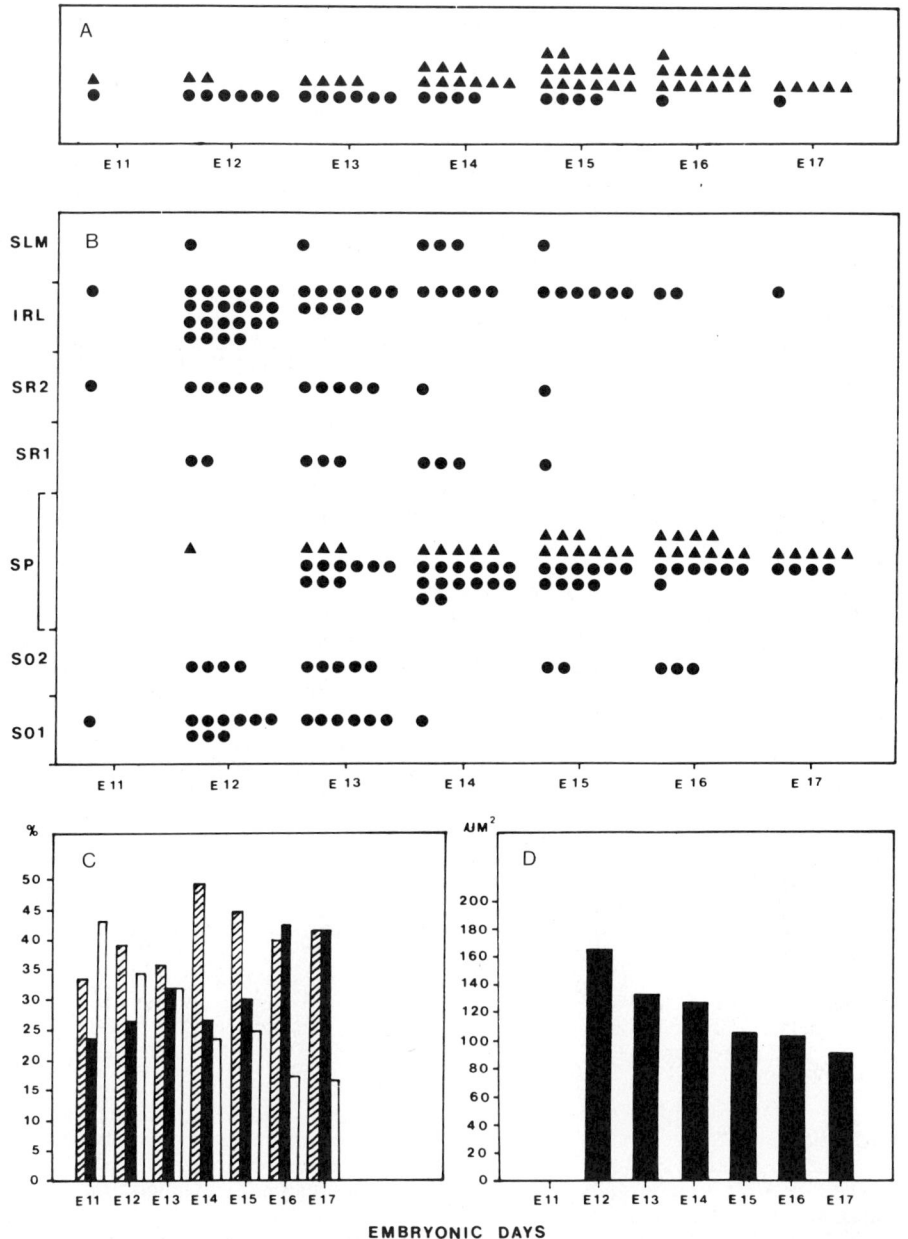

Figure 2. (a, b) Schematic representations illustrating aspects of the neurogenesis of GABAergic cells (dots) and pyramidal neurons (triangles) in the regio superior of the mouse hippocampus. Data modified from Soriano et al. (1986, 1989a). (a) Rates of neurogenesis in the regio superior. Each dot and triangle represents 5 and 20 neurons/mm², respectively. (b) Radial gradients of neurogenesis in the regio superior demonstrating the "sandwich" sequence for GABAergic cells. Each dot represents 1 GABAergic neuron/mm² in the plexiform layers, and 5 GABAergic neurons/mm² in the stratum pyramidale (sp); the onset of neurogenesis for the pyramidal neurons is represented by triangles (200 neurons/mm²). (c) Percentages of horizontal (white bars), vertical (black bars) and espheroidal (hatched bars) GABAergic neurons in the regio superior, after injections of 3H-thymidine at different embryonic days. (d) Mean size of GABAergic cell perikarya in the pyramidal layer of regio superior, labelled with autoradiographic silver grains at different embryonic ages. For abbreviations, see Fig. 1.

(Soriano et al., in preparation). Thus, we believe that the differences mentioned above are most probably related to the laminar peculiarities of the GABAergic neurons in the hippocampus and therefore to the "sandwich" radial sequence of neurogenesis.

Reduced rates of neuronal cell death?

Several studies have shown that the earliest-generated neurons in the neocortex are not destined to the cortical plate, but occupy layers just below and above it, i.e., the subplate and the marginal zone (Marin-Padilla, 1978; Luskin and Shatz, 1985; Shatz et al., 1989; see also De Carlos et al., this volume). The next neuronal generations reach the cortical plate which in the adult forms the neocortical gray matter, with the exception of layer I. It is also known that most of these early-generated cells have morphological and immunohistochemical features which are typical of nonpyramidal cells and that they cannot be detected in the adult neocortex because they disappear during the postnatal period, most probably by neuronal cell death (Marin-Padilla, 1978; Chun et al., 1987; Shatz et al., 1989; Valverde and Facal-Valverde, 1987, 1988; McConnell et al., 1989). These observations suggest that the early genesis of nonpyramidal cells following a "sandwich" spatio-temporal gradient may be a common principle in the early development of the hippocampus and neocortex. To see whether a similar process of neuronal cell death occurs in the hippocampal cortex, we have examined the ocurrence of pyknotic cells during the postnatal development of this region in the rat (Ferrer et al., 1989b). Whereas no neuronal cell death was found in the upper plexiform layers (stratum radiatum and lacunosum-moleculare), the stratum oriens displayed a moderate amount of pyknotic cells during the first postnatal week. These data confirm previous observations in the guinea pig hippocampus (Janowsky and Finlay, 1983) and suggest a less pronounced cell death when compared with the marginal and subplate zones of the neocortex. This reduced neuronal cell death may be responsible for the early neurogenesis of GABAergic cells and for the "sandwich" gradient they follow in the mature hippocampus (Amaral and Kurz, 1985; Lübbers et al., 1985; Soriano et al., 1986, 1989b). Several studies have in fact suggested that region-specific rates of neuronal cell death may play an important role during the differentation and cytoarchitectonic refinement of various cortical regions (Finlay and Slattery, 1983; Shatz et al., 1989; Ferrer et al., 1989a). We are now characterizing transitory populations of neurons in the neocortex and the hippocampus, in order to understand the reasons that determine the differential rates of neuronal cell death in the both cortices.

AKNOWLEDGEMENTS

The authors wish to thank A. Fairén and A. Cobas for their participation in part of the developmental studies, and the support of the Institute Cajal (CSIC, Madrid), INSERM U-106 (Paris), Institute of Anatomy of the University of Frankfurt, and Institute of Anatomy of the University of Lausanne, where parts of the work reviewed here were done. We specially thank Michael Frotscher for helpful comments on the manuscript. This work was supported by grants CAICYT/CSIC 85-154 to A. Fairén, and DGICYT PB87-139 and FIS 89-0317 to E. Soriano. J.A. del Rio is a fellow from FIS.

REFERENCES

Amaral, D. G. (1978) A Golgi study of cell types in the hilar region of the hippocampus in the rat. *J. Comp. Neurol.*, 182: 851.
Amaral, D. G., and Kurz, J. (1985) The time of originof cells demonstrating glutamic acid decarboxylase-like immunoreactivity in the hippocampal formation of the rat. *Neurosci. Lett.*, 59: 33.
Angevine, J. B., Jr. (1965) Time of neuron origin in the hippocampal region. An autoradiographic study in the mouse. *Exp. Neurol. Suppl.*, 2: 1
Blackstad, T. W. (1956) Commissural connections in the hippocampal region in the rat, with special reference to their mode of termination. *J. Comp. Neurol.*, 105: 417.
Cajal, S. R. (1911) *Histologie du Systéme Nerveux del' homme et des Vertebrés*. Vol. II. Paris: Maloine.
Chun, J. J. M., Nakamura, M. J., and Shatz, C. J. (1987) Transient cells of the developing mammaliam telencephalon are peptide-immunoreactive neurons. *Nature,* 325: 617.
Cobas, A., Welker, E., Fairén, A., Kraftsik, R., and Van der Loos, H. (1987) The GABAergic neurons in the barrel cortex of the mouse: an analysis using neuronal archetypes. *J. Neurocytol.*, 16: 843

Cobas, A., and Fairén, A. (1988) GABAergic neurons of different morphological classes are cogenerated in the mouse barrel cortex. *J. Neurocytol.*, 17: 511.

De Felipe, J., Hendry, S.H.C., and Jones, E. G. (1989) Visualization of chandelier cell axons by parvalbumin immunoreactivity in monkey cerebral cortex. *Proc. Nat. Acad. Sci. USA*, 86: 2093.

Fairén, A., De Felipe, J., and Regidor, J. (1984) Nonpyramidal cells: general account, In: *Cerebral Cortex*, Vol. I. Peters, A., and Jones, E.G., eds., Plenum press: New York.

Fairén, A., Cobas, A., and Fonseca, M. (1986) Times of generation of glutamic acid decarboxylase immunoreactive neurons in mouse somatosensory cortex. *J. Comp. Neurol.*, 251: 67.

Ferrer, I., Hernández-Martí, M., Bernet, E., Calopa, M. (1989a) Formation and growth of cerebral convolutions. II.- Cell death in the gyrus suprasylvius and adjoining sulci in the cat. *Dev. Brain Res.*, 45: 303.

Ferrer, I., Serrano, T., and Soriano, E. (1989b) Naturally occurring cell death in the subicular complex and hippocampus in the rat during development, Neurosci. Res., in press.

Finlay, B.L., and Slattery, M. (1983) Local differences in the amount of cell death in the neocortex predict adult locqal specializations. *Science*, 219: 1349.

Freund, T.F., and Antal, M. (1988) GABA-containing neurons in the septum control inhibitory interneurons in the hippocampus. *Nature*, 336: 170.

Freund, T.F., and Gulyás, A.I. (1989) Interneurons are the primary targets of GABAergic basal forebrain neurons innervating the neocortex. *European J. Neurosci., Suppl.* 2: 376.

Frotscher, M., Leranth, C., Lübbers, K. and Oertel, W.H. (1984) Commissural afferents innervate glutamate decarboxylase immunoreactive non-pyramidal neurons in the guinea pig hippocampus. *Neurosci. Lett.*, 46: 137.

Frotscher, M., Kraft, J., and Zorn, U. (1988) Fine structure of identified neurons in the primate hippocampus: A combined Golgi/EM study in the baboon. *J. Comp. Neurol.*, 275: 254

Frotscher, M. (1988b) Neuronal elements in the hippocampus and their synaptic connections, in: *Advances in Anatomy, Embryology and Cell Biology*, Vol. 111: Neurotransmission in the hippocampus, Frotsher, M., Kugler, P., Misgeld, U., and Zilles, K., eds., Springer-Verlag, Berlin.

Katsamaru, H., Kosaka, T., Heizmann, C.W., and Hama, H. (1988) Immunocytochemical study of GABAergic neurons containing the Calcium binding protein parvalbumin in the rat hippocampus. *Exp. Brain Res.*, 72: 347.

Kosaka, T., Hama, K. and Wu, J.Y. (1984) GABAergic synaptic boutons in the granule cell layer of the rat dentate gyrus. *Brain Res.*, 293: 353.

Kosaka, T., Kosaka, K., Tateishi, K., Hamaoka, I., Yanaihara, N., Wu, J-Y., and Hama, K. (1985) GABAergic neurons containing CCK-8-like and/or VIP-like immunoreactivities in the rat hippocampus and dentate gyrus, *J. Comp. Neurol.*, 239: 420.

Kosaka, T., Katsumaru, H., Hama, K., Wu, J.Y., and Heizmann, C. W. (1987) GABAergic neurons containing the Ca2+-binding protein parvalbumin in the rat hippocampus and dentategyrus. *Brain Res.*, 419: 119.

Lacaille, J.C., and Schwatzkroin, A. (1988) Stratum lacunosum-moleculare interneurons of hippocampal CA1 region. I: Intracellular response characteristics, synaptic responses and morphology. *J. Neurosci.*, 8: 1400.

Lorente de Nó, R. (1934) Studies on the structure of the cerebral cortex. II. Continuation of the study of the ammonic system. *J. Psychol. Neurol.* (Lpz), 46: 113.

Lübbers, K., and Frostcher, M. (1987) Fine structure and synaptic connections of identified neurons in the rat fascia dentada. *Anat. Embryol.*, 177:1.

Lübbers, K., Wolff, J.R., and Frotscher, M. (1985) Neurogenesis of GABAergic neurons in the rat dentate gyrus: A combined autoradiographic and immunocytochemical study. *Neurosci. Lett.*, 62: 317.

Luskin, M.B., and Shatz, C.J. (1985) Studies of the earliest generated cells of the cat's visual cortex: Cogeneration of subplate and marginal zones. *J. Neurosci.*, 5: 1062.

Marín-Padilla, M. (1978) Dual origin of the mammaliam neocortex and evolution of the cortical plate. *Anat. Embryol.*, 152: 109.

McConnell, S.K., Ghosh, A., and Shatz, C.J. (1989) Subplate neurons pioneer the 1st axon pathway from the cerebral cortex. *Science*, 245: 4921

Miller, M.W. (1985) Cogeneration of retrogradely labeled corticocortical projection and GABA-immunoreactive local circuit neurons in cerebral cortex. *Dev. Brain. Res.*, 23: 187.

Misgeld, U., and Frotscher, M. (1986) Postsynaptic-GABAergic inhibition of non-pyramidal neurons in the guinea-pig hippocampus. *Neuroscience*, 19: 193.

Nitsch, R., Soriano, E., and Frotscher, M. (1989) Parvalbumin immunoreactive cells in the rat hippocampus: electron microscopy and coexistence with GABA. *European J. Neurosci.*, Suppl. 2: 749.

Peters, A. (1984) Chandelier cells, In: Peters, A., and Jones, E.G., eds., *Cerebral Cortex*, Vol. 1. Plenum Press: New York.
Ribak, C.E., Vaughn, J.E., and Saito, K. (1978) Immunocytochemical localization of glutamic acid decarboxylase in neuronal somata following colchicine inhibition of axonal transport. *Brain Res.*, 140: 315.
Ribak, C.E., and Seress, L. (1983) Five types of basket cell in the hippocampal dentate gyrus: a combined Golgi and electron microscopic study. *J. Neurocytol.*, 12: 577.
Schaffer, K. (1892) Beitrag zu histologie der Ammonshornformation, Arch. Mikrosk. *Anat.*, 39: 611.
Schlander, M., and Frotscher, M. (1986) Non-pyramidal neurons in the guinea pig hippocampus. A combined Golgi-electron microscope study. *Anat. Embryol.*, 174: 35.
Seress, L., and Ribak, C.E. (1983) GABAergic cells in the dentate gyrus appear to be local circuit and projections neurons. *Exp. Brain Res.*, 50: 173.
Seress, L., and Ribak, C.E. (1985) A combined Golgi-electron microscopic study of non-pyramidal neurons in CA1 area of the hippocampus. *J. Neurocytol*, 14: 717.
Shatz, C.J., Chun, J.J.M., and Luskin, M.B. (1989) The role of the subplate in the development of the mammalian telencephalon, In: *Cerebral Cortex, Development and maturation of the cerebral cortex*, Vol. 7. Peters, A., and Jones, E.G., eds., Plenum Press: New York.
Sloviter, R.S., and Nilaver, G. (1987) Immunocytochemical localization of GABA-, cholecystokinin-, vasoactive intestinal polypeptide-, and somatostatin-like immunoreactivity in the area dentata and hippocampus of the rat. *J. Comp. Neurol.*, 256: 42.
Sloviter, R.S. (1989) Calcium-binding protein (Calbindin-D28K) and parvalbumin immunohistochemistry: Localization in the hippocampus with specific reference to the selective vulnerability of hippocampal neurons to seizure activity. *J. Comp. Neurol.*, 280: 183.
Somogyi, P., Nunzi, , M.G., Gorio, A., and Smith, A.D. (1983a) A new type of specific interneuron in the monkey hippocampus forming synapses exclusively with the axon initial segments of pyramidal cells. *Brain Res.*, 259: 137.
Somogyi, P., Smith, A.D., Nunzi, M.G., Gorio, A., Takagi, H., and Wu, J.Y., (1983b) Glutamate decarboxylase immunoreactivity in the hippocampus of the cat: Distribution of immunoreactive synaptic terminals with special reference to the axon initial segment of pyramidal neurons. *J. Neurosci*, 3: 1450.
Somogyi, P., Freund, T.P., Hodgson, A.J., Somogyi, J., Beroukas, D., and Chubb, I.W. (1985) Identified axo-axonic cells are immunoreactive for GABA in the hippocampus and visual cortex of the cat. *Brain Res.*, 332: 143.
Soriano, E., Cobas, A., and Fairén, A. (1986) Asynchronism in the neurogenesis of GABAergic and non-GABAergic neurons in the mouse hippocampus. *Dev. Brain Res.*, 30: 88.
Soriano, E., and Fariñas, I. (1987) Chandelier cells in the rat hippocampal region. *Neurosci. Suppl.*, 22: S124.
Soriano, E., and Frotscher, M. (1989) A GABAergic axo-axonic cell in the fascia dentata controls the main excitatory hippocampal pathway. *Brain Res.*, in press.
Soriano, E., Cobas, A., and Fairén, A. (1989a) Neurogenesis of glutamic acid decarboxylase immunoreactive cells in the hippocampus of the mouse. I: regio superior and regio inferior. *J. Comp. Neurol.*, 281: 586.
Soriano, E., Cobas, A., and Fairén, A. (1989b) Neurogenesis of glutamic acid decarboxylase immunoreactive cells in the hippocampus of the mouse. II: area dentata. *J. Comp. Neurol.*, 281: 603.
Soriano, E., Nitsch, R., and Frotscher, M. (1989c) Axo-axonic chandelier cells in the rat fascia dentata: Golgi-EM and immunocytochemical studies. *J. Comp. Neurol.*, 293: 1.
Swanson, L.W., Köhler, C., and Björklund, A. (1987) The limbic region. I. The septohippocampal system. In: *Handbook of Chemical Neuroanatomy, Integrated Systems of the CNS, Part I.*, Vol. 5: Björklund, A., Hökfelt, T., and Swanson, L.W., eds, Elsevier, Amsterdam.
Szentágothai, J., and Arbib, M.A. (1974) Conceptual models of neural organization. *Neurosci. Res. Program Bull.*, 12: 307.
Tömböl, T., Somogyi, G., and Hajdu, F. (1978) Golgi study on cat hippocampal formation. *Anat. Embryol.*, 153: 331.
Valverde, F. (1983) A comparative approach to neocortical organization based on the study of the brain of the hedgehog (Erinaceus europaeus). In: *Ramon y Cajal's Contribution to the Neurosciences*, Grisolia, S., Guerri, C., Samson, F., Norton S., and Reinoso-Suarez, F., eds., Elsevier, Amsterdam.
Valverde, F., and Facal-Valverde, M.V. (1987) Transitory populations of cells in the temporal cortex of kittens. *Dev. Brain Res.*, 32: 283.
Valverde, F., and Facal-Valverde, M.V. (1988) Postnatal development of interstitial (subplate) cells in the white matter of the temporal cortex of kittens. A correlated Golgi and electron microscopic study. *J. Comp. Neurol.*, 269: 168.

MORPHOLOGICAL CHARACTERIZATION OF ALZ-50 IMMUNOREACTIVE CELLS IN THE DEVELOPING NEOCORTEX OF KITTENS

De Carlos, J.A.*, López-Mascaraque, L.*, and Valverde, F.

Instituto de Neurobiología
Santiago Ramón y Cajal (CSIC)
Avenida del Doctor Arce 37
28002 Madrid, Spain

INTRODUCTION

Alz-50 is a monoclonal antibody isolated from brain tissue of patients with Alzheimer's disease (Wolozin et al., 1986). This antibody recognizes an antigenic protein of 68 kilodaltons in neurons involved in the formation of neuritic plaques and neurofibrillary tangles. It has been reported recently that Alz-50 immunoreactive (ir) neurons are found in normal fetal and neonatal human brains and in brain tissue from neonatal individuals with Down's syndrome (Wolozin et al., 1988).

In addition, Alz-50 ir neurons have been found in the subplate and cortical plate of neonatal rats (Al-Ghoul and Miller, 1989) in a pattern that changes with development (Hamre et al., 1989). Most subplate neurons die shortly after birth (Kostovic and Rakic, 1980; Luskin and Shatz, 1985; Valverde and Facal-Valverde, 1987, 1988). Different authors have suggested that these neurons may be the same as those showing Alz-50 immunoreactivity (Wolozin et al., 1988; Al-Ghoul and Miller, 1989).

Here we report preliminary observations on the morphological characteristics and distribution of Alz-50 ir neurons in the visual cortex of kittens of different postnatal ages. Morphological features of these neurons were compared with those of neurons previously described in the white matter of the temporal cortex of kittens (Valverde and Facal-Valverde, 1987, 1988).

MATERIAL AND METHODS

Material taken from the visual cortex of kittens sacrificed at postnatal days (P) 2, 4, 9, 12, 16 and 23 was used. The animals were perfused with either 4% paraformaldehyde or 10% formaldehyde in 0.1M phosphate buffer (pH 7.4). The brains were removed and immersed in the same fixative with 30% sucrose at 4°C. Frozen sections 30-40 μm thick were cut in the transverse plane and stored in 0.1M phosphate buffered saline (PBS). Free floating sections were incubated in 5% non-fat dry milk in PBS for 1 hour. Incubation in Alz-50 (IgM subtype), diluted from 1:10 to 1:100 in 0.5% non-fat dry milk in PBS, was carried out for 24-48 hours at room temperature under continuous gentle agitation. The sections were subsequently washed in PBS and incubated with rabbit anti-mouse IgG (BioMakor, 1:50) in PBS for 1 hour. Following three 5-min rinses in PBS, the sections were incubated in monoclonal mouse PAP (BioMakor, 1:1000) in PBS for 90 min. All sections were pretreated with 0.03% hydrogen peroxide plus 0.3% Triton X-100 in PBS.

---

* Present address: Department of Anatomy and Neurobiology, Washington University School of Medicine, 660 South Euclid Avenue, St. Louis, Mo. 63110, USA.

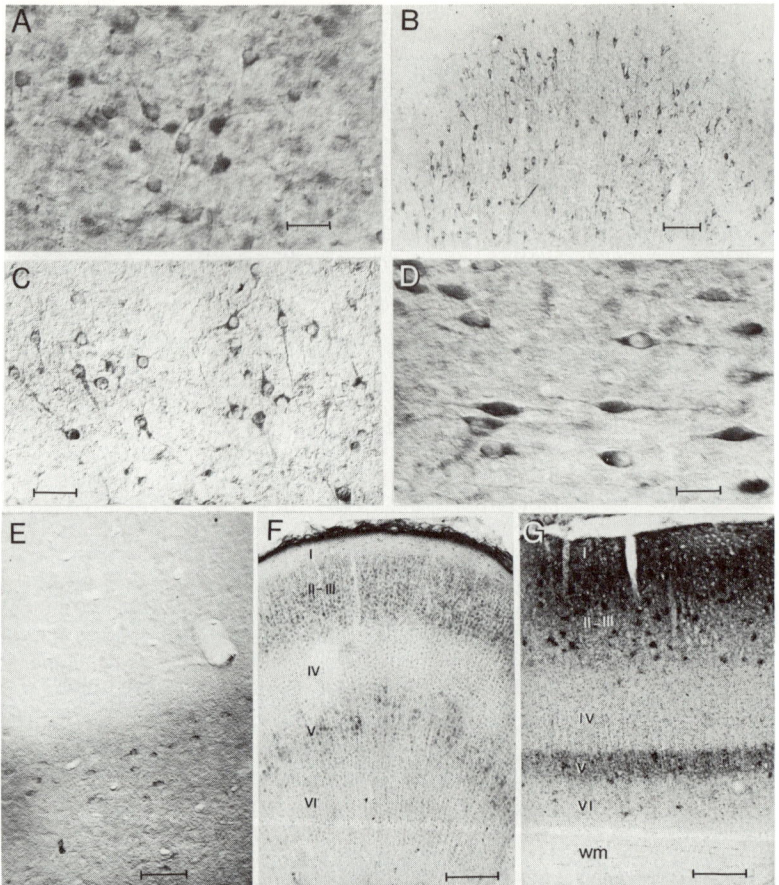

Figure 1. Morphology of Alz-50 immunoreactive cells at different postnatal ages (A-F) in the visual cortex of kittens. **A**, P2; **B**, P9; **C** and **D**, P23; **E**, P12; **F**, P4, counterstained section. **G**, Alz-50 immunoreactivity in a case of Alzheimer's disease in human visual cortex. Scale bars: A and C, 50 μm; D, 20 μm; B and E, 100 μm; F and G, 400 μm.

To visualize Alz-50 ir neurons, the sections were reacted for 4-5 min in a mixture of 0.03% 3'3 diaminobenzidine (DAB, Fluka) and 0.017% hydrogen peroxide in 50 mM Tris-ClH buffer (pH 7.6). After thorough rinsing in Tris-buffer, the sections were mounted on slides, dried, dehydrated and coverslipped with Dammar resin. Some sections were counterstained with cresyl violet to identify cortical layers. As control experiments, we incubated representative sections in the dilution buffer without primary antibody. In any case, no staining was found in these control experiments. Golgi sections from a large collection of brains of kittens from different ages were available for comparison.

We also tested the monoclonal antibody Alz-50 in neuropathologically confirmed cases of Alzheimer's disease of adult human visual cortex. Control adult human cerebral tissue (visual cortex) was obtained at autopsy from individuals with no clinical signs of impairment. The pieces were fixed by immersion in 10% formaldehyde in PBS for several months.

RESULTS

Labeling depended on the age of the kitten (Fig. 1, A-F) and on the area considered. Alz-50 ir cells on P2 were round multipolar and pyramid-like, restricted to the white matter (Fig. 1, A).

In some of them, we observed labeled dendritic processes. In the visual cortex at P4 the Alz-50 ir cells were localized mainly in layers II-III and V (Fig. 1, F). A small number of large, round, horizontally oriented fusiform cells were found in the subjacent white matter. In the anterior part of the visual cortex, we observed strong immunoreactivity including labeled cells in layer VI. Apical dendrites passing through layer IV were also observed. Interestingly, similar immunoreactivity is seen in P4 kittens and in Alzheimer's cases (compare Fig. 1, F and 1, G).

By P9 the processes become more differentiated (Fig. 1, B). Alz-50 ir cells were large and round, and sparsely distributed; others were multipolar and pyramid-like in form. At P12, a decrease in Alz-50 ir cells became apparent; the number was lower than in younger subjects (Fig.1, E). The morphology was similar to that reported previously. We observed fusiform cells in the depth of the sulci. At P16 the Alz-50 immunoreactivity was similar to that described in P12. However, at P23 an increase of the number of Alz-50 ir cells was noticed, and the cells exhibited clear immunoreactive somata and proximal segments of dendrites (Fig. 1, C). At this age, positive cells were two morphologically different populations. One population was found in the depth of the sulci, with fusiform morphology and horizontal distribution (Fig. 1, D), running parallel to fibers in the white matter. The other population was located in the deeper part of layer VI and underlying white matter at the crest of the gyri (Fig. 1, C). Here, the predominant cell morphology was round or pyramid-like and radially distributed. Furthermore, it is remarkable the striking morphological similarities of Alz-50 ir cells with the population of interstitial cells, which became manifest by the observation of pieces of kitten cortex impregnated by the Golgi method.

In human control tissue, Alz-50 ir cells were absent. However, in Alzheimer cases we observed a strong immunoreactivity in layers I-III and V, in neuritic plaques, neurofibrillary tangles and degenerated neuronal cell bodies (Fig. 1, G).

DISCUSSION

We selected the visual cortex to test Alz-50 immunoreactivity in kittens of different postnatal ages. In all cases, Alz-50 ir cells were found in the white matter. In addition, at P4, positive cells were also located in layers II-III and V. While at P4 the immunoreactivity is similar to that in human material with Alzheimer's disease, at all other ages, no label was found in any cortical layer, different from that described by Al-Ghoul and Miller (1989) in neonatal rat. In kittens from P2 to P16, the number of Alz-50 ir neurons decreased, as has been reported for the neonatal rat (Al-Ghoul and Miller, 1989; Hamre et al., 1989). However, there is a great increase in the number of Alz-50 ir cells at P23 in kittens.

Recent studies have shown that interstitial cells of the white matter below the cortical plate, the subplate cells, are the first neurons to differentiate. They correspond to certain cell types which have different transmitters and variable axonal morphologies (Wahle and Meyer, 1987, 1989; Valverde and Facal-Valverde, 1988; Chun and Shatz, 1989). The majority of them are transient and are probably eliminated by cell death (Kostovic and Rakic, 1980; Luskin and Shatz, 1985; Valverde and Facal-Valverde, 1987, 1988). Some persist in the adult forming the population of interstitial cells representing the remnants of the early generated subplate neurons (Chun and Shatz, 1989). Wolozin et al., (1988) and Al-Ghoul and Miller (1989) have suggested that Alz-50 may be a marker for neuronal death. Consistent with this hypothesis, we have found Alz-50 ir cells in the white matter in kittens which are likely to be subplate cells. However, the increase of the number of positive cells in P23 is difficult to explain. Although we have yet examined the brains of older kittens, it seems reasonable to suggest that Alz-50 may be expressed transiently in different developmental periods; first in a specific population of cells destined to disappear by cell death, and later in other cells for other reasons (Tsumoto et al., 1988). It is possible that Alz-50 may be re-expressed again in aged brains.

It has been reported that Alz-50 recognizes an abnormally phosphorylated protein Tau in Alzheimer disease (Grundke-Iqbal et al., 1986; Wood et al., 1986; Selkoe, 1989). Tau is a microtubule-associated protein implicated in microtubule assembly by phosphorylation-dephosphorylation mechanisms (Ksiezak-Reding et al., 1988). A likely function of this assembly is the active transport of different substances through the axon. Alz-50 might recognize either abnormally phosphorylated stages of Tau in Alzheimer disease or normal transitory stages of physiological phosphorylation of this protein, varying in activity throughout development and differentiation.

In human material, neuritic plaques and neurofibrillary tangles in Alzheimer's patients are preferentially located in cortical association areas and are more common in output cell-rich layers II, III and V (Pearson et al., 1985; Rogers and Morrison, 1985). However, we have found strong immunoreactivity in the visual cortex as well which might relate to the visual deficits in Alzheimer's patients. Visual cortical pathology had already been reported by Beach and McGeer (1988) with different methods. These authors have found astrocytic gliosis and senile plaques in layers II, III, IVa and IVc, while we observed heaviest Alz-50 immunoreactivity in layers I-III and V.

Further work will examine later postnatal days with convergent techniques to address the questions raised by this preliminary study.

ACKNOWLEDGEMENTS

This study has been supported by grant PB87-0412 (DGICYT) from the Ministerio de Educación y Ciencia. We thank Dr. P. Davies for the generous gift of Alz-50.

REFERENCES

Al-Ghoul, W.M., and Miller, M.W. (1989) Transient expression of Alz-50 immunoreactivity in developing rat neocortex: a marker for naturally occurring neuronal death? *Brain Res.*, 481: 361-367.
Beach, T.G., and McGeer, E.G. (1988) Lamina-specific arrangement of astrocytic gliosis and senile plaques in Alzheimer's disease visual cortex. *Brain Res.*, 463: 357 361.
Chun, J.J.M., and Shatz, C.J. (1989) Interstitial cells of the adult neocortical white matter are the remnant of the early generated subplate neuron population. *J. Comp. Neurol.*, 282: 555-569.
Grundke-Iqbal, I., Ibqal, K., Quinlan, M., Tung, Y.-C., Zaidi, M.S., and Wisniewski, H.M. 1986 Abnormal phosphorylation of the microtubule-associated protein tau in Alzheimer cytoskeletal pathology. *J. Biol. Chem.*, 261: 6084-6089.
Hamre, K.M., Hyman, B.T., Goodlett, C.R., West, J.R., and Van Hoesen, G.W. (1989) Alz-50 immunoreactivity in the neonatal rat: changes in development and co-distribution with MAP-2 immunoreactivity. *Neurosci. Lett.*, 98: 264-271.
Kostovic, I., and Rakic, P. (1980) Cytology and time of origin of interstitial neurons in the white matter in infant and adult human and monkey telencephalon. *J. Neurocytol.*, 9: 219-242.
Ksiezak-Reding, H., Davies, P., and Yen, S.-H. (1988) Alz-50, a monoclonal antibody to Alzheimer's disease antigen, cross-reacts with Tau proteins from bovine and normal human brain. *J. Biol. Chem.*, 263: 7943-7947.
Luskin, M.B., and Shatz, C.J. (1985) Studies of the earliest generated cells of the cat's visual cortex: cogeneration of subplate and marginal zones. *J. Neurosci.* 5: 1062-1075.
Pearson, R.C.A., Eisiri, M.M., Hiorns, R.W., Wilcock, G.K., and Powell, T.P.S. (1985) Anatomical correlates of the distribution of the pathological changes in the neocortex in Alzheimer's disease. *Proc. Natl. Acad. Sci. USA*, 82: 4531-4534.
Rogers, J., and Morrison, J.H. (1985) Quantitative morphology and regional laminar distribution of senile plaques in Alzheimer's disease. *J. Neurosci.*, 5: 2801-2808.
Selkoe, D.J. (1989) Biochemistry of altered brain proteins in Alzheimer's disease, *Ann. Rev. Neurosci.*, 12: 463-490.
Tsumoto, T., Sato, H., and Sobue, K. (1988) Immunohistochemical localization of a membrane-associated, 4.1-like protein in the rat visual cortex during postnatal development. *J. Comp. Neurol.*, 271: 30-43.
Valverde, F., and Facal-Valverde, M.V. (1987) Transitory population of cells in the temporal cortex of kittens. *Dev. Brain Res.*, 32: 283-288.
Valverde, F., and Facal-Valverde, M.V. (1988) Postnatal development of interstitial (subplate) cells in the white matter of the temporal cortex of kittens. A correlated Golgi and electron microscopic study. *J. Comp. Neurol.*, 269: 168-192.
Wahle, P., and Meyer, G. (1987) Morphology and postnatal changes of transient NPY-ir neuronal populations during early postnatal development of the cat visual cortex. *J. Comp. Neurol.*, 261: 165-195.
Wahle, P., and Meyer, G. (1989) Early postnatal development of vasoactive intestinal polypeptide- and peptide histidine isoleucine-immunoreactive structures in the cat visual cortex. *J. Comp. Neurol.*, 282: 215-248.

Wolozin, B.L., Pruchnicki, A., Dickson, D.W., and Davies, P. (1986) A neuronal antigen in the brains of Alzheimer patients. *Science*, 232: 648-650.

Wolozin, B.L., Scicutella, A., and Davies, P. (1988) Reexpression of a developmentally regulated antigen in Down syndrome and Alzheimer disease. *Proc. Natl. Acad. Sci., USA*, 85: 6202-6206.

Wood, J.G., Mirra, S.S., Pollock, N.J., and Binder, L.I. (1986) Neurofibrillary tangles of Alzheimer disease share antigenic determinants with the axonal microtubule-associated protein tau (t). *Proc. Natl. Acad. Sci. USA*, 83: 4040-4043.

GUIDANCE OF CHICK RETINAL AXONS IN VITRO

Jochen Walter and Friedrich Bonhoeffer

Max-Planck-Institut für Entwicklungsbiologie
Spemannstr. 35
D-7400 Tübingen, FRG

To understand the mechanism of axonal guidance, the behavior of growing axons has been studied in several *in vitro* systems (Gundersen and Barrett, 1979; Lumsden and Davies, 1983; Kapfhammer et al., 1986; for review see Bray and Hollenbeck, 1988).

In one such system, the so-called stripe assay, which has been described recently (Walter et al., 1987a), retinal axons grow on alternating stripes of two different but similar substrates. These consist of membrane vesicles prepared from anterior or posterior chick tectal tissue. The growing axons repeatedly encounter the situation where they have a choice between the anterior or the posterior substrate. The assay determines whether they grow preferentially on one of the two substrates. Chick temporal retinal axons (Fig.1a) but not nasal retinal axons (Fig.1b) show a distinct growth behavior at the border between the two substrates. They have a strong preference to grow on the lanes of anterior membranes and they do not cross the borderline between anterior and posterior substrates. This assay presents a very clear example of axonal guidance *in vitro*. The axons are guided by biological material which they also contact *in vivo*.

This position-specific difference of tectal membranes is developmentally regulated. Temporal axons are guided by material derived from tecta of embryonic day 5 to embryonic day 12 (Walter et al., 1987a). Membranes derived from older embryonic tecta are no longer active. This coincides well with the formation of the retinotectal projection in the chick where retinal axons invade the optic tectum from embryonic day 6 to 12 (Crossland et al., 1975; McLoon, 1985).

A similar preference of temporal retinal axons to grow on the anterior tectal membranes but not on posterior tectal membranes has been observed in two other species, in goldfish (Vielmetter and Stürmer, 1989) and mouse (Godement and Bonhoeffer, 1989). Whereas in mouse, membranes of the superior colliculus lose their guiding activity for temporal axons within the first postnatal week (Godement and Bonhoeffer, 1989) after an imprecise topographic map has been formed (O'Leary et al., 1986), goldfish tectal membranes of all ages guide temporal axons. This corresponds to a difference in the development of chick and mouse compared to goldfish; the goldfish continues to grow throughout its life and generates new ganglion cells which have to find their retinotopic target sites (Raymond and Easter, 1983).

An important indication for the biological relevance of the guidance observed in the stripe assay is the finding of Godement and Bonhoeffer (1989) that retinal axons of one species respond to tectal membranes derived from another one. For example chick temporal retinal axons choose between anterior and posterior mouse superior colliculus membranes. Thus, in both retina and tectum, positional markers exist which are probably similar in function and in their overall distribution in both species. This and the developmental regulation suggest a similarity of the *in vivo* and the observed *in vitro* guidance.

What makes the temporal axons grow only on one set of membranes and not on the other? The simplest explanation of the observed guidance phenomenon, i.e. that anterior material is permissive and posterior membranes are non-permissive for axonal growth, is probably wrong. As

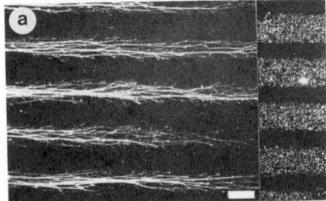 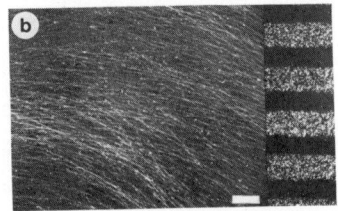

Figure 1. Growth pattern of temporal axons (a) and nasal axons (b) on a striped membrane carpet. The distribution of rhodamin-labeled axons is given on the left; the position of the posterior membrane stripes is given on the right, indicated by FITC-labeled beads which had been mixed among the posterior tectal membranes (Walter et al., 1987a). Anterior tectal membranes are filling the free space between the posterior tectal membranes. Temporal axons grow preferentially in the stripes of anterior tectal membranes and leave the stripes of posterior membranes free of neurites. Bar = 100μm.

shown in Fig. 2a, temporal axons do grow on posterior membranes (Walter et al., 1989). Outgrowth on posterior membranes seems to occur at a similar rate and with similar frequency as on the anterior membranes (Fig. 2b). Only when temporal retinal axons are offered the choice between anterior and posterior material simultaneously, then the axons do not grow onto the posterior membranes.

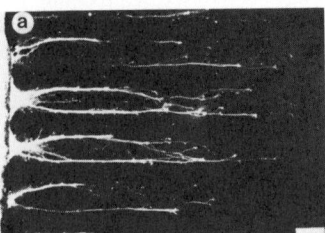 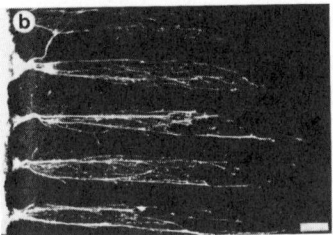

Figure 2. Growth of temporal retinal axons on posterior (a) and anterior (b) tectal membranes. Single stripes consisting of either anterior or posterior membrane fragments were separated by an interspace containing no substrate. Neurites have grown from left to right over a period of 28 hours. Temporal retinal axons reach the same length when growing on either type of substrate but tend to be more fasciculated when growing on posterior (a) than on anterior tectal membranes (b). Bar = 100μm.

Why then do temporal axons grow preferentially on anterior membranes as soon as they are given the choice of anterior and posterior membranes? To answer this question, one should know whether the anterior membranes show a specific attractivity or whether the posterior membranes contain a component which leads to the avoidance of the posterior lanes.

Experiments with heat inactivation of the membrane fragments have shown that a brief heat treatment of posterior membranes leads to a change in the growth pattern of temporal axons on the striped carpets so that they now cross the stripe borders freely and show no preference for the anterior stripes (Walter et al., 1987b). The result of such an experiment is depicted in Fig. 3a, b. A similar result is obtained when, instead of the unspecific heat inactivation, the posterior membranes are treated with PI-PLC, an enzyme which removes from the cell surface those proteins which are bound at the cell membrane by a phosphatidyl-inositol glycan anchor (Low and Saltiel, 1988). Treatment of posterior membrane vesicles with this enzyme does not affect their ability to support axonal growth, but it changes their properties in such a way that temporal axons cannot distinguish between anterior and posterior membranes (Walter et al., 1989). The avoidance reaction (Fig. 3c) is lost (Fig. 3d).

The conclusion of such experiments is that axons recognize at least two different components or sets of components in the stripe assay system. Molecules that are used as substratum for elongation seem to be equally distributed on both types of membranes. Posterior membranes contain in addition a component which is selectively avoided by temporal retinal axons.

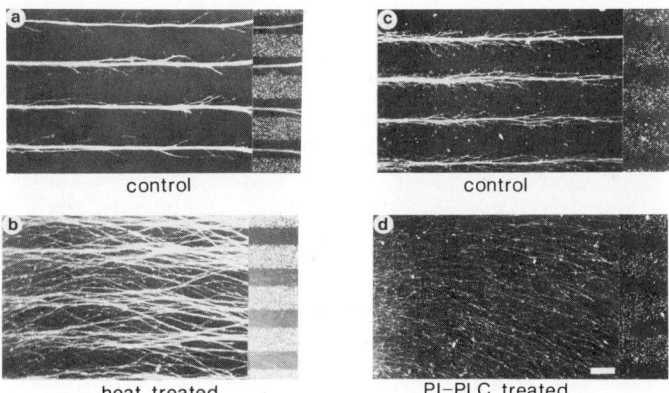

Figure 3. Loss of axonal guidance after inactivation of posterior tectal membranes by heat treatment (63°C, 8 min) (b) or PI-PLC treatment (1mU, 1 hour, 37°C) (d). Striped carpets were prepared of either untreated anterior and untreated posterior tectal membranes (a,c) or of untreated anterior membranes and treated posterior tectal membranes (b,d). The distribution of posterior material is indicated by FITC-labeled beads which have been mixed among the posterior membrane fragments. Axons from temporal retina grow only in the stripes of anterior tectal membranes in control carpets (a,c) but extend on both types of membranes in carpets consisting of anterior and treated posterior membranes (b,d). They show a slight preference for the heat-treated posterior membranes (b) as has been described earlier (Walter et al., 1987b). Bar = 100μm.

What is the mechanism of the avoidance reaction? A simple model of general growth inhibition is challenged by the observation that temporal retinal axons do grow on posterior membranes if they have no choice. In this case a mechanism based on inhibition would be possible if one assumes habituation so that growth cones being continuously exposed to the inhibitory component lose their responsiveness to it. However experiments performed by Müller (1988) indicate a mechanism which is not related to a change in the growth rate. He observed single growth cones growing along the border of anterior to posterior tectal membranes. The majority of temporal retinal growth cones are deflected from the lanes of posterior material without an obvious reduction of their growth rate.

Additional information on the interaction of growth cones with membrane components comes from the use of a new in vitro assay which has been introduced by Raper and Kapfhammer (1988, 1990). They found that growth cones of axons of dorsal root ganglion (DRG) explants growing on laminin react in a specific manner to membrane particles derived from the central nervous system. The growth cones lose their flat lamellipodial and filopodial structure, they "collapse" in a similar manner to the reaction of a DRG growth cone when it contacts a retinal neurite in vitro (Kapfhammer and Raper, 1987). Their "collapse assay" has been adapted to the retinotectal system of the chick by Cox et al. (1990). Retinal growth cones growing on laminin in vitro usually develop flat lamellipodia (Fig. 4a). The addition of posterior membranes to temporal retinal explants makes most axons retract. They lose their normal growth cone morphology and end in a blind tip (Fig. 4b).

Several lines of evidence indicate that the molecule active in the stripe assay and in the collapse assay is the same. Posterior but not anterior tectal membranes contain the collapsing activity; only temporal but not nasal retinal growth cones are affected (Cox et al., 1990). The collapsing activity is heat labile and developmentally regulated (Müller, personal communication) and both activities can be blocked by the same polyclonal antiserum (Stahl, personal communication).

Experiments were performed as described (Cox et al., 1990). The percentage of normal (Fig. 4a) and of collapsed (Fig. 4b) growth cones has been calculated. Numbers missing to 100% correspond to growth cones of an intermediate configuration. Membranes have been added (50μg

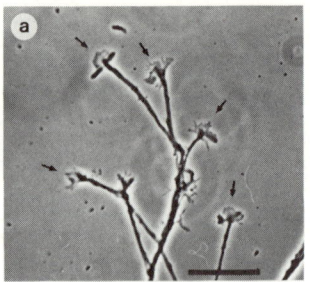 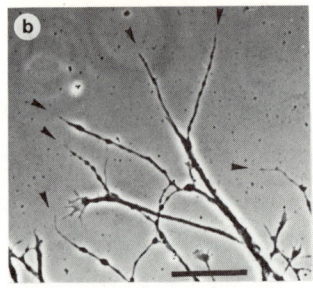

Figure 4. Morphology of normal (a) and collapsed growth cones (b). Retinal growth cones are growing on alaminin substratum. In control cultures (a) most axons end with lamillipodial growth cones (arrows). (b) Culture of temporal retinal axons 1 hour after the addition of posterior tectal membranes as described by Cox et al. (1990). Most axons end without a growth cone in a blind tip (arrow heads). Bar = 50 μm.

in 400μl culture medium) 1 hour before fixation. Inactivation of posterior membranes with PI-PLC (110 mU/ml for 1 hour at 37°C in culture medium) has been performed before the addition of membranes to temporal retinal explants.

The collapsing activity is also sensitive to the enzyme PI-PLC as summarized in Table I. Whereas temporal retinal axons in control cultures have predominantly lamellipodial growth cones (70%), the addition of posterior membranes results in only 10% normal growth cones. Pretreatment of the membranes in culture medium with the enzyme PI-PLC before the addition to the explants inactivates the collapsing activity of posterior membranes and the percentage of normal growth cones is similar to control values.

Table 1: PI-PLC inactivates the collapsing activity of posterior tectal membranes.

	Percentage of growth cones	
	Normal	Collapsed
Control	71.3 ± 7.1	10.2 ± 4.4
Posterior membranes	8.5 ± 5.5	79.9 ± 6.6
PI-PLC-inactivated posterior membranes	70.6 ± 7.7	15.5 ± 4.7

The component of posterior tectal membranes acting in the two assay systems seems to be identical relative to its biochemical characterization; the action to the growth cones seems not to be the same at the first glance.

In the collapse assay the growth cones stop growing when they are confronted with the posterior tectal material and this is accompanied by a dramatic change of the cyto-skeleton (Raper and Kapfhammer, 1990; Cox et al., 1990).

This can be interpreted in a way that the contact with the membranes leads to a destabilization of the lamellipodia. In the stripe assay, the local contact of only a small part of the growth cone leads to an avoidance reaction based on a partial lamellipodial destabilization, without an obvious influence on the growth rate. In the collapse assay, the larger amount of material in contact with all the growth cone leads to the complete destabilization of its lamellipodia - the growth cone becomes destabilized and collapses.

REFERENCES

Bray, D., and Hollenbeck, P.J. (1988) Growth cone motility and guidance. *Ann. Rev. Cell Biol.*, 4:43-61.
Cox, E.C., Müller, B., and Bonhoeffer, F.B. (1990) Axonal guidance in the chick visual system: Posterior tectal membranes induce collapse of growth cones from the temporal retina. Neuron 4: 31-37.

Crossland, W.J., Cowan, W.M. and Rogers, L.A. (1975) Studies on the development of the chick optic tectum. IV. An autoradiographic study of the development of retino-tectal connections. *Brain Res.*, 91:1-23.

Godement, P., and Bonhoeffer, F. (1989) Cross-species recognition of tectal cues by retinal fibers in vitro. *Development,* 106: 313-320.

Gundersen, R.W., and Barrett, J.N. (1979) Neuronal chemotaxis: Chick dorsal-root axons turn toward high concentrations of nerve growth factor. *Science,* 206: 1079-1080.

Kapfhammer, J.P., Grunewald, B.E. and Raper, J.A. (1986) The selective inhibition of growth cone extension by specific neurites in culture. *J. Neurosci.,* 6: 2527-2534.

Kapfhammer, J.P., and Raper, J.A. (1987) Collapse of growth cone structure on contact with specific neurites in culture. *J. Neurosci,* 7: 201-212.

Low, M.G., and Saltiel, A.R. (1988) Structural and functional roles of glycosyl-phosphatidylinositol in membranes. *Science*, 239: 268-275.

Lumsden, A.G.S., and Davies, A.M. (1983) Earliest sensory nerve fibres are guided to peripheral targets by attractants other than nerve growth factor. *Nature,* 306: 786-788.

McLoon, S.C. (1985) Evidence for shifting connections during development of the chick retinotectal projection. *J. Neurosci.*, 5: 2570-2580.

Müller, B (1988) Untersuchungen zum Wachstumsverhalten retinaler Ganglienzellaxone des Hühnchens auf Membranen des Zielgewebes. Diploma thesis, Tübingen, FRG.

O'Leary, D.D.M., Fawcett, J.W. and Cowan, W.M. (1986) Topographic targeting errors in the retinocollicular projection and their elimination by selective ganglion cell death. *J. Neurosci.,* 6: 3692-3705.

Raper, J.A. and Kapfhammer, J.P. (1990) A growth cone collapsing activity in embryonic chick brain membranes. *Soc. Neurosci. Abstr,.* 14: 596.

Raper, J.A., and Kapfhammer, J.P. (1989) The enrichment of a neuronal growth cone collapsing activity from embryonic chick brain. Neuron, 4, 21-29.

Raymond, P.A., and Easter, S.S. (1983) Postembryonic growth of the optic tectum in goldfish. I. Location of germinal cells and numbers of neurons produced. *J. Neurosci*, 3: 1077-1091.

Vielmetter, J., and Stuermer, C.A.O. (1989) Goldfish retinal axons respond to position-specific properties of tectal cell membranes in vitro. *Neuron,* 2: 1331-1339.

Walter, J., Kern-Veits, B., Huf, J., Stolze, B. and Bonhoeffer, F. (1987a) Recognition of position-specific properties of tectal cell membranes by retinal axons in vitro. *Development,* 101: 685-696.

Walter, J., Henke-Fahle, S. and Bonhoeffer, F. (1987b) Avoidance of posterior tectal membranes by temporal retinal axons. *Development,* 101: 909-913.

Walter, J., Müller, B. and Bonhoeffer, F. (1989) Axonal guidance by an avoidance mechanism. *J. Physiol.*, (Paris): in press.

CROSS-MODAL PLASTICITY IN SENSORY CORTEX
Visual Responses in Primary Auditory Cortex in Ferrets with Induced Retinal Projections to the Medial Geniculate Nucleus

Sarah L. Pallas

Department of Brain and Cognitive Sciences, E25-618
Massachusetts Institute of Technology
Cambridge, MA 02139
U.S.A.

INTRODUCTION

The evolution of the mammalian brain has involved marked degrees of encephalization, and this trend is particularly spectacular in the neocortex (see Jerison, Finlay, this volume). An important question in understanding neocortical evolution is how this expansion may be exploited by structures which form afferent connections with the expanded cortical populations. For example, what happens when additional cortical processing circuitry becomes available to sensory inputs as a result of mutation or duplication? Does the newly acquired circuitry replicate the existing mode(s) of information processing, or does it process sensory input in a new way? The latter change would be more likely to increase the animal's behavioral repertoire and hence reproductive"fitness". Certainly in the visual system, the number of separable visual cortical areas increases from hedgehogs to rats to cats and monkeys (Kaas et al., 1970; see Kaas, 1987 for review), and there is a large body of evidence that the different areas perform different transformations on their sensory input. Whether this segregation of function derives from differences inherent in cortical circuitry, or from parcellation of subtypes of afferent input (or both) is unknown from either an evolutionary or developmental perspective.

Another important aspect of cortical evolution involves the addition of new sensory modalities or sensory afferent subtypes (e.g., the addition of cones to an all-rod retina, or the subdivisions of auditory cortex in bats). Can a new afferent system impose a certain type of central processing on its cortical targets, or does it make use of cortical processing circuits already present? Rakic (this volume) presents evidence that excess cortical tissue (his area 'X') can be specified by its thalamic input into a new cytoarchitectonic area, but whether there are functional changes as well has not yet been addressed.

Cortical evolution has involved not only additive events, but also losses of afferent inputs. Examples include the deemphasis of olfactory structures in the human brain, or the loss of the lateral line system and electroreception in amniotes (see Fritzsch, this volume). Such regressive events potentially make brain space available that could be used for another, more adaptive purpose. Some recent investigations on blind mole rats are relevant in this regard. Blind mole rats (*Spalax ehrenbergi*) have only vestigial eyes, which lose their connections with the lateral geniculate nucleus (LGN) and the superior colliculus during development (Bronchti et al., 1989b). As a result, the LGN and occipital cortex receive auditory input (Bronchti et al., 1989a,c). It is unknown from where this input arises, or whether it represents a stabilization of early exuberant projections. The mole rats represent a natural evolutionary variation where afferents of one modality have 'colonized' vacant target tissue that is normally of a different modality.

What we need to know before we can address these evolutionary issues is whether different sensory cortices actually differ in their intrinsic circuitry, and if so, why. As sensory information passes from thalamus to cortex, it is usually transformed in its representation. For example, in the visual system, concentric receptive fields in lateral geniculate give way to elongated, oriented fields in cortex. The type of transformation appears quite different in cortices of different sensory modalities. Topography of representation also differs between different cortical areas, both within and between modalities, with different maps emphasizing different aspects of the stimulus or different parts of the sensory epithelium.

These differences could arise for three reasons. First, different primary sensory cortical areas may not differ in their processing circuitry, but rather perform the same operations on all their inputs, regardless of modality. In this case, the operations would only appear different because the afferent modality differs. Second, specific thalamic inputs could induce unique cortical processing circuitry. The elegant studies of Henrik Van der Loos and his colleagues (reviewed in this volume) have certainly shown that afferent input can have a profound effect on structural aspects of cortex. It would not be surprising if functional changes occur as well. Finally, sensory cortices might develop unique processing circuits independent of thalamic input. Some major features of visual system organization, such as LGN lamination, are not prevented by bilateral enucleation (Brunso-Bechtold & Casagrande, 1981). Similarly, mice that are congenitally anopthalmic have topographically normal projections from the lateral geniculate nucleus to area 17 of visual cortex (Kaiserman-Abramof et al., 1980). These results argue that at least some aspects of cortical circuitry are organized without afferent influence.

The evolutionary changes in brain organization that I have described require permissive developmental mechanisms. For this reason, it is appropriate to study brain development in extant species in order to understand brain evolution. In Mriganka Sur's lab, we have approached the question of sensory cortical specification directly by a developmental diversion of afferent information of one modality into cortex which normally processes a different modality. In the following, I will both review our previous work on this preparation and summarize some preliminary results from our recent work.

Retinal afferents can be induced to project to auditory thalamic targets during development

Schneider (1973) and Frost (1981; this volume) have shown in the hamster that by reducing the normal targets of the retina, and deafferenting either auditory or somatosensory thalamus, retinal afferents can be induced to project or to maintain exuberant projections to the deafferented thalamus. We have pursued this observation in ferrets. Ferrets are carnivores (Order Carnivora, Family Mustelidae) with a well-differentiated visual system quite similar to the cat's, but ferrets are born much earlier in gestation (ferrets have a 41 day gestation vs. 65 days in the cat). For this reason, they are useful for manipulations of the developing nervous system.

The neonatal manipulation is schematized in Fig. 1. On the day of birth, ferret pups are anesthetized by hypothermia, and the superior colliculus and visual cortex are ablated. The lateral geniculate nucleus largely degenerates as a result of the cortical lesion, as it does in cats and rodents (Perry & Cowey, 1979; Pearson et al., 1981; Raabe et al., 1986). The auditory thalamus (MGN) is deafferented by cutting the fiber pathway from the inferior colliculus. These manipulations allow the retinal afferents to grow into the MGN, which they do in a patchy fashion. These anomalous retino-MGN projections, in contrast to retinal inputs to somatosensory thalamus in hamsters (Frost, 1981), are not a result of the stabilization of early transient collaterals but represent sprouting of the axons into new territory (Linden et al., 1981; Hahm & Sur, unpubl.). This sprouting is undoubtedly facilitated by the proximity of the optic tract to the MGN as it travels on the lateral surface of the diencephalon. In these "rewired" ferrets, the MGN projects to the primary auditory cortex (AI) as it does in normal animals, but as a result of the lesions, it carries visual rather than auditory information.

Comparison of primary auditory and primary visual cortex

In order to generate hypotheses about how AI might handle visual information from MGN, it is necessary to examine the similarities and differences between normal visual and auditory cortex. The differences which are normally present will be especially important in determining whether afferents can influence processing circuitry in their target.

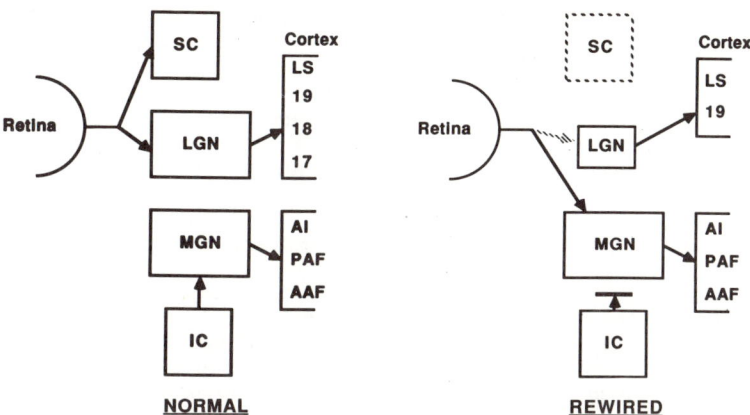

Figure 1. Neonatal manipulation to induce retinal projections to MGN. Ablation of areas 17 and 18 of visual cortex causes massive retrograde degeneration in the dorsal lateral geniculate nucleus. This, coupled with lesion of the superior colliculus, removes or reduces severely the normal target space for the retinal ganglion cell axons. Deafferentation of the medial geniculate nucleus by section of the brachium from the inferior colliculus makes terminal space available there. These manipulations allow the retinal afferents to grow into the MGN, which then carries visual information to the auditory cortex.

There are a number of similarities in the extrinsic connectivity patterns of primary visual and primary auditory cortex (Swadlow, 1983; Gilbert & Wiesel, 1979; Ferster & Lindstrom, 1983; Imig & Reale, 1980; Mitani & Shimokouchi, 1985; Mitani et al., 1985; Winguth & Winer, 1986). Thalamic input terminates primarily in layer IV, subcortical efferents arise from layers V and VI, and intercortical and callosal efferents arise from layer III in both cases. Also, intrinsic connectivity in visual cortex resembles that in auditory cortex, in that layer II/III and layer IV are reciprocally connected, and layer II/III also projects to layer V. These similarities would suggest that processing circuitry in the two areas is also similar.

In contrast to similarities in radial connectivity, afferent input is mapped quite differently between primary visual (area 17) and primary auditory cortex (AI). In both cases the map reflects the sensory epithelium, but the cochlea is a unidimensional surface, in contrast to the two-dimensional photoreceptor array. As a result most visual cortical areas contain a two-dimensional retinotopic map whereas AI has a one-dimensional cochleotopic map. The second dimension in AI is an isofrequency representation. The isofrequency axis is subdivided into two to three sets of binaural bands (groups of cells which get either excitatory input from both ears, or excitatory from one and inhibitory input from the other ear), each of which receives input from each of two to three binaural bands in MGN (Middlebrooks & Zook, 1983; Middlebrooks et al., 1980). This projection is highly convergent and divergent and is not a map of space. Numerous attempts to demonstrate a synthetic space map in auditory cortex have failed, though such maps have been demonstrated at lower levels (Knudsen & Konishi, 1978; Middlebrooks & Knudsen, 1984; King & Hutchings, 1987).

Particularly interesting for the purposes of this study are the transformations in stimulus representation that occur between thalamus and cortex. Some receptive field properties in visual cortex have obvious correlates in auditory cortex, and others do not. Orientation selectivity is very characteristic of primary visual cortex, but because it depends on a two-dimensional sensory epithelium, it has no strict correlate in AI. Direction selectivity, on the other hand, is analogous to sensitivity to the direction of an FM sweep, which has been demonstrated in AI (Mendelson & Cynader, 1985). Velocity selectivity in the visual domain would correspond to sensitivity to the rate of an FM sweep, which has been reported in the cat (Mendelson & Cynader, 1985) and gerbil auditory cortex (Scheich, this volume). End-inhibition, the decline in responsiveness of some cortical cells when an oriented bar is extended beyond a certain length, is similar in some respects to 2-tone inhibition or bandwidth sharpening in auditory cortex. As a last example, binocularity combines input from two eyes at the cortical level. AI has binaural information, but input from the two ears is first combined in the brainstem and not in cortex (Goldberg & Brown, 1969). The information is not spatially mapped in AI, but is highly divergent and convergent (Middlebrooks & Zook, 1983).

Given these differences and similarities between AI and visual cortex, we can predict how visual input might be handled by AI. Again, what we are ultimately interested in is whether the properties of sensory neocortex are determined by its intrinsic organization or by specific inputs during development.

A particular subset of retinal ganglion cell axons project to the deafferented MGN

One major advantage of the ferret for this study is the similarity of its visual system to that of the cat. The well-defined subclasses of retinal ganglion cells in cat and ferret allow us to establish which group(s) are responsible for the plasticity we observe, and then to compare their normal properties with those in the new target tissue. We have several lines of evidence that the retinal input to MGN in the rewired ferrets derives largely from the W class of retinal ganglion cells. First, it has been shown in the cat that most X cells die as a result of extensive lesions of area 17 in visual cortex (Pearson et al., 1981; Tong et al., 1982; Sur et al., 1987). The LGN is markedly shrunken in our lesioned ferrets, and the number of medium sized somata in the retina declines (Roe et al., 1987). The X cells probably die because they are singly-targeted; they have only one axon collateral which projects to the LGN, whereas Y and W cells also project to the superior colliculus (see Sherman & Spear, 1982, for review). Second, HRP injections in MGN of the lesioned ferrets indicate that only ganglion cells with small somata are backfilled and thus project to the MGN (Roe et al., 1987). The class of ganglion cells with the smallest somata has been correlated with W cell physiological characteristics. Third, recordings of postsynaptic responses of visual cells in MGN to optic chiasm stimulation show that they have long latency responses, large receptive fields, and they are sluggish and readily habituating. All of these properties are characteristic of W cells (Sur & Sherman, 1982; Sur et al., 1988). Finally, recent evidence from bulk-filling of retinal axons in the optic tract of normal and neonatally lesioned ferrets (Pallas et al., 1989) shows that retinal terminal arbors in the MGN do not resemble X or Y cells. The axon trunks in the MGN are often very thin (1-5 μ), and the arbors are sparse and small (Fig. 2) with many enpassant swellings and crenulated endings. Many fibers look quite immature. Some do not terminate in the MGN but course through it, (probably on their way to the remaining fragment of LGN or superior colliculus) forming many swellings along the way. The retinal axon arbors that do terminate in MGN often occur in clusters, as predicted by the patchiness of the terminal label following eye injections. It is difficult to demonstrate that the axon arbors are W-like, because of the lack of data in the literature concerning morphology of arbors in cells identified physiologically as W in character. One characteristic of the axons in MGN that is typical of W axons, however, is the fine caliber of the axon trunks (Mason & Robson, 1979; Fukuda et al., 1984; Stanford, 1987). We are now studying arbor morphology of retinal ganglion cells in normal ferret superior colliculus and the C-lamina of the LGN, areas known to receive largely W-cell input (Sur & Sherman, 1982; Leventhal et al., 1985), so that we can compare them with the ganglion cell arbors in MGN.

Response properties of visual cells in AI

Our recordings of response latencies in AI to optic chiasm stimulation are consistent with our interpretation that W cells provide the major input to AI in rewired ferrets. The average latency in AI is 6.6 ms, compared to 4 ms in area 17. This is short enough to suggest a monosynaptic input from MGN. Our anatomical data show that the MGN provides the major input to AI in the rewired ferrets (Pallas et al., 1988 and in press).

We have recently tested the responses of single visually responsive cells in AI to various visual stimuli (Roe, Pallas, Kwon & Sur, unpublished; see also Sur et al., 1988). As in the MGN, responses to visual stimuli in AI are sluggish and rapidly habituating. Receptive fields are much larger than those in area 17 in normal ferrets, and often have diffuse edges. We find many receptive field properties which are normally characteristic of visual cortex (Table 1). Cells have either non-oriented, circular receptive fields, or oriented, rectangular fields. Some also show direction selectivity. In most of the oriented cells the tuning is broader than that seen in area 17 of visual cortex, but some cells are well-tuned to orientation and/or direction. The orientation-selective cells resemble complex cells in that there are usually no subfields of ON or OFF type response; rather the field is uniform. Many cells will also respond to flashing lights.

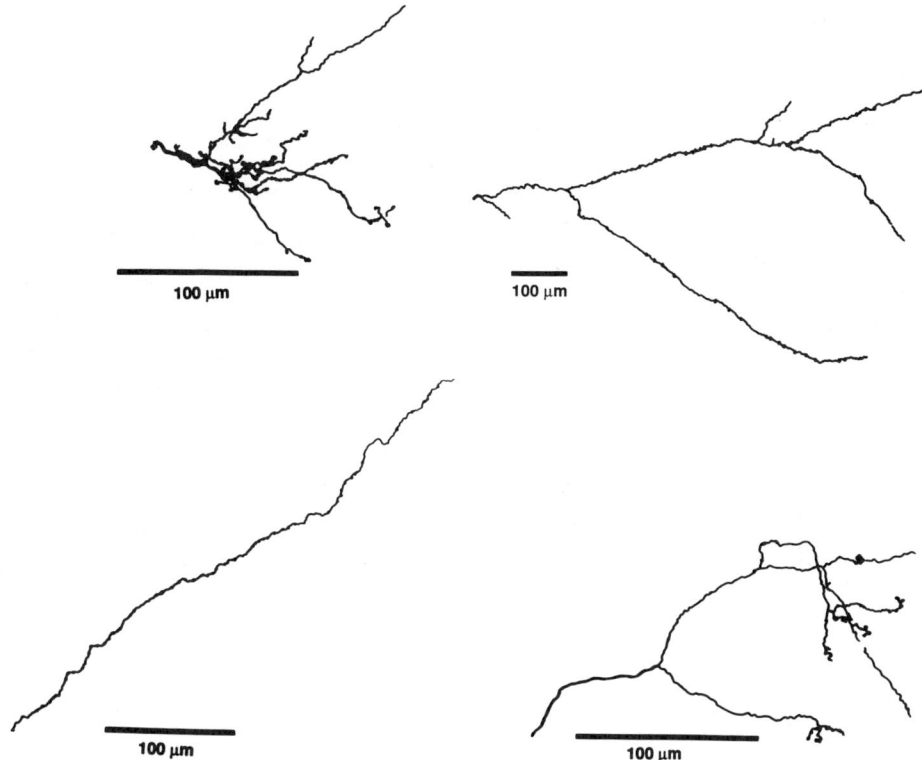

Figure 2. Camera lucida reconstructions of retinal axons in the medial geniculate nucleus. In these sagittal sections, dorsal is up and caudal is to the right. In all cases the optic tract is to the left. Three of the axons shown have their terminal arbors in MGN. These arbors are sparse and small with *en passant* swellings and crenulated endings. Most of the axons course through the MGN without making terminal arbors, as the one shown in the lower left. Note the differing scales.

Velocity tuning is usually marked, but there is a large range of preferred velocities between different cells in AI. Occasionally we see binocular cells, and a few cells are driven only by the ipsilateral eye. This is particularly interesting because the lesions are unilateral. We have investigated the source of the ipsilateral input in only one animal to date. Although the data are preliminary, injections of different tracers in each of the eyes in this animal clearly showed that the ipsilateral eye also projects to MGN (unpub. results). This raises the possibility that ocular dominance columns could form in AI in the lesioned ferrets.

TABLE 1. Summary of receptive field properties of visually-driven cells in primary auditory cortex.

Property	*Incidence*	*Percent*
Oriented	33/94	35%
Directional	7/17	41%
Binocular	13/67	19%
Ipsilateral	3/67	4%
End-inhibited	3/17	18%

The responses in AI of the rewired ferrets are different in some respects from visual responses in normal area 17. However, to compare the responses of visual cells in AI to those in 17 is somewhat inappropriate. Area 17 gets input from both X and Y classes of retinal ganglion cells, and receives little if any W input, whereas AI in the rewired ferrets apparently gets most of its input from W cells. Therefore, the appropriate comparison would be with a cortical area which receives pure W cell input directly from the LGN. Unfortunately, little information is available in the literature about cortical cells with W input. One response property which has been correlated with W cell input is end-inhibition. Area 19 receives much of its input from W cells (Dreher et al., 1980; Kimura et al., 1980) and end-inhibition is very common in Area 19 (Duysens et al., 1982a). Although we have not yet systematically tested for this receptive field property, we have recorded from end-inhibited visual cells in AI on a couple of occasions. The large receptive fields and long latencies we see in the visual cells in AI are also characteristic of Area 19, as is the reduced selectivity for direction and orientation (Dreher et al., 1980; Duysens et al., 1982a, b).

These results demonstrate that some of the transformations in stimulus representation that occur in normal thalamocortical pathways in the visual system can also occur in AI of the rewired ferrets. The presence of orientation selectivity and complex receptive fields in AI suggests that afferent input can influence the function of cortical cells in at least some respects. However, the possibility that all sensory neocortices handle afferent input in a similar way, regardless of modality, is not ruled out. Direction selectivity, velocity selectivity, and binocularity all have potential correlates in the auditory cortex.

Mapping of visual input in AI

Because the mapping of the sensory epithelium is one-dimensional in normal AI and two-dimensional in visual cortex, we were interested to find out how visual input would map in AI. Kelly and colleagues (Kelly et al., 1986; Phillips et al., 1988) have shown that the frequency map in AI of ferrets is oriented such that high frequencies are represented medially and low frequencies laterally. The orthogonal dimension is the isofrequency axis. Because of this arrangement, it seemed likely to us that the visual input to AI would be specific in only one dimension, leading to a one-dimensional map of space in AI. To test this hypothesis, we systematically mapped the surface of AI in the rewired ferrets with visual stimuli.

We have found, contrary to our expectations, that the neonatal lesions result in a two-dimensional map of visual space in AI (Roe et al., 1988). Figure 3 is a cartoon showing the form of the retinotopic map in AI. In the AI map shown in Fig. 3A, visual field elevation increased from caudal to rostral and azimuth increased from medial to lateral. The maps tend to be much more regular along the tonotopic axis (medial to lateral) than along the isofrequency axis, but in most animals there is nevertheless a general progression of visual field locations along this axis as one maps anterior to posterior. The maps can show local discontinuities, and even local reverses in receptive field sequence. In another ferret, the map of azimuth still increased in a roughly medial to lateral direction, but the map of elevation was completely reversed (Fig. 3B). I will return to this point below.

Closer analysis of the maps has shown that they are fairly linear. As one moves roughly along an isoelevation line, azimuths change in a linear manner. The same is true for elevations. This would be expected for a W cell map, because the distribution of W cells in the retina is fairly flat. Y and particularly X cells map in a more non-linear fashion, emphasizing area centralis. This idea is supported by an analysis of magnification factors. Magnification (mm of cortex/degree of visual field) remains fairly constant as the map is traversed along each isoelevation line, again supporting the idea that W cells provide the major input.

Anatomical basis for the visual map in AI

The fact that a two-dimensional map can be imposed on a cortical surface that normally maps in only one dimension is intriguing. How does this map arise? It is possible that it comes indirectly from the retina via a retinotopic map in MGN. It is also possible that the map is derived from some other visual structure which projects directly to AI as a result of the early lesions. A third possibility is that the map (or at least its second dimension) arises *de novo* in AI.

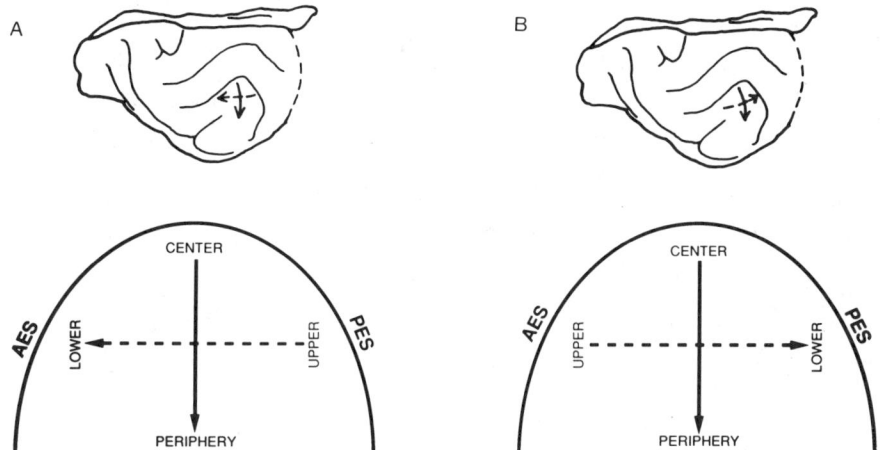

Figure 3. Visual field maps in adult ferrets with retinal inputs to MGN. The surface of AI was mapped with an electrode in these two animals to determine the nature of the retinotopic map. A). In this ferret, as shown by the solid arrows and lines, azimuth (central taperipheral visual field) increased from medial to lateral. Elevation, as shown by the dashed arrows and lines, changed from lower to upper visual field as the electrode was advanced rostrally. B). In the second ferret shown here, azimuth still increased in a roughly medial to lateral direction, but the elevation map was reversed as compared to A. AES and PES are the anterior and posterior ectosylvian sulci, respectively.

In consideration of the first possibility, it is important to remember that the nature of the MGN to AI projection is distinctly different from the LGN to area 17 projection. The highly divergent and convergent MGN to AI projection pattern is in contrast to the more point-to-point type projection from LGN to area 17. Merzenich, Middlebrooks and their colleagues (Middlebrooks and Zook, 1983; also see Merzenich et al., 1984, for review) have strong evidence in the cat that each point in MGN projects in a divergent way to an isofrequency lamina in AI, and neighboring points along an isofrequency lamina in MGN project in an overlapped fashion to the same isofrequency lamina in AI. In this way, a small area of MGN can influence large areas of cortex. Cells in two different isofrequency laminae in MGN, in contrast, project to non-overlapping laminae in AI.

We have demonstrated (Pallas et al., 1988 and in press) that the normal thalamocortical projections in the auditory system of the ferret are similar to those in the cat. Injections of neuroanatomical tracers (HRP, Fluoro-Gold, Fast Blue, and Rhodamine-labelled microspheres) in AI were made to describe the pattern of the geniculocortical projection. In most animals, more than one tracer was used in order to determine the topography of afferent inputs to AI.

In the normal brain schematized in Fig. 4, two different dyes were injected in different locations in AI. Retrograde label was heavy in MGN and the posterior thalamic group (PO). As in the cat, input to AI arises from slabs of cells in MGN extending rostrocaudally through the nucleus (not shown; see Andersen et al., 1980). The fact that the two labels backfilled non-overlapping slabs of cells in MGN suggests that the injection sites were in different frequency laminae.

When we did the same experiment in the rewired ferrets, we found that, as in normal ferrets, input to AI arises from rostrocaudally oriented slabs in MGN (Fig. 4). Also, as in normal ferrets, injections in two different AI locations can produce segregated slabs of label in MGN depending on where the injections are located with respect to the isofrequency axis. Note, however, that we cannot identify isofrequency bands in an AI that lost its auditory input at birth.

These results suggest that the nature of the MGN-AI projection changes little or not at all as a result of the early lesions, and that the MGN projects in its usual slab-to-slab pattern to AI. As a result, it is unlikely that the second dimension of the visual map in AI in the rewired ferrets arises from the thalamocortical projection from MGN. This leaves two possibilities for the source of the map: mapped visual input to AI from other sources, or a map that arises within AI.

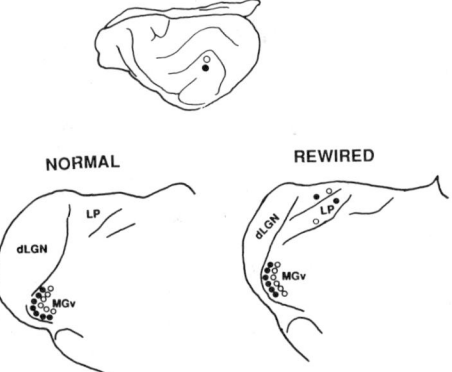

Figure 4. Pattern of thalamocortical projections shown by tracer injections in AI. In normal ferrets, as in cats, slabs or laminae of cells in the MGN, which extend in the rostrocaudal direction, project to single loci in AI. Injections in different sites, if they are located in different isofrequency laminae in AI, label non-overlapping laminae in MGN. In the rewired ferrets, this organization is unchanged. However, there are sparse additional inputs to AI in the rewired ferrets from the LP/Pulvinar region of the dorsal thalamus.

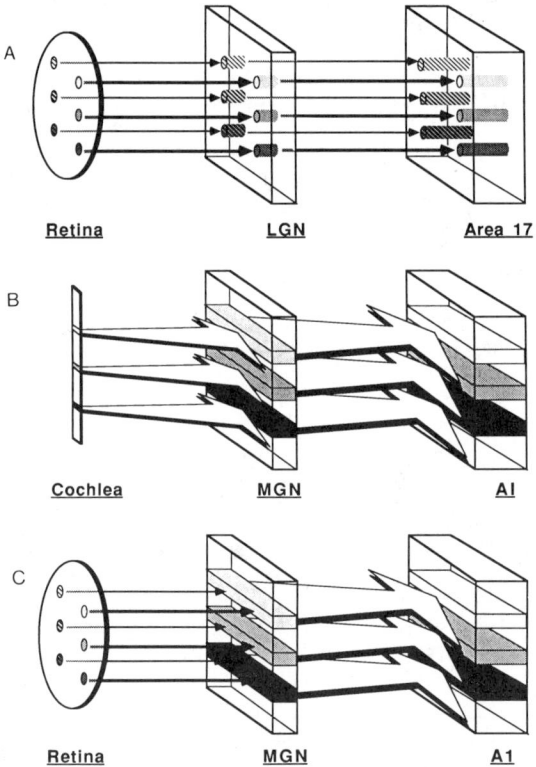

Figure 5. Summary of thalamocortical projections in normal visual system (A), auditory system (B), and the retina-MGN-AI projection in the rewired ferrets (C). The normal visual system has a roughly point-to-point type projection pattern, while the normal auditory system, reflecting the one-dimensional nature of the cochlea, projects in a slab-to-slab pattern. In the rewired ferrets, a two-dimensional retinal input is imposed on MGN, but MGN's projection to AI remains highly overlapped. We do not know the nature of the retina to MGN projecitons at this time, but in the schematic, we have assumed that the retinal terminals project to MGN in register. The nature of the thalamocortical projection in the rewired ferrets suggests that the second dimension of the visual map is created in AI (see text). Note that in the schematic we have simplified the connection from the cochlea to MGN, which is actually indirect.

Effect of the early lesions on thalamocortical and corticocortical connectivity patterns of AI

In the rewired ferrets in which tracers were injected in AI, there were cells backfilled in the dorsal part of the thalamus that are not present in normal ferrets (Fig. 4). These cells are in the lateral posterior (LP) nucleus as well as the pulvinar and lateromedial-suprageniculate complex. This result is particularly interesting because thalamocortical connectivity shows little exuberance or sensitivity to target removal during normal development (Dawson & Killackey, 1985; Rakic, 1976; Crandall & Caviness, 1984). Patterns of terminal label from Fluoro-Gold and HRP injections showed that these projections are reciprocal. The dorsal part of the thalamus also receives anomalous inputs from the retina in these animals, so it is possible that it could impose a visual map on AI. The LP/Pulvinar complex does contain several maps of visual space (Hutchins & Updyke, 1989). However, the input is much sparser than that from MGN and PO, and the retinal terminals in LP/Pul are mainly dorsal to the somata which project to AI (Pallas et al., 1988 and in press). Because of this, it seems unlikely that these dorsal thalamic nuclei influence the map in AI to any significant extent. In contrast, there is heavy overlap between retinal terminals in MGN and MGN cells which project to AI. We saw no evidence of other anomalous subcortical inputs or outputs of AI in the rewired ferrets.

We have also looked for sources of mapped visual input to AI in other cortical structures. We find that, although there is an expansion of a weak projection from medial cortex to AI (which is reciprocal) in the rewired ferrets, there is no input from visual cortical areas (Pallas et al., 1988). By the process of elimination, it seems probable that the visual map in AI arises from circuitry within AI itself. In our lesioned ferrets, even though two-dimensional input is imposed on MGN, the inputs to AI from MGN remain highly divergent (Fig. 5). Because of this, creation of the second dimension of the AI visual map must occur by functional selection of a subset of the anatomical inputs. This selection process could occur by means of activity-dependent selection via correlated activity in neighboring retinal locations. This type of mechanism has been suggested by Merzenich et al. (1984) and others (see also Edelman & Finkel, 1984) on the basis of experiments in the somatosensory cortex, where a large number of anatomical inputs seem to be suppressed under normal circumstances, but can be unmasked by appropriate stimulation. This argument is supported by the fact that the retinotopic map in AI can sometimes switch polarity from animal to animal along the isofrequency axis. This switching is suggestive of a dynamic process underlying map formation.

As I have described, we have now studied three aspects of primary auditory cortex in the rewired ferrets: receptive field properties, topography, and extrinsic connectivity. In contrast to the changes we have described in response properties and topography, the thalamocortical, corticocortical, and subcortical connections of AI in the rewired animals are quite similar to the connections in normal animals.

SUMMARY AND DISCUSSION OF FINDINGS

The findings from our experiments can be summarized as follows:

1) In ferrets, retinal afferents can be induced to make projections to auditory thalamus (MGN and PO) by removing normal visual targets and deafferenting MGN at birth.
2) These afferents appear to be composed primarily of W cell axons, and are composed of at least two arbor types; those that make *en passant* boutons on their way through MGN, and those that have very sparse arbors in MGN.
3) Cells in MGN acquire visual response properties as a result of the early lesions that are characteristic of LGN-W cells (long latency, sluggish responses).
4) We have studied the differences and similarities between normal visual cortex and visually-responsive auditory cortex along three dimensions: response properties, topography, and connectivity.

   a) Cells in primary auditory cortex acquire some new properties as a result of the lesions– such as orientation/direction selectivity, velocity selectivity, binocularity, and complex re ceptive fields– which are normally characteristic of the visual system.
   b) The neonatal lesions result in a two-dimensional map of visual space in AI. The MGN to AI thalamocortical projection retains its normal slab-to-slab, one-dimensional pattern, suggesting that the second dimension of the visual map in AI arises from circuitry intrinsic to AI. This could occur by functional selection of a subset of inputs, as suggested by Edelman & Finkel (1984).

c) Changes in extrinsic connectivity of AI as a result of the lesions are minor. There is an expansion of normal weak inputs to AI from medial cortex, and there are anomalous inputs to AI from the dorsal part of the thalamus, but these inputs are extremely sparse.

As I have described above, visual input to AI provides this cortex with properties normally characteristic of visual cortex. What do these results suggest about the development of sensory cortices of different modalities? Are primary auditory cortex and primary visual cortex fundamentally different, and if so, how do they become different? The fact that we can create receptive field properties typical of visual cortex in AI of the rewired ferrets seems incompatible with the idea that there are inherent differences between different primary sensory cortices, at least along the dimensions we have examined. This means that either afferents induce differences during development, or there are in fact no differences in processing circuitry between these two areas.

How is it possible to distinguish between these two hypotheses? It is necessary to reexamine the basic similarities and differences in auditory versus visual cortex. In some respects, auditory cortex is not different enough from visual cortex to make this distinction. Parallels can be drawn between the two for most response properties. The experiments of Frost (this volume) in diverting retinal afferents into somatosensory thalamus also suffer from this difficulty, perhaps even more so because both visual and somatosensory cortex contain two-dimensional space maps. However, there are some properties which seem not to have parallels in auditory cortex, such as orientation selectivity and simple/complex-type receptive fields. It is difficult to imagine a correlate for orientation selectivity in the auditory system, because orientation requires a two-dimensional sensory surface. There is evidence that the isofrequency axis in AI contains mapped information on bandwidth and FM rate (Schreiner and Cynader, 1984; Scheich, this volume), but these properties are not strict analogs of a second spatial dimension. Increases in bandwidth are equivalent to an elongation along the sensory epithelium, but because the cochlea is one-dimensional, this elongation can occur only along one orientation (up or down the frequency scale). Simple and complex-type receptive fields have not been demonstrated in AI, perhaps because they also depend on mapping of a single parameter (space) in two dimensions. The creation of these properties in AI of the lesioned ferrets may depend simply on the developmental induction of a two-dimensional visual map.

CONCLUSIONS

We can conclude that afferents are capable of imposing dimensionality on their sensory cortical targets, and that visual receptive field properties that appear in primary auditory cortex are a result of either 1) afferent induction of the cortical circuitry responsible for transformations in stimulus representation, or 2) afferent utilization of existing cortical circuitry to produce modality-specific transformations. What implications do these conclusions have for cortical development and evolution?

If all sensory cortices perform similar transformations on their input, then evolutionary expansion of total cortical area or addition/subtraction of modalities poses little difficulty. Afferent availability itself is not a limiting factor in cortical cell survival beyond a certain minimal requirement (see Finlay & Pallas, 1989, for review). Genetic specification would not be required beyond its identification as a sensory cortex. Simple duplication of a cortical area or an extra round of cell division, coupled with subsequent differentiation of afferent input into submodalities or a new modality entirely, would provide the substrate for a new cortical area. This interpretation is supported by the fact that there are commonalities between visual and auditory receptive field properties, and by the fact that intrinsic cortical circuitry is quite similar between the two areas.

If our other hypothesis is correct, and afferents can induce cortical circuitry, then at what level, and how is this achieved? If the function of cortex is simply to enhance differences between incoming stimuli, then this possibility is not substantially different from a situation where all sensory cortices have the same intrinsic circuitry. Afferents might influence cortex to the extent that each modality (and each environment) provides cortex with a different set of stimuli. The cortex would become wired up on the basis of that set of stimuli alone, detecting similarities and differences between those inputs. Afferent influence is supported by our finding that the dimensionality of the sensory epithelium can be imposed on cortex. Additional indirect support comes from the profound effects of monocular deprivation or exposure to restricted visual stimuli on cortical response properties (Wiesel & Hubel, 1963, 1965; Singer, 1977; Wilson & Sherman, 1977; Shatz & Stryker, 1978). One way to test this idea further would be to look for subtle differences in intrinsic cortical connectivity in normal visual and auditory cortex. If afferent input can influence cortical circuitry to a significant degree, then we should be able to detect changes in intracortical projection patterns in the rewired ferrets.

A critical point in the interpretation of our results is the timing of the lesions. While we were able to induce changes in topographic mapping and receptive field properties in auditory cortex by switching the input modality, we were not able to change the extrinsic connectivity pattern of AI to a significant extent. This could indicate that there are hard-wired inherent differences in extrinsic circuitry between AI and visual cortex. Alternatively, it is possible that the lesions were made too late for afferents to have an influence over extrinsic connectivity. There is evidence that callosal, subcortical and thalamocortical connections are established before intracortical connections. Shatz and colleagues have shown that a population of cells in the cortical subplate (Luskin & Shatz, 1985a, b; Chun et al., 1987), which appear before the generation of the six cortical layers, make connections with callosal and thalamic targets at very early developmental stages (Chun et al., 1987; Chun & Shatz, 1988, 1989). McConnell et al. (1989) have suggested that these early projections pioneer the efferent cortical pathways, even before the thalamocortical axons elongate to their position below the cortical plate (Lund & Mustari, 1977; Rakic, 1977; Wise et al., 1977; Innocenti, 1981; Shatz & Luskin, 1986), and long before the generation of the cortical layers. The thalamocortical afferents and cortical efferents would then follow this pathway as they elongate. This hypothesis is supported by recent evidence that ablation of the subplate cells prevents thalamic axons from reaching the appropriate cortical target (Ghosh et al., 1989). It is not known how specific these early pioneer pathways are. In addition to the pathways characteristic of adult cortex, they may also pioneer the early, exuberant callosal and subcortical projections. Perhaps they provide the cortex with a limited 'menu' of possible projections, which can be pruned back later in development by collateral elimination (Innocenti et al., 1977; Innocenti & Caminiti, 1980; Innocenti, 1980; Ivy & Killacky, 1982; Stanfield & O'Leary, 1985; Stanfield et al., 1982), but outside of which projections are prevented.

On the day of birth in the ferret (comparable to E42 in the cat), when we make our lesions, the infragranular layers of cortex are migrating into position (Jackson et al., 1989), and thalamocortical afferents have not quite reached the cortical plate. Thus, although the thalamocortical and intracortical connections are not yet formed, intercortical and subcortical connectivity patterns may already be specified by the subplate pioneers. This could explain why the modality switch in our experiments apparently causes changes in intracortical circuitry (and minor changes in thalamocortical circuitry) while failing to affect cortical efferent projections. The changes in corticocortical connectivity which do occur (the expansion of the projections to and from medial cortex) appear to arise from a weak normal projection. The substrate for the expansion could be the stabilization of an early exuberant projection that is normally lost by collateral elimination.

Ultimately, I think it is quite possible that both hypotheses, afferent induction of cortical circuitry and uniformity of different sensory cortical areas, are valid. Some operations, such as difference detection and enhancement, may be performed by all sensory cortices, and the important evolutionary event is the differentiation of afferent modalities or submodalities that can take advantage of these commonalities. Other operations, such as topographic mapping, ocular dominance column formation, or the distribution of orientation selectivity, may require afferent control because of their unpredictability. Neither mechanism poses constraints for cortical expansion and differentiation, but in fact acts as a potentiating force.

ACKNOWLEDGEMENTS

I am indebted to Anna Roe, Jong-on Hahm, Mriganka Sur, and Young Kwon for the use of their unpublished data. Paul Katz, Anna Roe, Jong-on Hahm, and Mriganka Sur provided insightful comments on the manuscript. Thanks also to Teresa Sullivan for technical assistance with histology and photography, and to Paul Katz for help with illustrations. The work reported here was supported by an NIH postdoctoral grant EY 06121 to Sarah Pallas and NIH grant EY 07719 and a McKnight Foundation grant to Mriganka Sur.

REFERENCES

Andersen, R.A., Knight, P.L. and Merzenich, M.M. (1980) The thalamocortical and corticothalamic connections of AI, AII, and the anterior auditory field (AAF) in the cat: Evidence for two largely segregated systems of connections. *J. Comp. Neurol.*, 194: 663-701.

Bronchti, G., Heil, P., Scheich, H. and Wollberg, Z. (1989a) Auditory pathway and auditory activation of primary visual targets in the blind mole rat (*Spalax ehrenbergi*): I. 2-deoxyglucose study of subcortical centers. *J. Comp. Neurol.*, 284: 253-274.

Bronchti, G., Heil, P., Scheich, H. and Wollberg, Z. (1989c) Auditory activation of cortical visual fields in the blind mole rat. *Proc. 2nd Intl. Cong. Neuroethol.*, Abstract # 147.

Bronchti, G., Rado, R., Terkel, J. and Wollberg, Z. (1989b) Ontogenetic degeneration of retinal projections in the blind mole rat (*Spalax ehrenbergi*). *Proc. 2nd Intl. Cong. Neuroethol.*, Abstract # 201.

Brunso-Bechtold, J.K. and Casagrande, V.A. (1981) Effect of bilateral enucleation on the development of layers in the dorsal lateral geniculate nucleus. *Neuroscience*, 6: 2579-2586.

Chun, J.J.M., Nakamura, M.J. and Shatz, C.J. (1987) Transient cells of the developing mammalian cerebral telencephalon are peptide-immunoreactive neurons. *Nature*, 325: 617-620.

Chun, J.J.M. and Shatz, C.J. (1988) Redistribution of synaptic vesicle antigens is correlated with the disappearance of a transient synaptic zone in the developing cerebral cortex. *Neuron*, 1: 297-310.

Chun, J.J.M. and Shatz, C.J. (1989) The earliest-generated neurons of the cerebral cortex: Characterization by MAP2 and neurotransmitter immunohistochemistry during fetal life. *J. Neurosci.*, 9: 1648-1667.

Crandall, J.E. and Caviness, V.S. (1984) Thalamocortical connections in newborn mice. *J. Comp. Neurol.*, 228: 542-556.

Dawson, D.R. and Killackey, H.P. (1985) Distinguishing topography and somatotopy in the thalamocortical projections of the developing rat. *Dev. Brain Res.*, 17: 309-313.

Dreher, B., Leventhal, A.G. and Hale, P.T. (1980) Geniculate input to cat visual cortex: a comparison of area 19 with areas 17 and 18. *J. Neurophysiol.*, 44: 804-826.

Duysens, J., Orban, G.A., van der Glas, H.W. and de Zegher, F.E. (1982a) Functional properties of Area 19 as compared to Area 17 of the cat. *Brain Res.*, 231: 279-291.

Duysens, J., Orban, G.A., van der Glas, H.W. and Maes, H. (1982b) Receptive field structure of Area 19 as compared to Area 17 of the cat. *Brain Res.*, 231: 293-308.

Edelman, G.M. and Finkel, L.H. (1984) Neuronal group selection in the cerebral cortex. In G.M. Edelman, W.E. Gall, and W.M. Cowan (eds.): *Dynamic Aspects of Neocortical Function*. New York, NY: Neurosciences Research Foundation, pp. 653-695.

Ferster, D. and Lindstrom, S. (1983) An intracellular analysis of geniculo-cortical connectivity in Area 17 of the cat. *J. Physiol.*, (Lond.) 342: 181-215.

Finlay, B.L. and Pallas, S.L. (1989) Control of cell number in the developing mammalian visual system. *Prog. Neurobiol.*, 32: 207-234.

Frost, D.O. (1981) Orderly anomalous retinal projections to the medial geniculate, ventrobasal, and lateral posterior nuclei of the hamster. *J. Comp. Neurol.*, 203: 227-256.

Fukuda, Y., Hsiao, C.-F., Watanabe, M. and Ito, H. (1984) Morphological correlates of physiologically identified Y-, X-, and W-cells in cat retina. *J. Neurophysiol.*, 52: 999-1013.

Ghosh, A., Antonini, A., McConnell, S.K. and Shatz, C.J. (1989) Ablation of subplate neurons alters the development of geniculocortical axons. *Soc. Neurosci. Abstr.*, 15: 960.

Gilbert, C.D. and Wiesel, T.N. (1979) Morphology and intracortical projections of functionally characterized neurones in the cat visual cortex. *Nature*, (Lond.) 280: 120-125.

Goldberg, J.M. and Brown, P.B. (1969) Response of binaural neurons of dog superior olivary complex to dichotic tonal stimuli: Some physiological mechanisms of sound localization. *J. Neurophysiol.*, 32: 613-636.

Hahm, J. and Sur, M. (1988) The development of individual retinogeniculate axons during laminar and sublaminar segregation in the ferret LGN. *Soc. Neurosci. Abstr.*, 14: 460.

Hutchins, B. and Updyke, B.V. (1989) Retinotopic organization within the lateral posterior complex of the cat. *J. Comp. Neurol.*, 285: 350-398.

Imig, T.J. and Reale, R.A. (1980) Patterns of cortico-cortical connections related to tonotopic maps in cat auditory cortex. *J. Comp. Neurol.*, 192: 293-332.

Innocenti, G.M. (1981) Growth and reshaping of axons in the establishment of visual callosal connections. *Science*, 212: 824-827.

Innocenti, G.M. and Caminiti, R. (1980) Postnatal shaping of callosal connections from sensory areas. *Exptl. Brain Res.*, 38: 381-394.

Innocenti, G.M., Fiore, L. and Caminiti, R. (1977) Exuberant projection into the corpus callosum from the visual cortex of newborn cats. *Neurosci. Lett.*, 4: 237-242.

Ivy, G.O. and Killackey, H.P. (1982) Ontogenetic changes in the projections of neocortical neurons. *J. Neurosci.*, 2: 735-743.

Jackson, C.A., Peduzzi, J.D. and Hickey, T.L. (1989) Visual cortex development in the ferret. I. Genesis and migration of visual cortical neurons. *J. Neurosci.*, 9: 1242-1253.

Kaas, J.H. (1987) The organization of neocortex in mammals: Implications for theories of brain function. *Ann. Rev. Psychol.*, 38: 129-151.

Kaas, J.H., Hall, W.C. and Diamond, I.T. (1970) Cortical visual areas I and II in the hedgehog: Relation between evoked potential maps and architectonic subdivisions. *J. Neurophysiol.*, 33: 595-615.

Kaiserman-Abramof, I.R., Graybiel, A.M. and Nauta, W.J.H. (1980) The thalamic projection to cortical area 17 in a congenitally anopthalmic mouse strain. *Neuroscience*, 5: 41-52.

Kelly, J.B., Judge, P.W. and Phillips, D.P. (1986) Representation of the cochlea in primary auditory cortex of the ferret (*Mustela putorius*). *Hearing Res.*, 24: 111-115.

Kimura, M., Shiida, T., Tanaka, K. and Toyama, K. (1980) Three classes of area 19 cortical cells characterized by their neuronal connectivity and photic responsiveness. *Vision Res.*, 20: 69-77.

King, A.J. and Hutchings, M.E. (1987) Spatial response properties of acoustically responsive neurons in the superior colliculus of the ferret: a map of auditory space. *J. Neurophysiol.*, 57: 596-624.

Knudsen, E.I. and Konishi, M. (1978) A neural map of auditory space in the owl. *Science*, 200: 795-797.

Leventhal, A.G., Rodieck, R.W. and Dreher, B. (1985) Central projections of cat retinal ganglion cells. *J. Comp. Neurol.*, 237: 216-226.

Linden, D.C., Guillery, R.W. and Cucchiaro, J. (1981) The dorsal lateral geniculate nucleus of the normal ferret and its postnatal development. *J. Comp. Neurol.*, 203: 189-211.

Lund, R.D. and Mustari, M.J. (1977) Development of the geniculocortical pathway in rats. *J. Comp. Neurol.*, 173: 289-306.

Luskin, M.B. and Shatz, C.J. (1985a) Studies of the earliest generated cells of the cat's visual cortex: cogeneration of subplate and marginal zones. *J. Neurosci.*, 5: 1062-1075.

Luskin, M.B. and Shatz, C.J. (1985b) Neurogenesis of the cat's primary visual cortex. *J. Comp. Neurol.*, 242: 611-631.

Mason, C.A. and Robson, J.A. (1979) Morphology of retino-geniculate axons in the cat. *Neuroscience*, 4: 79-97.

McConnell, S.K., Ghosh, A. and Shatz, C.J. (1989) Subplate neurons pioneer the first axon pathway from the cerebral cortex. *Science*, 245: 978-982.

Mendelson, J.R. and Cynader, M.S. (1985) Sensitivity of cat primary auditory cortex (AI) neurons to the direction and rate of frequency modulation. *Brain Res.*, 327: 331-335.

Merzenich, M.M., Jenkins, W.M. and Middlebrooks, J.C. (1984) Observations and hypotheses on special organizational features of the central auditory nervous system. In G.M. Edelman, W.E. Gall, and W.M. Cowan (eds.): *Dynamic Aspects of Neocortical Function.* New York, NY: Neurosciences Research Foundation, pp. 397-424.

Merzenich, M.M., Knight, P.L. and Roth, G.L. (1975) Representation of cochlea within primary auditory cortex in the cat. *J. Neurophysiol.*, 38: 231-249.

Middlebrooks, J.C., Dykes, R.W. and Merzenich, M.M. (1980) Binaural response-specific bands in primary auditory cortex (AI) of the cat: Topographical organization orthogonal to isofrequency contours. *Brain Res.*, 181: 31-48.

Middlebrooks, J.C. and Knudsen, E.I. (1984) A neural code for auditory space in the cat's superior colliculus. *J. Neurosci.*, 4: 2621-2634.

Middlebrooks, J.C. and Zook, J.M. (1983) Intrinsic organization of the cat's medial geniculate body identified by projections to binaural response-specific bands in the primary auditory cortex. *J. Neurosci.*, 3: 203-224.

Mitani, A. and Shimokouchi, M. (1985) Neuronal connections in the primary auditory cortex: An electrophysiological study in the cat. *J. Comp. Neurol.*, 235: 417-429.

Mitani, A., Shimokouchi, M., Itoh, K., Nomura, S., Kudo, M. and Mizuno, N. (1985) Morphology and laminar organization of electrophysiologically identified neurons in the primary auditory cortex in the cat. *J. Comp. Neurol.*, 235: 430-447.

Pallas, S.L., Hahm, J.-O. and Sur, M. (1989) Retinal axon arbors in a novel target: Morphology of ganglion cell axons induced to arborize in the medial geniculate nucleus of ferrets. *Soc. Neurosci. Abstr.*, 15: 495.

Pallas, S.L., Roe, A.W. and Sur, M. (1988) Retinal projections induced into auditory thalamus in ferrets: Changes in inputs and outputs of primary auditory cortex. *Soc. Neurosci. Abstr.*, 14: 460.

Pallas, S.L., Roe, A.W. and Sur, M. (in press) Visual projections induced into the auditory pathway of ferrets: I. Novel inputs to primary auditory cortex (AI) from the LP/Pulvinar complex and the topography of the MGN-AI projection. *J. Comp. Neurol.*

Pearson, H.E., Labar, D.R., Payne, B.R., Cornwaell, P. and Aggarwal, N. (1981) Transneuronal retrograde degeneration in the cat following neonatal ablation of visual cortex. *Brain Res.*, 212:470-475.

Perry, V.H. and Cowey, A. (1979) The effects of unilateral cortical and tectal lesions on retinal ganglion cells in rats. *Exp. Brain Res.*, 35: 97-108.

Phillips, D.P., Judge, P.W. and Kelly, J.B. (1988) Primary auditory cortex in the ferret (*Mustela putorius*): neural response properties and topographic organization. *Brain Res.*, 443: 281-294.

Raabe, J.I., Windrem, M.S. and Finlay, B.L. (1986) Control of cell number in the developing visual system. II. Visual cortex ablation. *Devel. Brain Res.*, 28: 1-11.

Rakic, P. (1976) Prenatal genesis of connections subserving ocular dominance in the rhesus monkey. *Nature,* 261: 467-471.

Rakic, P. (1977) Prenatal development of the visual system in rhesus monkey. *Phil. Trans. Roy. Soc.*, (Lond.) B 278: 245-260.

Roe, A.W., Garraghty, P.E. and Sur, M. (1987) Retinotectal W-cell plasticity: Experimentally induced retinal projections to auditory thalamus in ferrets. *Soc. Neurosci. Abstr.*, 13: 1023.

Roe A.W., Pallas, S.L., Hahm, J., Kwon, Y.H. and Sur, M. (1988) Retinal projections induced into auditory thalamus in ferrets: Visual topography in primary auditory cortex. *Soc. Neurosci. Abstr.*, 14: 460.

Schneider, G.E. (1973) Early lesions of the superior colliculus: Factors affecting the formation of abnormal retinal projections. *Brain, Behav. Evol.*, 8: 73-109.

Schreiner, C.E. and Cynader, M.S. (1984) Basic functional organization of second auditory cortical field (AII) of the cat. *J. Neurophysiol.*, 51: 1284-1305.

Shatz, C.J. and Luskin, M.B. (1986) The relationship between the geniculocortical afferents and their cortical target cells during development of the cat's primary visual cortex. *J. Neurosci.*, 6: 3655-3668.

Shatz, C.J. and Stryker, M.P. (1978) Ocular dominance in layer IV of the cat's visual cortex and the effects of monocular deprivation. *J. Physiol.*, (Lond.) 281: 267-283.

Sherman, S.M. and Spear, P.D. (1982) Organization of visual pathways in normal and visually deprived cats. *Physiol. Rev.*, 62: 738-855.

Singer, W. (1977) Effects of monocular deprivation on excitatory and inhibitory pathways in cat striate cortex. *Exptl. Brain Res.*, 134: 568-572.

Stanfield, B.B. and O'Leary, D.D.M. (1985) The transient corticospinal projection from the occipital cortex during the postnatal development of the rat. *J. Comp. Neurol.*, 238: 236-248.

Stanfield, B.B., O'Leary, D.D.M. and Fricks, C. (1982) Selective collateral elimination in early postnatal development restricts cortical distribution of rat pyramidal tract neurones. *Nature*, 298: 371-373.

Stanford, L.R. (1987) W-cells in the cat retina: Correlated morphological and physiological evidence for two distinct classes. *J. Neurophysiol.*, 57: 218-244.

Sur, M., Garraghty, P.E. and Roe, A.W. (1988) Experimentally induced visual projections into auditory thalamus and cortex. *Science*, 242: 1437-1441.

Sur, M., Roe, A.W. and Garraghty, P.E. (1987) Evidence for early specificity of the retinogeniculate X cell pathway. *Soc. Neurosci. Abstr.*, 13: 590.

Sur, M. and Sherman, S.M. (1982) Linear and nonlinear W cells in C-laminae of the cat's lateral geniculate nucleus. *J. Neurophysiol.*, 47: 869-884.

Swadlow, H.A. (1983) Efferent systems of primary visual cortex: A review of structure and function. *Brain Res. Rev.*, 6: 1-24.

Tong, L., Spear, P.D., Kalil, R.E. and Callahan, E.C. (1982) Loss of retinal X-cells in cats with neonatal or adult visual cortex damage. *Science*, 217: 72-75.

Wiesel, T.N. and Hubel, D.H. (1963) Single cell responses in striate cortex of kittens deprived of vision in one eye. *J. Neurophysiol.*, 26: 1003-1017.

Wiesel, T.N. and Hubel, D.H. (1965) Comparison of the effects of unilateral and bilateral eye closure on cortical unti responses in kittens. *J. Neurophysiol.*, 28: 1029-1040.

Wilson, J.R. and Sherman, S.M. (1977) Differential effects of early monocular deprivation on binocular and monocular segments of the cat striate cortex. *J. Neurophysiol.*, 40: 892-903.

Winguth, S.D. and Winer, J.A. (1986) Corticocortical connections of cat primary auditory cortex (AI): Laminar organization and identification of supragranular neurons projecting to Area AII. *J. Comp. Neurol.*, 248: 36-56.

Wise, S.P., Hendry, S.H.C. and Jones, E.G. (1977) Prenatal development of sensory-motor cortical projections in cats. *Brain Res.*, 138: 538-544.

VISUAL RESPONSES OF NEURONS IN SOMATOSENSORY CORTEX OF HAMSTERS WITH EXPERIMENTALLY INDUCED RETINAL PROJECTIONS TO SOMATOSENSORY THALAMUS

Christine Métin* and Douglas O. Frost**

*Laboratoire des Neurosciences de la Vision
Université de Paris
Paris, France

**Dept. of Neurology
Massachusetts General Hospital
Boston, MA., USA

INTRODUCTION

In thalamic nuclei and cortical areas of the visual and somatosensory systems, information about peripheral stimuli is abstracted by single neurons that respond preferentially to particular values of one or more stimulus parameters. To what extent is information processing in the two systems similar and how do these systems differentiate during ontogeny? To study these questions, we exploited the fact that in newborn hamsters, retinal ganglion cell (RGC) axons can be surgically induced to form permanent, retinotopic projections to the primary somatosensory (ventrobasal, VB) thalamic nucleus (Campbell and Frost, 1988; Frost, 1981; 1982; 1986; Frost and Metin, 1985). We made neurophysiological recordings from single neurons in the principal targets of VB, the first and second somatosensory cortices (SI and SII, respectively), of neonatally operated, adult hamsters. We quantitatively compared the visually evoked responses of these neurons with those of single neurons in the primary visual cortex (VI, area 17) of normal, adult hamsters. We found that in operated hamsters, SI/SII neurons normally associated with somatic sensation have visual response properties that resemble those of neurons in VI of normal animals.

METHODS

Permanent retinal projections to VB were induced in anesthetized, newborn Syrian hamsters, as described previously (Campbell and Frost, 1988; Frost, 1981; 1982; 1986; Frost and Metin, 1985): two of the principal targets of RGC axons, the superior colliculus (SC) and dorsal lateral geniculate nucleus (LGd), were ablated. Heat lesions of SC were made bilaterally; unilateral, retrograde degeneration of LGd was induced by making a heat lesion of the ipsilateral occipital cortex. The VB ipsilateral to the cortical lesion was made an alternative target for RGC axons by making a midbrain hemisection to cut its ascending somatosensory afferents.

For recording experiments, adult hamsters were anesthetized with urethane and prepared as described previously (Frost and Metin, 1985). Physiological status and anethesia level were assessed by continuous monitoring of the electrocardiogram. Recording micropipettes (4-6MΩ impedance) containing 5% NaCl and 4% pontamine blue penetrated the dura, perpendicular to the cortical surface. Soma/dendrite recordings were distinguished from axon recordings by established criteria (Hubel, 1960). Visual receptive fields (RF's) were plotted on a screen with a hand-held projector, then studied quantitatively using computer generated, stationary or moving bars or spots, and drifting or alternating phase gratings of variable spatial and temporal frequencies, presented on a CRT. Single unit responses to visual stimuli were recorded and analysed by computer; in some units, we qualitatively evaluated responses to light cutaneous stimulation.

After recording, the anesthetized animals were intracardially perfused with 10% formalin. Brains were sectioned frozen at 50 μm and stained with cresyl violet. Visually responsive neurons in operated hamsters were shown to lie in SI and SII by two independent criteria (Frost and Metin, 1985): i) Some recording sites were marked by iontophoresis of pontamine blue. Subsequent histological examination revealed that these and other sites were in SI or SII as cytoarchitectonically defined (Caviness, 1975; Caviness and Frost, 1980). ii) The neurons were within regions whose somatic representations had polarities characteristic of SI or SII (Frost and Metin, 1985).

We studied 35 visually responsive cells in SI/SII of 7 operated animals aged 9-18 months, and, as controls, 48 visually responsive cells in VI of 12 normal animals aged 6-18 months. Of these cells, 26 in SI/SII and 41 in VI were fully characterized; partial data were obtained from the rest.

RESULTS

Spatial organization of receptive fields (RF's)

We recorded units with distinct visual RF's, in VI of normal- and in SI/SII of operated hamsters. (Visually evoked responses cannot be obtained in SI/SII of normal hamsters [Frost and Metin, 1985]). These RF's showed zones from which responses were elicited by turning luminous spots or bars ON, OFF, or both ON and OFF (Fig. 1, top frames). In VI of normal- and SI/SII of operated hamsters, RF's had the same three types of spatial organization: 1) "unizone" RF's had one ON, OFF, or ON/OFF zone which could be homogeneous (Fig. 1, RF's 2-3b) or, occasionally, heterogeneous (Fig. 1, RF's 2a, 1b) with respect to the intensity of the response evoked by stimulating different subregions of the RF; 2) "concentric" RF's had ON or OFF centers and antagonistic surrounds (Fig. 1, RF 3a); 3) "multizone" RF's had adjacent, non-concentric zones of ON and OFF, or ON and ON/OFF response (Fig. 1, RF 1a).

The spatial organization of unit RF's in SI/SII of operated hamsters differed from that of unit RF's in VI of normal hamsters in two respects. First, unit RF's in VI consist of a single responsive region; 57% of the visual RF's in SI/SII of operated animals consisted of two responsive regions, 20°-40° apart, either completely separated (Fig. 1, RF 3b), or, less often, linked by a region of weak, irregular response (Fig. 1, RF 2b). Usually, one responsive region gave much more robust visual responses than the other; unit response properties were tested only in the former. Second, units in SI/SII had visual RF's that were larger than those of units in VI. The mean RF area in VI was 93^{o2}. In SI/SII, the mean was 274^{o2}, considering only the more responsive region and 416^{o2}, summing the two responsive regions, for units with more than one.

Functional categories of RF's

We distinguished three functional categories of neurons in VI of normal hamsters (Fig. 1, lower frames 1-3a); the same three types of visually responsive neurons were present in SI/SII of operated hamsters (Fig. 1, lower frames 1-3b), in proportions not significantly different from those in VI (chi^2 test, p=0.7; Table 1): i) *Orientation selective* (Fig.1, 1a, 1b). These units had a preferred orientation when stimulated with a stationary, flashed bar. There was no significant difference in orientation bias (defined in Fig. 1 legend) between orientation selective units in VI and in SI/SII (2-tailed Mann-Whitney U-test; p>0.1). The best response to a moving bar was obtained when a bar of the preferred orientation moved along one direction (for unidirectional units, eg., 1a) or both directions (for bidirectional units, eg., 1b) orthogonal to the preferred orientation. These units gave stronger responses to optimally oriented bars than to spots with the same area, intensity and velocity of motion (not shown) but preferred directions were similar for bars and spots. ii) *Non-oriented, direction selective* (Fig. 1, 2a, 2b). These units had no orientation preference for stationary bars (2b) or gratings (not shown), nor did they give stronger responses to a moving bar than to a moving spot of the same area, intensity and velocity (2a). However, these units preferred movement in one (2a, 2b) or both (not shown) directions along a particular axis. iii) *Non-oriented, non-direction selective* (Fig. 1, 3a, 3b). These units had no preferred orientation when stimulated with stationary gratings (3a) or bars (3b), and no preferred direction of movement when stimulated with moving gratings (3a), bars (3b) or spots (not shown).

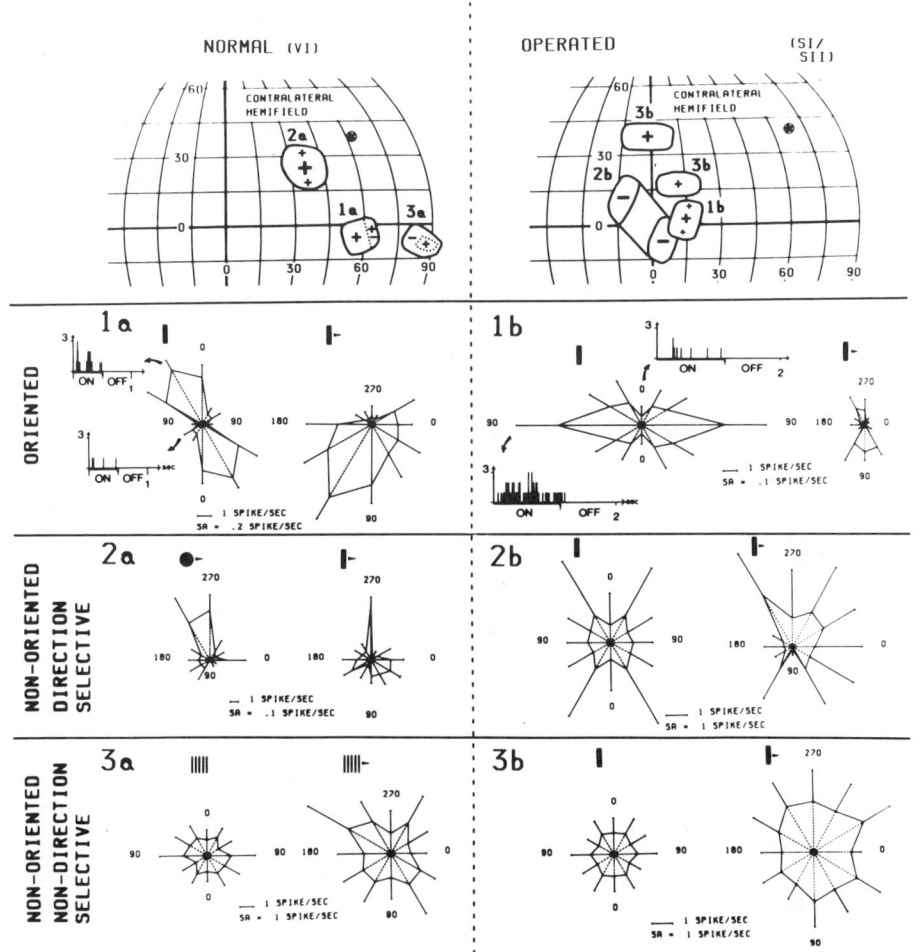

Figure 1. Visual RF's and response properties of 3 neurons in VI of normal hamsters and 3 neurons in SI/SII of operated hamsters, stimulated through the contralateral eye. RF's and polar histograms with the same number were obtained from the same neuron.

Top frames: spatial organization of visual RF's. Spherical representations of part of the visual field; vertical meridian (0) is the projection of the body plane of symmetry; horizontal meridian (0) is the projection of the horizontal plane containing the eyes. "30", "60" and "90" are eccentricities of meridia and elevation lines in degrees. Circled star is the projection of the optic disk. RF borders were determined using small, stationary, flashed stimuli. "+", "-" and "+/-" indicate zones in which responses were evoked by luminous stimuli turned on, off or on and off, respectively. Response intensity in unizone RF's occasionally varied with position in the RF, as indicated in 2a and 1b by large and small symbols.

Six lower frames: 3 functional categories of visual RF's (see text). Same neurons as above in VI (1-3a) and SI/II (1-3b). Symbols above each polar histogram indicate stimulus used to evoke responses illustrated. "[ ]" and "III" correspond, respectively, to *stationary*, flashed bars and to *stationary*, alternating phase gratings of 6 orientations separated by 30° (0° = vertical, 90° = horizontal) and presented randomly in a sequence repeated ≥5 times. Orientation of each dotted radius on polar plot indicates stimulus orientation; length of dotted radius gives mean response rate (spikes sec^{-1}) at that orientation. Solid line segments show standard deviation of each response. "[]>", "•>", and "III>" correspond, respectively, to bars, spots, and sinusoidal gratings, *moving* in 12 directions separated by 30° (0° = nasal to temporal, 90° = superior to inferior) and repeatedly presented in a random sequence. Bars and gratings moved in directions perpendicular to their long axes. Orientations of radii on polar plots indicate direction of stimulus movement; lengths of radii give mean response rate; solid line segments indicate standard deviation. SA indicates mean spontaneous activity rate.

1a and 1b - Orientation selective units. Units were considered to be orientation selective if their orientation bias, B, was ≥ 0.7. ($B=1-R_{min}/R_{max}$; R_{min} and R_{max} are the mean firing rates evoked by stationary bars at orientations producing the weakest and strongest responses, respectively. For units responding equally at all orientations, B=0, while for those giving no response at the least effective orientation, B=1). Post stimulus time histograms (PSTH's) are for responses to stationary flashed bars with the preferred orientations (150° and 90°, respectively) and with orthogonal orientations (60° and 0°, respectively). PSTH's show that the ON response to an optimally oriented, flashed bar was phasic in unit 1a and more tonic in unit 1b. In both cases, the ON response strongly decreased when the units were stimulated with bars orthogonal to the preferred orientation.

2a and 2b - Non-oriented, direction selective units. Both were unidirectional. 2a preferred 270° movement while 2b prefered upward nasal movement.

3a and 3b - Non-oriented, non-direction selective units. They showed no preferred orientation when stimulated with stationary gratings (3a) or bars (3b), and no preferred direction of movement when stimulated with moving gratings (3a) or bars (3b).

In both normal and operated hamsters RF category and RF spatial organization were correlated: all oriented and non-oriented, direction selective units were either unizone or multizone; 27 of 29 non-oriented, non-direction selective units were either unizone or concentric, while 2 were multizone. In normal and operated hamsters, the distributions of preferred orientations (Fig. 2A) and directions of movement (Fig. 2B) were both random. The depth distributions of the different types of visually responsive neurons were similar in normal and operated hamsters: orientation selective neurons predominated in the supragranular cortical layers of both VI and SI/SII (6/10=60% and 5/10=50%, respectively, of neurons both physiologically characterized and histologicaly localized in layers II-III) but were less common in deeper layers (4/24=17% and 3/13=23%, respectively, of neurons both characterized and localized in layers IV-VI).

The mean latency of response to a visual stimulus (Fig. 3A) was 198 msec in VI and 209 msec in SI/SII; a t-test showed no significant latency difference between neurons in VI and in SI/SII (p=0.56). Neurons were divided into three groups according to their preferred stimulus velocity. Slowly moving stimuli (0-15° sec^{-1}) were preferred by relatively fewer neurons in SI/SII than in VI, while stimuli of medium velocity (15-60° sec^{-1}) were relatively more often preferred in SI/SII than in VI; neurons preferring high velocities (>60° sec^{-1}) were about equally common in VI and SI/SII (Fig. 3B). In order to perform a chi^2 test on the preferred velocity distributions, we had to combine the medium and high velocity categories because of the small number of units in the latter; the distributions for VI and SI/SII were significantly different (p = 0.025) when tested this way.

Subsequent to neonatal midbrain hemisections, ascending somatosensory afferents grow rostral to the cut and reinvade VB (DOF, unpublished data). Thus, 22 visually responsive neurons in SI/SII were tested for somatosensory responsiveness. Eight (36%) responded to somatosensory stimulation; their response properties were qualitatively similar (unpublished data) to those of neurons in SI/SII of normal animals. One of the bimodal neurons was oriented, 1 was non-oriented, direction selective, 5 were non-oriented, non-direction selective and 1 was uncategorized.

Neurons in SI/SII often responded less intensely to visual stimuli than did neurons in VI. Furthermore, for units that responded to both visual and somatosensory stimulation, the somatosensory response was generally more robust. There was no correlation, however, between the intensity of a unit's visual response and its selectivity for various stimulus parameters. For example, in Fig. 1, although unit 1b in SI/SII does not respond to moving bar stimuli as well as unit 1a in VI, it is more sharply tuned for the direction of stimulus movement; in addition, unit 1b responds more robustly to stationary, optimally oriented bars.

DISCUSSION

In hamsters with abnormal retino-VB projections, somatosensory cortical neurons have visual response properties that resemble, in several characteristic features, those of normal visual cortical neurons. These data support the hypothesis that at the thalamic and cortical levels, the

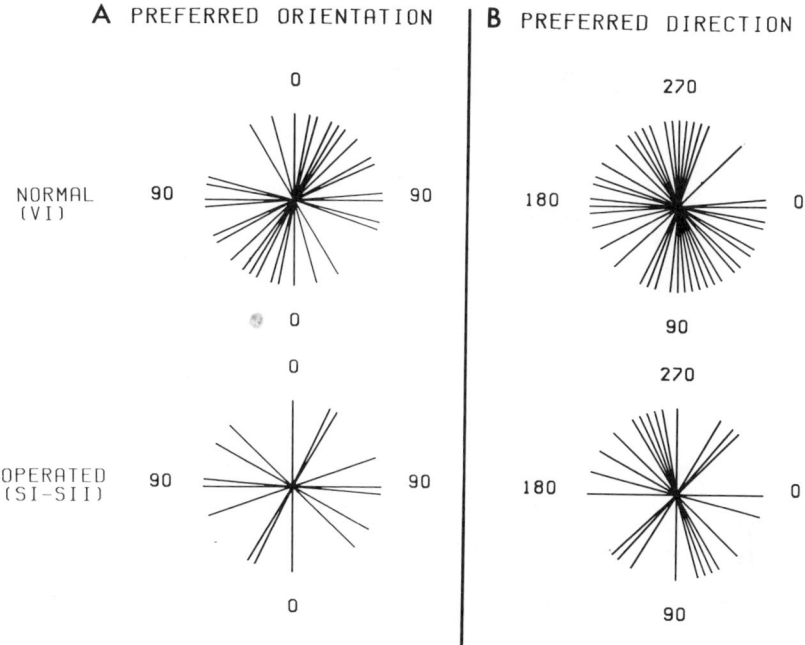

Figure 2. A. Preferred stimulus orientations of 15 cells in VI of normal hamsters and 8 cells in SI/SII of operated hamsters. Each diameter represents the preferred orientation of one cell. B. Preferred directions of movement for both orientation selective and non-oriented, direction selective cells. For each unidirectional neuron, preferred direction is represented by a radius pointing in that direction. For each bidirectional neuron, preferred axis of movement is represented by a diameter parallel to that axis. Conventions as in Fig. 1.

somatosensory and visual systems use similar neuronal circuits to perform similar transformations on their inputs[1], a possibility raised on the basis of a common organizational plan of the thalamus (Jones, 1985) and neocortex (Mountcastle, 1978; Rakic and Singer, 1988). We first consider several issues related to alternative interpretations.

In operated hamsters visual information almost certainly reaches SI/SII via the retino-VB projection: i) Visually evoked responses cannot be obtained in SI/SII of normal hamsters (Frost and Metin, 1985). ii) RGC axons make synapses in VB (Campbell and Frost, 1988). iii) Intraocular injections of ^{3}H amino acids label parts of VB and topographically corresponding regions of SI/SII; in SI/SII, the laminar distribution of label corresponds to the distribution of thalamocortical axons (Frost, 1982). iv) The topography of the visual field representations in SI/SII is predicted by the topographies of the retino-VB and VB to SI/SII projections (Frost and Metin, 1985). v) The somatosensory RF's of bimodal units were always on the mystacial vibrissae, head or ears; these regions are represented on the lateral aspect of VB (Waite, 1973), which is also the site of termination of the retino-VB projection.

Our data from single neurons in VI are consistent with those of previous studies in rodents (Tiao and Blakemore, 1976; Metin et al., 1988; Montero et al., 1973). There are qualitative and quantitative differences between the visual response properties of neurons in VI of normal hamsters and neurons in SI/SII of operated hamsters. Although some of these may reflect differences in the intrinsic circuitry of the visual and somatosensory thalamic nuclei or cortices, some probably have other causes: i) The multiple responsive regions and supranormal areas of the visual RF's of some neurons in SI/SII arise because unlike LGd and VB of normal animals, in

---

[1]"Transformation" denotes the relationship between the information flowing into and out of a neural structure.

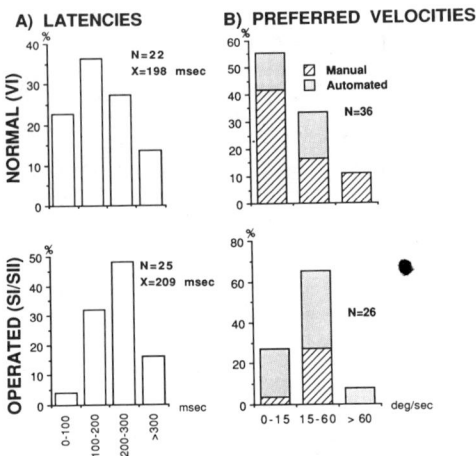

Figure 3. Response latencies (A) and preferred stimulus velocities (B) for neurons in VI of normal- and in SI/SII of operated hamsters. Latencies were measured from onset of optimal stimuli to the first peak in a unit's response. Preferred velocities were determined manually or using the computer. In manual tests (cross hatching), an otherwise optimal stimulus was swept across the RF at speeds in the ranges of 0-15°/sec, 15-60°/sec or >60°/sec; in automated tests (stippling), a grating of optimum orientation and spatial frequency was swept across the RF at one of 16 different velocities, separated equally on a logarithmic scale from 0.5-25 °/sec and presented repeatedly in pseudorandom order; the preferred stimulus velocity was that which evoked the greatest mean firing rate. Vertical axes: percentage of neurons tested of a given category; horizontal axes: latency or velocity category. N = number of neurons tested; X = mean latency in msec.

which relay neurons that project to the same cortical loci get input from the same point on the receptor surface, in VB of operated animals neurons projecting to the same cortical locus get input from multiple retinal loci[2]. ii) The differences in latency and velocity preference between neurons in SI/SII and those in VI may reflect differences in the proportions of various RGC types that project to VB and LGd, respectively. In carnivores and primates, there are distinct classes of RGC's that differ with respect to multiple response parameters, including conduction velocity and preferred stimulus velocity (Stone, 1983). A transient retinal projection to VB in normal neonatal hamsters (Frost, 1984) contributes to the permanent projection in operated animals (Frost, 1986) and arises from RGC's that do not project to the thalamus in normal, adult animals (Langdon et al., 1987).

---

[2]In normal mammals, orderly maps of the retina and body in VI and SI/SII, respectively, arise because in LGd and VB, "lines of projection" (the zones of termination of afferents representing a restricted region of the receptor surface) are congruent with the set of relay neurons that project beneath a single corical locus are distributed in arcs that lie in a plane approximately orthogonal to the lines of projection defined by retino-VB afferents in operated hamsters (2) and, therefore, intercept multiple, but not necessarily continuous, lines of projection.

Neonatal lesions of SC and the brachium of the inferior colliculus produce abnormal retinal projections to the medial geniculate nucleus (MG), the principal thalamic auditory nucleus (Campbell and Frost, 1988; Frost, 1981; 1982; 1986; Schneider, 1977; Kalil and Schneider, 1975; Sur and Garraghty, 1986). In operated ferrets with retino-MG projections, some neurons in the auditory cortex (AC) respond weakly to visual stimulation from large, poorly defined RF's (Sur and Garraghty, 1986). The reasons for the differences between our results and those obtained in ferret AC are unclear.

In sensory thalamic nuclei and cortical areas, the differentiation of some biochemical and morphological features underlying normal function may reflect the modality of the sensory input. While sensory input of the appropriate *modality* clearly influences the development of sensory systems (Fregnac and Imbert, 1984; Sherman and Spear, 1982), there are few data on how the differentiation of sensory systems depends on the *modality* of their input. Available evidence, namely our electron microscopic studies of the retinal projections to VB and MG, has not demonstrated such a dependence: retino-VB and retino-MG axons participate in synaptic complexes that morphologicly resemble those of normal, somatosensory and auditory thalamic afferents, respectively, rather than visual ones (Campbell and Frost, 1988). (This datum also suggests that the morphological features of thalamic synaptic complexes may not determine the parameters of cortical neuronal responses assayed in our visual RF studies). However, computer models demonstrate the plausibility of neuronal networks whose connection strengths are modifiable in such a way that the response properties of their constituent neurons develop particular features as a result of their sensory input (Lehky and Sejnowski, 1988; Linsker, 1988). Furthermore, it has been demonstrated both for lagomorphs (Levik, 1967) and birds (Maturana and Frenk, 1963) that there are small populations of RGC's that have strong preferences for stimulus orientation; if similar RGC's are present during development in rodents, carnivores and primates, it is possible that during development in normal or neonatally operated animals, the cortically evoked activities of these RGC's could instruct the cortex (via an activity-dependent mechanism [eg., Bear et al., 1987]) to form circuits appropriate for the analysis of contour orientation.

Three lines of evidence support the hypothesis that thalamic nuclei and cortical areas at corresponding levels in the visual and somatosensory systems perform similar transformations on their inputs:

a) The visual and somatosensory systems use similar information processing strategies, based on similar morphological substrates. In these systems, the internal anatomical and functional organization of thalamic nuclei (Jones, 1985) and cortical areas (Mountcastle, 1978; Rakic and Singer, 1988) are similar: Each cortical area consists of uniform, multiply replicated modules that are basic units of information processing. In both systems, there are i) orderly maps of the receptor surface at the thalamic (Jones, 1985) and cortical (Daniel and Whitteridge, 1961; Merzenich et al., 1978) levels, ii) multiple, hierarchically organized, reciprocally connected cortical areas (Van Essen, 1985; Jones et al., 1978), iii) similar laminar segregations within the cortex of various classes of afferent axons and efferent neurons (Van Essen, 1985; Jones et al., 1978; Gilbert and Kelly, 1975; Jones and Wise, 1977) and iv) parallel pathways for processing information concerning distinct submodalities (Van Essen, 1985; Kaas, 1983).

b) Orientation selective neurons occur with equal frequency and are equally sharply tuned in VI of normal- and SI/SII of operated hamsters[3]. The similar depth distributions of orientation selective units in VI and SI/SII, give further evidence of the similarity of circuitry in these cortical regions.

---

[3]This datum supports the hypothesis if either of two explanations of cortical orientation selectivity is correct. i) It was originally suggested that the orientation preference of visual cortical neurons is an emergent property of thalamocortical connectivity or cortical circuitry (32). ii). It is now known that in carnivores, RGC's and LGd neurons show weak orientation biases, although the contribution of these biases to the orientation preferences of cortical neurons is controversial (33). There has been no systematic study in rodents to determine where in the visual pathway different stimulus features are first abstracted. Thus, in rodents, carnivores and other orders, the response preferences of RGC's or thalamic neurons may be sharpened by the cortex.

c) The similarity of the visual and somatosensory response properties of bimodal neurons in SI/SII of operated hamsters to those of single neurons in VI and SI/SII, respectively, of normal hamsters suggests that similar circuits in the visual and somatosensory thalamic nuclei and cortices can generate both visual and somatosensory responses.

It is not known if the visual and somatosensory systems use similar strategies to accomplish similar tasks. While neurons in somatosensory cortex of normal animals are selective for the direction and velocity of stimulus movement (Essick and Whitsel, 1985), orientation selectivity is rare (Hyvarinen and Poranen, 1978) and these features may not be abstracted by the same mechanisms as in the visual system. Even if these stimulus parameters are not analysed similarly in the two systems, the visual and somatosensory cortices may perform a common operation, eg., selectively filtering their inputs so as to emphasize changes in the spatial or temporal domains. The hypothesis that visual and somatosensory forebrain structures perform similar transformations on their inputs implies that differences in information processing strategy between the two systems occur principally at prethalamic levels.

ACKNOWLEDGEMENTS

Supported by grants EY03465 and NS22807 from NIH (DOF), 1-1060 from March of Dimes Birth Defects Foundation (DOF), UA1199 from CNRS (to M. Imbert), 0212/87 from NATO (DOF & MI) and from La Fondation de France (CM). We thank M. Imbert, G. Innocenti, D. McCormick, P. Rakic, M. Schwartz and C.F. Stevens for stimulating discussions.

REFERENCES

Bear, M.F., Cooper, L.N. and Ebner, F.F. (1987) A physiological basis for a theory of synapse modification. *Science*, 237: 42-48.
Campbell, G., and D.O. Frost (1988) Synaptic organization of anomalous retinal projections to the somatosensory and auditory thalamus: target-cotrolled morphogenesis of axon terminals and synaptic glomeruli. *J. Comp. Neurol.*, 272: 383-408.
Caviness, V.S (1975) Architectonic map of neocortex of the normal mouse. *J. Comp. Neurol.*, 164: 247-264.
Caviness, V.S. , and D.O. Frost (1980) Tangential organization of thalamic projections to the neocortex in the mouse. *J. Comp. Neurol.*, 194: 335-367.
Daniel, P.M. and Whitteridge, D. (1961) The representation of the visual field on the cerebral cortex in monkeys. *J. Physiol. (Lond.)* 159: 203-221.
Essick, G.K. and Whitsel, B.L. (1985) Factors influencing cutaneous direction sensitivity: A correlative psychophysical and neurophysiological investigation. *Br. Res. Rev.* 10: 213-230.
Fregnac, Y. and Imbert, M. (1984) Development of neuronal selectivity in primary visual cortex of cat. *Physiol. Rev.* 64: 325-434.
Frost, D.O. and Caviness, V.S., Jr. (1980) Radial organization of thalamic projections to the neocortex in the mouse. *J. Comp. Neurol.* 194: 369-393.
Frost, D.O. (1981) Orderly anomalous retinal projections to the medial geniculate, ventrobasal, and lateral posterior nuclei of the hamster. *J. Comp. Neurol.*, 203: 227-256.
Frost, D.O. (1982) Anomalous visual connections to somatosensory and auditory systems following brain lesions in early life. *Dev. Brain Res.*, 3: 627-635.
Frost, D.O. (1984) Axonal growth and target selection during development: retinal projections to the ventrobasal complex and other "nonvisual" structures in neonatal Syrian hamsters. *J. Comp. Neurol.*, 230: 576-592.
Frost, D.O. (1986) Development of anomalous retinal projections to nonvisual thalamic nuclei in Syrian hamsters: A quantitative study. *J. Comp. Neurol.*, 252: 95-105.
Frost, D.O., and C. Metin (1985) Induction of functional retinal projections to the somatosensory system. *Nature*, 317 (162-164).
Gilbert, C.D. and Kelly, J.P. (1975) The projections of cells in different layers of the cat's visual cortex. *J. Comp. Neurol.* 163: 81-106.
Hubel, D.H. and Wiesel, T.N. (1962) Receptive fields, binocular interaction and functional architecture in the cat's visual cortex. *J. Physiol. (Lond.)* 160: 106-154.
Hyvarinen, J. and Poranen, A. (1978) Movement-sensitive and direction and orientation-selective cutaneous receptive fields in the hand area of the post-central gyrus in monkeys. J. Physiol. (Lond.) 283: 523-537.
Jones, E.G., J.D. Coulter, and S.H.C. Hendry (1978) Intracortical connectivity of architectonic fields in the somatic sensory , motor and parietal cortex of monkeys. *J. Comp. Neurol.*, 181: 291-348.

Jones, E.G., and S.P. Wise (1977) Size, laminar and columnar distribution of efferent cells in thesensory-motor cortex of primates. *J. Comp. Neurol.*, 175: 391-438.

Kaas, J.H. (1983) What, if anything, is S-I? Organization of first somatosensory area of cortex. *Physiol. Rev.*, 63: 206-231.

Kalil, R.E. and Schneider, G.E. (1975) Abnormal synaptic connections of the optic tract in the thalamus after midbrain lesions in newborn hamsters. *Br. Res.* 100: 690-698.

Langdon, R.B., J.M. Freeman, and D.O. Frost (1987) Trajectories and branching patterns of optic tract axons that project transiently to somatosensory thalamus in the neonatal hamster. *Neurosci. Abs.*, 13 (1023)

Lehky, S.R. and Sejnowski, T.J. (1988) Network model of shape-from-shading: neural function arises from both receptive and projective fields. *Nature* 333: 452-454.

Levick, W.R. (1967) Receptive fields and trigger features of ganglion cells in the visual streak of the rabbit's retina. *J. Physiol.* (Lond.) 188: 285-307.

Linsker, R. (1988) Self-organization in a perceptual network. *Computer* 21: 105-117.

Maturana, H.R. and Frenk, S. (1963) Directional movement and horizontal edge detectors in the pigeon retina. *Science* 142:977-979.

Merzenich, M.M., J.H. Kaas, M. Sur, and C.-S. Lin (1978) Double representation of the body surface within cytoarchitectonic areas 3b and 1 in S1 in the owl monkey (Aotus trivirgatus). *J. Comp. Neurol.*, 181: 41-74.

Metin, C., P. Godement, and . Imbert (1988) The primary visual cortex of the mouse: receptive field properties and functional organization. *Exp. Br. Res.*, 69: 594-612.

Montero, V.M., A. Rojas, and F. Torrealba (1973) Retinotopic organization of stirate and peristriate visual cortex in the albino rat. *Brain Res.*, 53: 197-201.

Mountcastle, V.B. (1978) An organizing principle for cerebral function: The unit module and the distributed system. In *The Mindful Brain* V.B. Mountcastle and G.M. Edelman, eds. pp 7-50, MIT Press, Cambridge, MA.

Rakic, P. and Singer, W. (1988) *Neurobiology of neocortex*, John Wiley & Sons, New York.

Schneider, G.E. (1973) Early lesions of the superior colliculus: factors affecting the formation of abnormal retinal projections. *Brain Behav. Evol.*, 8 (73-109).

Sherman, S.M., and P.D. Spear (1982) Organization of visual pathways in normal and visually deprived cats. *Physiol. Rev.*, 62: 738-855.

Stone, J. (1983) *Parallel processing in the visual system*, Plenum Press, New York.

Sur, M. and Garraghty, P.E. (1986) Experimentally induced visual responses from auditory thalamus and cortex. *Neurosci. Abs.* 12: 592.

Tiao, Y.-C., and C. Blakemore (1976) Functional organization of the superior colliculus of the golden hamster. *J. Comp. Neurol.*, 168: 459-482.

Van Essen, D.C. (1985) Functional organization of the primate visual cortex. In The Cerebral Cortex. Edited by A. Peters and E. G. Jones. 259-329. New York: Plenum Press.

Vidyasager, T.R. (1987) A model of striate response properties based on geniculate ani sotropies. *Biol. Cybern.* 57: 11-23.

Waite, P.M.E. (1973) Somatotopic organization of vibrissal responses in the ventro-basal complex of the rat thalamus. *J. Physiol. (Lond.)* 223: 527-540.

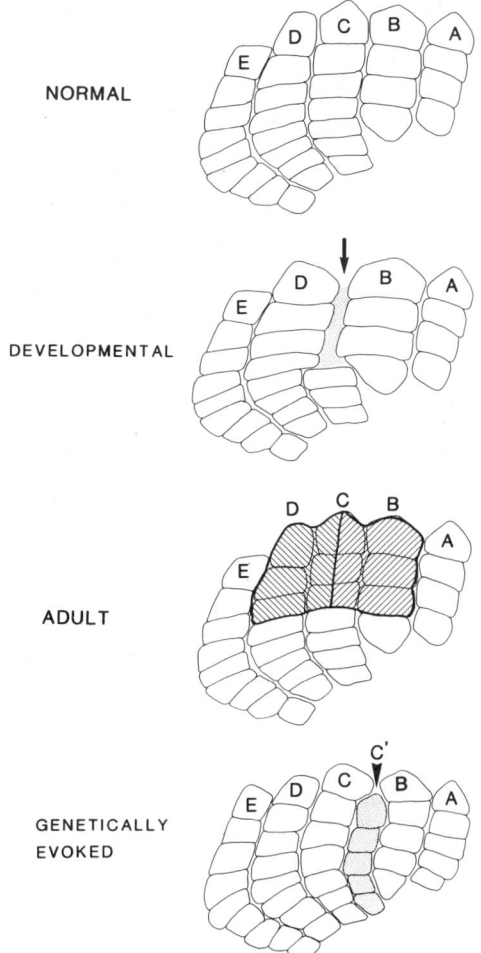

Figure 4. Cartoons representing the *NORMAL* barrelfield and the three types of plasticity of the barrel cortex as discussed in the text: *DEVELOPMENTAL*, as observed after postnatal lesions of the four posterior follicles of row C (here illustrated for lesions performed at postnatal day 2); *ADULT*, reorganization of the functional map (indicated by hatched zones, outlining the enlarged representation of follicles of rows B and D), as studied using the deoxyglucose technique and electrophysiological recordings, performed 60 days after denervation of the three posterior follicles of row C in the adult mouse; *GENETICALLY EVOKED*, the appearance of a supernumerary row of barrels (C') as a result of selective breeding for as many as possible supernumerary follicles between rows B and C.

strains were investigated in detail (Welker and Van der Loos, 1986b) and it was found that while there is great stability within each strain as to follicle and barrel size, there exists, between strains, considerable variation with respect to these parameters.

So far, this chapter has dealt with signs of plasticity in the whisker-to-barrel pathway, notably in its end-station, the barrel cortex. The question arises whether this cortical plasticity has consequences for its projections towards other brain areas (Fig. 2, *right*). We have begun approaching this point by making minute injections of the anterograde tracer Phaseolus vulgaris-leucoagglutinin in different parts of the barrel cortex, and by analyzing the topological organization of efferents of this cortical field in normal mice (Hoogland et al., 1987; Hoogland et al., 1988 and Welker et al., 1988a). We observed interesting differences between the organization of the cortical efferents towards different brain areas. The distribution of terminal labeling varies from

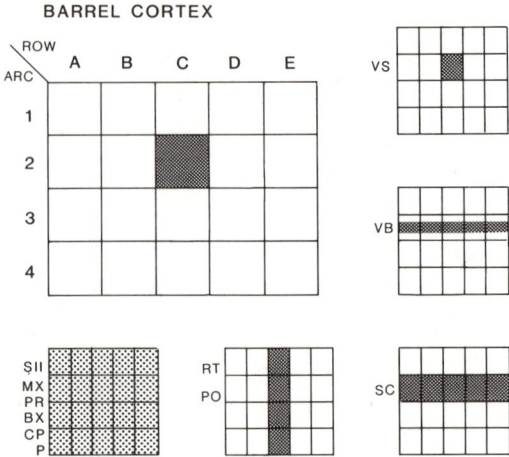

Figure 5. Schematic representation of the topographic organization of the efferents from the barrel cortex as studied by Phaseolus vulgaris-leucoagglutinin (PHA-L) injections in individual barrel columns of the mouse. The large block stands for the barrelfield in which letters indicate the representations of rows of vibrissae; numbers, those of arcs of vibrissae. The same organization holds for the five small blocks. The organization of the efferents is shown for those arising from barrel column C2. Sites of PHA-L injection and of labelling of terminal branches are stippled. Note the intriguing variation in the organization of the feedback projections (towards *VS*, the trigeminal sensory complex and *VB*, the ventrobasal thalamic nucleus) in which topographic precision is maintained and that of the feedforward projections in which various degrees of "map-dissolution" were observed (towards *SC*, the superior colliculus; *RT*, the reticular thalamic nucleus; *PO*, the posterior thalamic nucleus; *SII*, the secondary somatosensory cortex; *MX*, the motor cortex; *PR*, the perirhinal cortex; *BX*, the contralateral barrel cortex; *CP*, the ipsi- and contralateral caudo-putamen; and to *P*, the pons). From Welker et al. (1988a), with permission.

punctiform (i.e. each barrel column[2] having a separate projection area within a given brain area), via an overlap of projections arising from one row or one arc of barrel columns, towards completely overlapping projections within restricted zones of the target regions (Fig. 5 summarizes these findings). These differences in organization of the efferents of the barrel cortex demonstrate a contrast between feedback- and feedforward projections and sollicit an investigation of the consequences - if any - of the altered cortical circuitry for the organization of the efferent connections from the area of cortex in question.

In summary, we hope to have demonstrated that, besides the existence of developmental and adult plasticity of an important bit of neocortex, there is a plasticity that works on the brain by way of the genome acting on the periphery. Understanding this latter mode of making brain-maps may help finding clues to the evolution of homologous receptor sheets and of those parts of the brain to which they project directly, as well as of parts of the brain beyond the primary sensory cortical fields.

ACKNOWLEDGEMENTS

We thank Mr. S. Daldoss for assistance in preparing the figures, and the Swiss National Science Foundation for support (grant 3100-009468).

---

[2] A barrel column is defined as a part of cortex consisting of a barrel and the cortical tissue lying above and below it.

REFERENCES

Armstrong-James, M., Fox, K. (1987) Spatiotemporal convergence and divergence in the rat S1 "barrel" cortex. *J. Comp. Neurol.* 263: 265-281.

Andrés, F.L., Van der Loos, H. (1983) Cultured embryonic non-innervated mouse muzzle is capable of generating a whisker pattern. *Int. J. Devel. Neurosci. 1*: 319-338.

Andres, K.H. (1966) Ueber die Feinstruktur der Rezeptoren an Sinushaaren. *Z. Zellforsch*, 75: 339-365.

Belford, G.R., Killackey, H.P. (1979) The development of vibrissae representation in subcortical trigeminal centers of the neonatal rat. *J. Comp. Neurol.* 188: 63-74.

Blue, M.E., Molliver, M.E. (1989) Serotonin influences barrel formation in developing somatosensory cortex of the rat. *Soc. Neurosci. Abstr.* 15: 419.

Christensen, J., Woolsey, T.A. (1988) Peanut lectin staining in the mouse whisker-barrel pathway and its modification by peripheral lesions at different postnatal ages. *Soc. Neurosci. Abstr.* 14: 1273.

Cooper, N.G., Steindler, D.A. (1986) Lectins demarcate the barrel subfield in the somatosensory cortex of the early postnatal mouse. *J. Comp. Neurol*. 249: 157-169.

Dörfl, J. (1985) The innervation of the mystacial region of the white mouse. A topographical study. *J. Anat. (Lond.)* 142: 173-184.

Hoogland, P.V., Welker, E., Van der Loos, H. (1987) Organization of the projections from barrel cortex to thalamus in mice studied with Phaseolusvulgaris-leucoagglutinin and HRP. *Exp. Brain Res.*, 68: 73-78.

Hoogland, P.V., Welker, E., Van der Loos, H., Wouterlood, F.G. (1988) The organization and structure of the thalamic afferents from the barrel cortex in the mouse; a PHA-L study. In: *Cellular Thalamic Mechanisms.* Bentivoglio, M. and Spreafico, R. (eds.), Elsevier, Amsterdam, pp. 151-162.

Jeanmonod, D., Rice, F.L., Van der Loos, H. (1981) Mouse somatosensory cortex: alterations in the barrelfield following receptor injury at different early postnatal ages. *Neurosci.*, 6: 1503-1535.

Jensen, K.F. (1987) The development of the vibrissae related neocortical afferents to somatosensory cortex of the rat. *Soc. Neurosci. Abstr.* 13: 77

Lee, K.J., Woosley, T.A. (1975) A proportional relationship between peripheral innervation and cortical neuron number in the somatosensory system of the mouse. *Brain Res.*, 99: 349-353.

Melzer, P., Van der Loos, H., Dörfl, J., Welker, E., Robert, P., Emery, D., Berrini, J.-C. (1985) A magnetic device to stimulate whiskers of freely moving or restrained small rodents: its application in a deoxyglucose study. *Brain Res.*, 348: 229-240.

Melzer, P., Yamakado, M., Van der Loos, H., Welker, E., Dörfl, J. (1988) Plasticity in the barrel cortex of adult mouse: effects of peripheral deprivation on the functional map; a deoxyglucose study. *Soc. Neurosci. Abstr.* 14: 844.

Nussbaumer, J.C., Van der Loos, H. (1985) An electro-physiological and anatomical study of projections to the mouse cortical barrelfield and its surroundings. *J. Neurophysiol. 53*: 686-698.

Rakic, P. (1988) Specification of cerebral cortical areas. *Science*, 241: 170-176.

Rice, F.L., Van der Loos, H. (1977) Development of the barrels and barrelfield in the somatosensory cortex of the mouse. *J. Comp. Neurol.*, 171: 545-560.

Simons, D.J. (1978) Response properties of vibrissa units in rat S1 somatosensory neocortex. *J. Neurophysiol.*, 41: 798-820.

Steindler, D.A., Cooper, N.G., Faissner, A., Schachner, M. (1989) Boundaries defined by adhesion molecules during development of the cerebral cortex: the J1/tenascin glycoprotein in the mouse somatosensory cortical barrel field. *Dev. Biol.*, 131: 243-260.

Van der Loos, H., Woolsey, T.A. (1973) Somato-sensory cortex: structural alterations following early injury to sense organs. *Science*, 179: 395-398.

Van der Loos, H. (1976) Barreloids in mouse somatosensory thalamus. *Neurosci. Lett.* 2: 1-6.

Van der Loos, H., Dörfl, J. (1978) Does the skin tell the somatosensory cortex how to construct a map of the periphery? *Neurosci. Lett.* 7: 23-30.

Van der Loos, H. (1979) The development of topological equivalences in the brain. In: *Neuronal growth and differentiation.* Meisami, E. and Brazier, M.A.B. (eds.), Raven Press: New York, pp. 331-336.

Van der Loos, H., Dörfl, J., Welker, E. (1984) Variation in pattern of mystacial vibrissae in mice; a quantitative study of ICR stock and of several inbred strains. *J. Hered.*, 75: 326-336.

Van der Loos, H., Welker, E. (1985) Development and plasticity of somatosensory brain maps. In: *Development, Organization and Processing in Somatosensory Pathways.* Rowe, M.J. and Willis, W.D., Jr. (eds.), Alan Liss: New York, pp. 53-67.

Van der Loos, H., Welker, E., Dörfl, J., Rumo, G. (1986) Selective breeding for variations in pattern of mystacial vibrissae of mice; bilaterally symmetrical strains derived from ICR-stock. *J. Hered.*, 77: 66-82.

Waite, P.M.E. (1977) Normal nerve fibers in the barrel region of developing and adult mouse cortex. *J. Comp. Neurol.*, 173: 165-174.

Welker, C. (1976) Receptive fields of barrels in the somatosensory neocortex of the rat. *J. Comp. Neurol.*, 166: 173-190.

Welker, E., Van der Loos, H. (1986a) Is areal extent in cerebral cortex determined by peripheral innervation density? *Exp. Brain Res.*, 63: 652-654.

Welker, E., Van der Loos, H. (1986b) Quantitative correlation between barrelfield size and the sensory innervation of the whiskerpad: a comparative study in six strains of mice, bred for different patterns of mystacial vibrissae. *J. Neurosci.*, 6: 3355-3373.

Welker, E., Hoogland, P.V., Van der Loos, H. (1988a) Organization of feedback and feedforward projections of the barrel cortex: a PHA-L study in the mouse. *Exp. Brain Res.*, 73: 411-435.

Welker, E. Leclerc, S.S., Van der Loos, H., Yamakado, M., Dykes, R.W. (1988b) Plasticity in the barrel cortex of adult mouse: effects of peripheral deprivation on the functional map; an electrophysiological recording study. *Soc. Neurosci. Abstr.* 14: 843.

Welker, E., Soriano, E., Van der Loos, H. (1989) Plasticity in the barrel cortex of the adult mouse: effects of peripheral deprivation on GAD-immunoreactivity. Exp. Brain Res., 74: 441-452.

Welker, E., Van der Loos, H., Dörfl, J., Soriano, E. (1990) The possible role of GABAergic innervation in plasticity of adult cerebral cortex. In: *The Neocortex: Ontogeny and Phylogeny*, Finlay, B.L., Innocenti, G., Scheich, H. (eds), Plenum Press: New York.

Woolsey, T.A., Van der Loos, H. (1970) The structural organization of layer IV in the somatosensory region (S1) of mouse cerebral cortex. The description of a cortical field composed of discrete cytoarchitectonic units. *Brain Res.*, 17: 205-242.

Yarris, L.M., Chiaia, N.L., Bennett-Clarke, C.A., Jacquin, M.F., Haring, J.H. Macdonald, G.J., Rhoades, R.W. (1989) Serotonin immunoreactive fibers yield a complete map of the body surface in somatosensory cortex of perinatal rats. *Soc. Neurosci. Abstr.* 15: 874.

THE POSSIBLE ROLE OF GABA-ERGIC INNERVATION IN PLASTICITY OF ADULT CEREBRAL CORTEX

Egbert Welker, Hendrik Van der Loos, Josef Dörfl, Eduardo Soriano*

Institute of Anatomy
University of Lausanne
1005 Lausanne, Switzerland

*Unitat de Biologia Celular
Facultat de Biologia
Universitat de Barcelona
08071 Barcelona, Spain

INTRODUCTION

The configuration of the map of the body surface in the primary somatosensory cortex (SI) has been shown to be dependent on the configuration of the sensory peripheral sheet. Not only does the periphery have an important role in the ontogenetic establishment of its representation in SI, but it also plays a crucial role in its maintenance during adulthood (for references see Van der Loos et al., this volume). This latter aspect has been first shown for the hand representation of the adult monkey (Merzenich et al., 1983). Since then, the peripheral somatosensory system has been modified in adults of a variety of species. The chosen paradigm in most of these cases has been either surgical removal of a part of the sensory sheet, or a partial denervation through division of a sensory nerve. The common observation in these experiments was that, after a brief period in which the zone of the cortex that corresponds with the removed or denervated part of the periphery was silent, neuronal responses could be evoked, in that same zone, from skin areas neighboring the damaged periphery. The first sign of plasticity[1] in the functionally defined maps may occur within only a few hours after the division of a nerve (Calford & Tweedale, 1988). This very short time-lapse makes it unlikely that the cortical adaptations are due to axonal sprouting, peripheral or central. Therefore, it has been proposed that these adaptations may be due to alterations in the chemical environment of cortical neurons. The inhibitory neurotransmitter GABA (γ-aminobutyric acid) seemed to be a potential candidate since Dykes et al. (1984) have shown that the iontophoretic application of the GABA-antagonist bicuculline methiodide in the environment of neurons in the primary somatosensory cortex of the cat led to an increase of the receptive field of these neurons. Therefore, GABA does seem to be a good candidate to play a role in the expression of adult plasticity in SI. If so, one may expect its presence in the somatosensory cortex to be under the influence of peripheral sensory activity.

Following removal of one eye or unilateral eyelid suture in the adult monkey, Hendry and Jones (1986) demonstrated a reduction in GABA- and GAD-immunoreactivity[2] in the primary visual cortex. The effects were restricted to those parts of the visual cortex that were under the control of the affected eye. They were present after the shortest survival time used by these authors, i.e. two weeks after eye removal. Subsequently, the same authors demonstrated that the

---

[1] We define plasticity as the capacity of the nervous system to react in a specific, creative manner to factors that influence its normal development or its maintenance.
[2] GAD = Glutamic acid decarboxylase, the rate-limiting enzyme in the synthesis of GABA.

level of GABA-immunoreactivity returned to normal after reopening the eye (Hendry and Jones, 1988). They concluded that the GABA-level in the visual cortex is regulated by sensory experience.

Apart from those mechanisms that operate within the central nervous system (perhaps best alluded to as "psychological" mechanisms), variations in sensory experience may be considered as relative differences in the intensity of sensory stimulation between parts of the peripheral sheet. Sensory *deprivation* of a part of the sheet creates such a relative difference. Sensory *stimulation* would be its natural counterpoint. It is important to realize that most of our knowledge on adult plasticity comes from experiments using *deprivation* as the initiator of the differential sensory input.

Using the whisker-to-barrel pathway of the mouse (see Van der Loos et al. this volume) we investigated the effects of *decreased* and *increased* sensory stimulation upon GAD-immunoreactivity at the level of the barrel cortex[3]. In what follows we shall review these two sets of experiments (see for details and full documentation Welker et al., 1989a, b).

GAD-immunohistochemistry in the barrel cortex

GAD-immunoreactivity[4] is present in all layers of the barrel cortex. Layer IV is the most densely GAD-labeled layer, whereas layer Va is almost devoid of labeling. Variation between layers is mainly due to differences in density of "puncta", (a term we use as a synonym for buttons).

In tangential sections through the barrelfield GAD-immunoreactivity reveals a clear barrel pattern. This is due to the near-absence of immunoreactive elements in the septa between barrels, whereas the hollows and sides of the barrels are densely visited by GAD-positive cell bodies and puncta. At mid-height of a barrel, i.e. in the middle of layer IV, the GAD-positive puncta are concentrated in a ring comprising the central rim of the barrel side and the peripheral part of the barrel hollow. This ring surrounds the center of the hollow which is less densely packed with labeled boutons. GAD-positive cell bodies and puncta vary much in size and shape and in intensity of staining. The puncta form baskets around labeled and unlabeled somata, but also occur "free", i.e. unrelated to other visible structures.

GAD-immunoreactivity after peripheral deprivation

After surgical removal of the four caudal-most follicles of row C and of row D in adult mice, GAD-immunoreactivity in the corresponding barrels drops significantly (Fig. 1). This decrease is relative to the GAD-immunoreactivity observed in the neighboring barrels corresponding to intact follicles and also to that in barrels of control mice (Welker et al., 1989a). This effect was due to a diminution in the estimated numerical density of labeled puncta, as well as to a decrease in the intensity of staining of the GAD-positive somata and the remaining puncta. The density of GAD-positive cell bodies does not seem to have changed. The decrease in GAD-staining was present throughout the thickness of layer IV; no effects were observed in other layers. The decrease occurred as early as three days after the peripheral intervention[5]. The decrease in immunostaining seemed to reach a maximum after 2-4 weeks of peripheral deprivation, after which the reactivity returned gradually to normal levels (the longest survival time we tested was 7 months). Inspection of silver-impregnated sections of the whiskerpad revealed the presence of regenerating nerve fibers as early as two weeks after follicle removal. Most of the nerves formed neuromas just distal to the site where they had been transected.

---

[3] Barrel cortex is the part of the somatosensory cortex of the mouse that contains in layer IV barrels corresponding to the large, posterior mystacial vibrissal follicles (roughly equivalent to the "PMBSF" of Woolsey and Van der Loos, 1970).
[4] Our protocol is based on that of Mugnaini and Dahl (1983); and Baer et al. (1985). We used the GAD-antibody of Oertel et al. (1981).
[5] In deprived barrels of a mouse that was killed two days after surgery, there was only a slight decrease of GAD-immunoreactivity. Unpublished observation.

Careful single-unit analyses (Armstrong-James and Fox, 1987) have shown the existence of a surround receptive field[6] for neurons in all cortical layers of the rat barrel cortex. Assuming that on this score mouse and rat do not differ much, it seems that in the case of prolonged peripheral stimulation this surround receptive field is not powerful enough to spread the GAD-effect to neighboring barrel columns: in no case did we find a significant "spill-over" of the increased GAD-activity into the barrels adjacent to the stimulated one.

Recent studies in our laboratory have shown that following denervation of a set of follicles the representation of the neighboring untouched follicles expands into the deprived cortical zone (Melzer et al., 1988; Welker et al., 1988). Here we propose that the initial drop in GAD-levels may well be the "permissive" factor that allows this reorganization of the functional map to occur. The expansion can be interpreted as an increase in the relative impact of the surround receptive field upon neurons in the deprived cortical zone. The appearance of the new functional arrangement within the deprived zone could well be the driving force behind the return of normal GAD-levels. Interestingly, the return of normal GAD-levels 6 weeks after the peripheral intervention does not seem to coincide with the end of the reorganization of the map which continues up till at least 160 days after the lesion (Melzer et al., 1988)[7].

We have shown that GAD-immunoreactivity in layer IV of the barrel cortex of the adult mouse can be influenced by modifications in the sensory periphery. We argued that the GAD-level in layer IV reflects peripheral sensory activity and can be down-regulated or up-regulated after sensory deprivation or sensory stimulation, respectively. Assuming that this regulation has immediate and parallel effects on GABA-levels, more or less inhibitory activity would occur in the barrel column in question. We consider the experimental conditions under which these results were obtained as extreme cases of what may happen in a natural setting, and propose that GAD-regulation is a powerful mechanism of the adult cerebral cortex to adapt its organization to changes in the animal's experience in a world contained in its whisker-space.

Consideration of our results together with those of Hendry and Jones (1986) in the visual cortex of the monkey tends to suggest that the relationship between peripheral activity and cortical GAD-levels may exist in all primary sensory cortices. However, visual deprivation in the adult cat does not modify GAD-levels in area 17 (Bear et al., 1985; Benson et al., 1989). Although we should be flattered by the notion that the barrel cortex of the mouse resembles more the visual cortex of the monkey than that of the cat, we are disappointed that a generalization of the relationship mentioned above can not be made. It would be of interest to know which factors may influence GAD-levels in *non*-primary sensory cortical areas.

Interestingly, Artola and Singer (1987) were able to elicit long term potentiation (LTP) in slices of rat visual cortex only after reduction of GABAergic inhibition using bicuculline. Our GAD-decrease obtained after peripheral sensory loss may well constitute a "natural" means by which LTP may be made to occur in the neocortex, although a biological parallel for the synchronous tetanic stimulus of 10-50 Hz is yet to be found.

While we can offer some suggestions about the role of GABA in the reorganization of the peripheral map in SI after sensory deprivation, we still lack data on the physiological effects of an increased GABA-level. However, preliminary results of a deoxyglucose study in behaving (i.e. whisking) mice indicate that prolonged peripheral stimulation results in cortical habituation: the animal's use (by whisking) of the follicles submitted to prolonged stimulation, along with all other whiskers, resulted in a relatively low deoxyglucose uptake in the corresponding barrels as compared to the uptake in the neighboring not pre-stimulated barrels (Rao et al., 1989). That GABA may be involved in cortical habituation was suggested by Creutzfeldt and Heggelund (1975) who based their conclusion on experiments on the visual system of adult cats. The GAD-

---

[6] The surround receptive field of a barrel neuron is defined as the ensemble of responses of this neuron upon stimulation of the whisker follicles that surround the principal (central) follicle.

[7] However, we have to be prudent in the juxtaposition of the results derived from the two experiments. Although both were performed on adult mice, they were not identical: the reorganization of the map has been studied after nerve division, whereas the GAD-results were based on follicle removal.

increase that we reported here may well be the mechanism that underlies cortical habituation. Future work will be directed towards a characterization of the physiological effects of prolonged sensory stimulation on the response characteristics of single neurons in the barrel cortex of the mouse.

ACKNOWLEDGMENTS

We thank Dr. M.L. Tappaz (Lyons, France) for the GAD-antibody, Mr. S. Daldoss for assistance in preparing the figure, and the Swiss National Science Foundation for support (grant 3100-009468).

REFERENCES

Armstrong-James M., Fox K. (1987) Spatiotemporal convergence and divergence in the rat S1 "barrel" cortex. *J. Comp. Neurol.*, 263: 265-281
Artola, A., Singer, W. (1987) Long-term potentiation and NMDA receptors in rat visual cortex. *Nature*, 330: 649-652.
Bear, M.F., Schmechel, D.E., Ebner, F.F. (1985) Glutamic acid decarboxylase in the striate cortex of normal monocularly deprived kittens. *J. Neurosci.*, 5: 1262-1275.
Benson, D.L., Isackson, P.J., Hendry, S.H.C., Jones, E.G. (1989) Expression of glutamic acid decarboxylase mRNA in normal and monocularly deprived cat visual cortex. *Mol. Brain. Res.*, 5: 279-287.
Calford, M.B., Tweedale, R. (1988) Immediate and chronic changes in responses of somatosensory cortex in adult flying-fox after digit amputation. *Nature*, 332: 446-448.
Creutzfeldt, O.D., Heggelund, P. (1975) Neural plasticity in visual cortex of adult cats after exposure to visual patterns. *Science*, 188: 1025-1027.
Dykes, R.W., Landry, P., Metherate, R., Hicks, T.P. (1984) Functional role of GABA in cat primary somatosensory cortex: shaping receptive fields of cortical neurons. *J. Neurophysiol.*, 52: 1066-1093.
Frost, D.O., Caviness, V.S. Jr. (1980) Radial organization of thalamic projections to the neocortex in the mouse. *J. Comp. Neurol.*, 194: 369-393.
Hendry, S.H.C., Jones, E.G. (1986) Reduction in number of immunostained GABAergic neurones in deprived-eye dominance columns of monkey area 17. *Nature*, 320: 750-753.
Hendry, S.H.C., Jones, E.G. (1988) Activity-dependent regulation of GABA expression in the visual cortex of adult monkeys. *Neuron*, 1: 701-712.
Melzer, P., Van der Loos, H., Dörfl, J., Welker, E., Robert, P., Emery, D., Berrini, J.C. (1985) A magnetic device to stimulate selected whiskers of freely moving or restrained small rodents: its application in a deoxyglucose study. *Brain Res.*, 348: 229-240.
Melzer, P., Yamakado, M., Van der Loos, H., Welker, E., Dörfl, J. (1988) Plasticity in the barrel cortex of adult mouse: effects of peripheral deprivation on the functional map; a deoxyglucose study. *Soc. Neurosci. Abstr.*, 14: 844.
Merzenich, M.M., Kaas, J.H., Wall, J., Nelson, R.J., Sur, M., Felleman, D. (1983) Topographic reorganization of somatosensory cortical areas 3b and 1 in adult monkeys following restricted deafferentation. *Neuroscience*, 8: 33-55.
Mugnaini, E., Dahl, A.L. (1983) Zinc-aldehyde fixation for light-microscopic immunocytochemistry of nervous tissues. *J. Histochem. Cytochem.*, 31: 1435-1438.
Oertel, W.H., Schmechel, D.E., Tappaz, M.L., Kopin, I.J. (1981) Production of a specific antiserum to cat brain glutamic acid decarboxylase by injection of an antigen-antibody complex. *Neuroscience*, 6: 2689-2700.
Rao, S.B., Welker, E., Dörfl, J., Melzer, P., Van der Loos, H. (1989) Plasticity in the barrel cortex of the adult mouse: effects of prolonged whisker stimulation on acute stimulus evoked DG-uptake. *Soc. Neurosci. Abstr.*, 15: 1222.
Van der Loos, H., Woolsey, T.A. (1973) Somatosensory cortex: structural alterations following early injury to sense organs. *Science*, 179: 395-398.
Van der Loos, H., Welker, E., Dörfl, J., Hoogland, P.V. (1990) Brain maps: development, plasticity and distribution of signals beyond. In: The Neocortex: Ontogeny and Phylogeny. Finlay, B.L., Innocenti, G., Scheich, H. (Eds.), Plenum Press, NY.
Wall, P.D., Gutnick, M. (1974) Ongoing activity in peripheral nerves: the physiology and pharmacology of impulses originating from a neuroma. *Exp. Neurol.*, 43: 580-593.

Welker, C. (1976) Receptive fields of barrels in the somatosensory neocortex of the rat. *J. Comp. Neurol.*, 166: 173-189.

Welker, E., LeClerc, S.S., Van der Loos, H., Yamakado, M., Dykes, R.W. (1988) Plasticity in the barrel cortex of adult mouse: effects of peripheral deprivation on the functional map; an electrophysiological recording study. *Soc. Neurosci. Abstr.*, 14: 843.

Welker, E., Soriano, E., Van der Loos, H. (1989a) Plasticity in the barrel cortex of the adult mouse: effects of peripheral deprivation on GAD-immunoreactivity. *Exp. Brain Res.*, 74: 441-452.

Welker, E., Soriano, E., Dörfl, J., Van der Loos, H. (1989b) Plasticity in the barrel cortex of the adult mouse: transient increased GAD-immunoreactivity following sensory stimulation. *Exp. Brain Res.*, 78: 659-664..

White, E.L. (1979) Thalamocortical synaptic relations: a review with emphasis on the projections of specific thalamic nuclei to the primary sensory areas of the neocortex. *Brain Res.*, 180: 275-311.

Woolsey, T.A., Van der Loos, H. (1970) The structural organization of layer IV in the somatosensory region (SI) of mouse cerebral cortex. The description of a cortical field composed of discrete cytoarchitectonic units. *Brain Res.*, 17: 205-242.

Woolsey, T.A., Dierker, M.L., Wann, D.F. (1975) Mouse SmI cortex: qualitative and quantative classification of Golgi-impregnated barrel neurons. *Proc. Natl. Acad. Sci. (USA)* 72: 2165-2169.

TRANSIENT RECEPTOR EXPRESSION IN VISUAL CORTEX DEVELOPMENT AND THE MECHANISMS OF CORTICAL PLASTICITY

M. Cynader*, C. Shaw*, F. van Huizen** and G. Prusky***

*Department of Ophthalmology
University of British Columbia Vancouver
Vancouver, British Columbia, Canada

**Department of Neuropharmacology
Organon International bv
The Netherlands

***Department of Biology
Yale University
New Haven, Connecticut, U.S.A.

During the first few months of postnatal life, the kitten visual cortex exhibits a remarkable plasticity, enabling it to match its organization to the visual environment with which it is confronted. If the visual environment is abnormal, as occurs with early monocular cataract or refractive error, then binocular competition results, with the eye that sees less well permanently losing cortical territory to the non-deprived eye (Cynader and Chernenko, 1976; Wiesel and Hubel, 1963a; Wiesel and Hubel, 1965). This binocular competition occurs only during a well-defined critical period which begins about three weeks after birth, peaks sharply between four and six weeks of age, and then declines slowly until the animal is about six months of age (Cynader et al., 1980; Hubel and Wiesel, 1970). We have investigated the mechanisms underlying binocular competition and the critical period that underlies it. In a series of electrophysiological experiments that have been described in detail elsewhere, the following features of binocular competition are now well established: (a) Binocular competition involves changes at the level of the lateral geniculate nucleus (LGN) (Wiesel and Hubel, 1963b), the terminals of LGN afferents in the cortex (Hubel et al., 1977) and in the responses of cortical cells (Wiesel and Hubel, 1963a), but the primary event in binocular competition appears to involve the responses of the postsynaptic cells in Layer IV. This has been established by investigations showing that binocular competition can be made to be completely orientation dependent (Cynader and Mitchell, 1977), a property of cortical cells and not their afferents (Hubel and Wiesel, 1962), and by experiments which involve selective excitation (Shaw and Cynader, 1984) or inhibition (Reiter and Stryker, 1988) of cortical cells. In these latter experiments it has been shown that disturbing the responsivity of cortical cells without affecting the asymmetric input from the LGN that occurs with monocular suture is sufficient to prevent the effects of monocular deprivation. (b) The duration of the critical period described above can be influenced by the visual history of the immature organism. If kittens are kept in the dark throughout the naturally occurring critical period (Cynader and Mitchell, 1980) then it is possible to extend the developmental period over which binocular competition can occur. In subsequent experiments (Cynader, 1983) it has been shown that maintaining kittens in the dark for a period of up to two years still results in a cortex with significant plasticity as assessed using the monocular deprivation paradigm. (c) The importance of extra-visual factors in determining the sensitivity of the visual system to binocular competition has become clear over the years. After the first demonstrations that it was possible to modify cortical ocular dominance by closing one eye for very brief periods, several different investigators attempted to record from the visual cortex of lightly anesthetized paralyzed kittens, to monitor the alterations of cortical ocular dominance in individual cells while they were being recorded. In general, these experiments failed to convincingly demonstrate effective modifica-

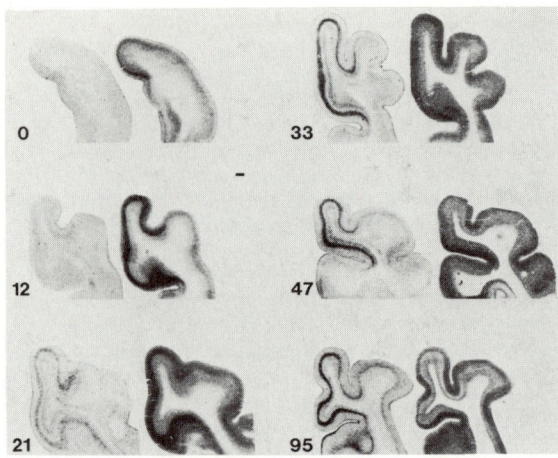

Figure 1. The distribution of [³H]nicotine and [³H]QNB (muscarinic) binding sites in the visual cortex of kittens of different ages. The photographs represent kittens of ages 0, 12, 21, 33, 47 and 95 days of age. The number to the left of each pair of sections represents the ages of the kittens. The left hand member of each pair of sections illustrates nicotine binding sites. Note the age-related increase in the binding in Layer IV and to a lesser extent in Layers I and VI in the primary visual cortex. There are few laminar specific binding sites in young animals but binding sites begin to concentrate in Layer IV at 21 days and reach their full adult density by postnatal day 47. The right hand member of each pair of sections represents [³H]QNB binding. Note the high density in Layer IV in 0 and 12 day old kittens. By 21 days of age, muscarinic receptors are also found in superficial layers and in the 33 day old kittens Layers I through IV are a zone of concentration. At 47 days of age, the middle layers of the cortex are showing lower levels of binding than the superficial and deep layers and this trend continues and becomes more pronounced in the 95 day old. The pattern of binding in adult cats is very similar to that observed in the 95 day old kitten. The scale bar is one millimeter.

tion of neuronal responses despite the fact that neurons responded well to the visual stimuli presented through the non-deprived eye at the time of recording. The reasons for the inability of visual stimuli to modify cortical neuronal organization in acute conditioning experiments remain unclear. It has been suggested (Singer and Rauschecker, 1982) that signals from ocular proprioception are necessary, and that it is essential that the eyes move about for ocular dominance to shift. This finding may be related to evidence that modifiability of cortical ocular dominance requires attentional mechanisms mediated by monoaminergic and/or cholinergic systems which terminate heavily within the cortex (Bear and Singer, 1986; Daw et al., 1984; Kasamatsu and Pettigrew, 1979). Perhaps the most neutral statement we can make in the face of our current ignorance is that some sort of enabling input seems to be important to allow cortical modifiability to occur during the critical period. The nature of this enabling signal is not known, but it may well be linked to the substantial innervation of neocortex by neuromodulatory substances.

Cellular mechanisms of visual cortex plasticity: developmental alterations of neurotransmitter receptors

Neurotransmitter receptors are protein moieties located on the surfaces of neurons or glial elements, which bind specific hormones or neurotransmitters and hence evoke cellular responses. We have used *in vitro* autoradiographic methods to localize and characterize receptors associated with neurotransmitters and neuromodulators in the developing visual system. Figure 1 illustrates the distribution and development of two classes of acetylcholine-related receptors in the visual cortex of kittens of various ages. The left-hand member of each pair represents nicotinic binding sites, and the right-hand member of each pair represents muscarinic binding sites, studied in an adjacent section of the same kitten.

The ages of the kittens are noted below the autoradiograms. The nicotinic sites show a gradual increase during development with low concentrations in the zero- and 12-day-old animals, and high levels of binding in Layer IV by 33 and 47 days of age. Note the remarkable concentration in the 33 and 47-day-old animals of receptors in the visual cortex as opposed to the subadjacent cingulate cortex. In other experiments, which are reported elsewhere (Prusky and Cynader, 1988a; Prusky and Cynader, 1988b; Prusky et al., 1987) we have localized these receptors to LGN terminals within the striate cortex. Note that, regardless of the animal's age, the greatest concentration of nicotine receptors is found in Layer IV of the cortex. The constancy of laminar distribution for the nicotinic binding sites contrasts with the developmental pattern observed for the muscarinic binding sites (the right-hand member of each pair in Figure 1). In the neonatal animal, muscarinic binding sites are concentrated in the middle layers of the cortex, and are present in far greater numbers than are nicotinic receptors. Note the concentration within the visual cortex, and near absence in neighboring cortical fields, in neonatal kittens. At 12 days of age, the increased preponderance of muscarinic receptors is still observed and they are still located in Layer IV, but by 21 days of age, as *nicotinic* receptors begin to appear in Layer IV, one notices muscarinic receptors located in the superficial cortical layers as well. This trend is even more pronounced at 33 days of age, by which time muscarinic receptors appear to be located in nearly all of the superficial (Layers I-IV) cortical layers. At 47 days of age, muscarinic receptors are concentrated in Layers I-III and VI. This pattern is retained, with some further thinning out of muscarinic receptors in Layer IV during the next several weeks, and the 95-day-old pattern is similar to that observed in adults. The adult pattern emphasises inner and outer cortical layers while Layer IV is sparsely labelled. Thus muscarinic acetylcholine receptors, unlike their nicotinic counterparts, appear to change their *laminar distribution* during the critical period. From an initial concentration in Layer IV (the input layer), these binding sites essentially *reverse* in distribution to favour all cortical layers except Layer IV by adulthood (Shaw et al., 1984).

We have examined over twenty different binding sites in the kitten visual cortex as a function of the animal's age. We find in fact that the type of receptor redistribution illustrated by the muscarinic binding sites is a common pattern. Nearly two-thirds of all the receptor populations that we have studied show some form of redistribution in visual cortex during postnatal development (Cynader and Shaw, 1986; Shaw et al., 1986). Figure 2 is a summary diagram illustrating the distribution of neurotransmitter receptors of various types in the cortex of neonatal vs. adult animals. (For references see Cynader and Shaw, 1986; Shaw et al., 1986.) At the top of each column the name of the particular neurotransmitter-neuromodulator and the associated ligand is displayed. The layers of the cortex are illustrated on the left of each figure. The density of receptors within each layer relative to other cortical layers is indicated by the density of shading in each cell of Figure 2. Darker shading indicates greater receptor density. For some neurotransmitter receptors, including $GABA_A$, and some of the subsets of glutamate receptors, the relative distribution of binding sites appears to remain constant throughout post-natal development, with the former sites concentrated in Layer IV throughout development, and the latter favouring superficial and deep layers throughout. Other binding sites, including muscarinic receptors, ß-adrenergic receptors, calcium channel binding sites, opiate receptors, and cholecystokinin-5 receptors, show a pattern in which Layer IV has the densest binding early in life, with this pattern reversing to involve a preferential distribution within superficial and deep layers later in development. Individual receptors may show highly individual and idiosyncratic patterns. The A_1-adenosine receptor, labelled with [^{3}H]cyclohexyladenosine (CHA), favours the deep cortical layers early in life and superficial and deep layers in adulthood. Other receptors, including those for nicotine and $GABA_B$, are not observed at birth, but are clearly present in adulthood. Certain patterns are notable by their absence. We have yet to find any binding sites that concentrate in superficial cortical layers early in life and then favour deeper layers in adulthood. This may reflect the ontogenetic history of the cortex, with cells of the superficial layers being generated later in life than those of the deep layers. It would be far too simple, however, to suggest that any or all of the receptor redistributions that are depicted in Figure 2 simply follow the ontogenetic history of cortical neurons.

Thus different transmitter receptors appear to undergo specific changes in distribution during postnatal development. In all cases the receptor redistributions take place during the physiologically-defined critical period for alteration of cortical unit responses. The ontogenetic time courses of the alterations in distribution also vary from receptor to receptor. This may even apply to receptors which show the same neonatal and adult pattern of redistribution. For example the muscarinic receptors, labelled with [^{3}H]quinuclidinylbenzilate (QNB) and the calcium binding sites (labeled with PN200) are both found in Layer IV in young animals and superficial and deep layers of old animals, but the calcium binding sites appear to vacate Layer IV by 20 days of age, while muscarinic receptors can still be found in this layer at 30 days of age. In some

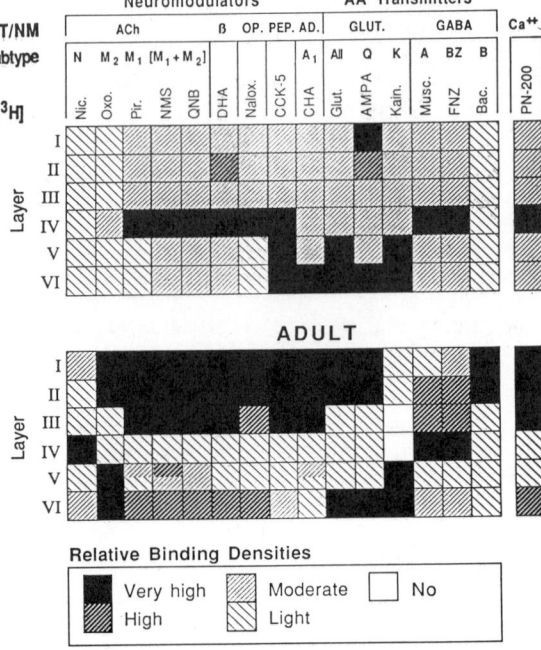

Figure 2. A schematic distribution of the laminar binding patterns for 16 binding sites studied in neonatal (1-3 day old) kittens and in adult cats. The top row of the figure lists the neurotransmitters (NT) and neuromodulators (NM) whose receptors have been examined. These include those for acetylcholine (ACH), beta-adrenergic (BETA), opiates (OP), peptides (PEP), and for A_1-adenosine (AD) within the broad class of neuromodulatory substances. Binding sites for glutamate and GABA represent the amino acid transmitters. The various receptor subtypes are illustrated on the next lower row. The tritiated ligand employed is illustrated in the subsequent row. The cortex is divided into 6 layers and the density of binding is illustrated using the shading scale given at the bottom of the figure.

In neonatal kittens the most common patterns of binding observed are either little or no discernible binding (nicotine, oxytremorine M_2 muscarinic, baclofen ($GABA_B$)), or a pronounced peak in Layer IV. The glutamate-related binding sites appear concentrated in Layer VI in the neonatal kitten. In adult animals, the pattern is very different, with relatively few binding sites (nicotine, muscimol and flunitrazepam) concentrated in Layer IV. Most receptors show a distribution emphasizing Layers I through III and VI.

cases, pairs of receptors may show the same laminar distribution in both neonatal and adult animals (e.g., pirenzepine and oxotremorine, muscarinic M_1 and M_2 subtypes respectively), but may show very different patterns at *intermediate* ages.

Effects of input on postnatal development of neurotransmitter receptors

We have noted earlier that the critical period for cortical plasticity can itself be altered by preventing the visual system from obtaining a normal pattern of input during postnatal development. We have found that the redistribution of neurotransmitter receptors that is normally observed during the critical period is also dependent on normal input to the cortex. Figure 3 illustrates the effects of surgically undercutting a portion of the visual cortex in a young animal, 24 days of age, and allowing the animal to survive throughout the period over which muscarinic acetylcholine receptors would normally have altered their laminar distribution (until 49 days of age). This manipulation effectively removes input from the undercut zone during the developmental process. Figure 3 is comprised of several panels illustrating: (A) The effect of the surgical undercut on Nissl staining in the cortex; little effect is observed indicating the normality of the the tissue; (B) This conlcusion is reinforced by the relative lack of alteration of cytochrome

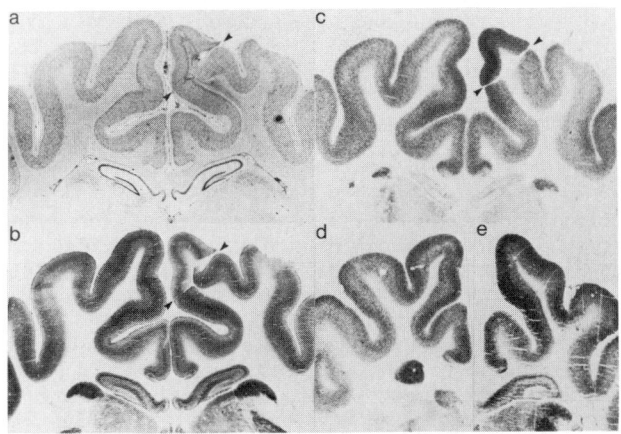

Figure 3. The effects of subjecting a kitten to surgical undercutting of visual cortex at 24 days postnatal and allowing the animal to survive until 49 days of age. a: A section of the control and undercut cortex was stained for Nissl substance with cresyl violet. The undercut cortex appears normal except for gliosis along the cut. Control cortex is on the left. b: Alternate sections of both cortices were processed for cytochrome C oxidase histochemistry. The reaction products, especially the dense band in Layer IV, are less well-defined but still present in the undercut zone as in the normal cortex. This difference from normal cortex may reflect maturational differences in the undercut zone. c-e: Autoradiograms illustrate the distribution of muscarinic ACh receptors (left). An adjacent section to those of Fig. 3a and 3b has been incubated with [^{3}H]QNB to visualize the distribution of muscarinic cholinergic receptors. The resulting autoradiogram shows that receptors in the control hemisphere (left hand side of Figure 3c) are concentrated in Layers I-III and VI. In the undercut zone of the opposite hemisphere (right hand side of Figure 3c), receptors are instead concentrated in cortical Layer IV. This pattern closely resembles that of a normal 27-day-old kitten shown in Figure 3e for comparison. Figure 3d illustrates muscarinic receptor binding in an unoperated kitten of 46 days of age. Note the similarity to the pattern observed in the control hemisphere of Figure 3c. Muscarinic ACh receptors were labelled with 5-7nM [^{3}H]QNB for 60 min at 20°C. Nonspecific binding was determined by coincubation of alternate sections with 10^{-4}M atropine sulfate. Ultrofilm exposure: c, 14 days; d, 12 days; e, 16 days. All autoradiograms have been photographed for highest contrast. For all panels, calibration bar = 1 mm. For further details, see Shaw et al., 1988.

oxidase activity in the undercut zone; (C) The distribution of muscarinic receptors in the undercut zone is compared with that outside the undercut zone. In the control cortex (Figure 3c left) muscarinic ACh receptor distribution is typical of that for an animal of this postnatal age (49 days) (Figure 1), i.e., Layers I-III and VI show the densest binding. These results are in marked contrast to the binding observed in the undercut zone of the opposite cortex which more closely resembles that typically seen in the cortex of a much younger kitten (i.e., the age of this kitten at the time of surgery). Within the isolated zone, the binding is densest in Layer IV and appears denser overall than outside the undercut area. Immediately outside the undercut zone, the binding pattern resembles that of the control cortex. Figure 3d illustrates [^{3}H]QNB binding in a normal 46 day old kitten for comparison to the control cortex in Figure 3c. Figure 3e illustrates [^{3}H]QNB binding in a normal 27 day old kitten for comparison with the undercut zone of Figure 3c.

The failure of receptors to redistribute following isolation of part of the visual cortex early in life appears to be a general phenomenon. We have observed effects like those illustrated in Figure 3 for several other neurotransmitter receptors which change their distribution (including those for cholecystokinin, adenosine, and various muscarinic subtypes). We have yet to observe the normal developmental pattern of redistribution of any neurotransmitter receptor in animals in which the surgical isolation was performed early in life. The data thus lend support to the notion that basic receptor characteristics in cat visual cortex are under the influence of extracortical factors. The mechanisms by which external inputs regulate the receptor redistribution normally observed as the animal ages remain uncertain. Our undercutting procedure removes many sources of extracortical input from the affected zone, including those representing visual influences from the thalamus, and also modulatory influences conveyed via cholinergic, noradrenergic, and serotonergic inputs. The role of alterations in cortical *electrical* activity (which may be especially disrupted in Layer IV, the major input layer) in the failure of the receptors to redistribute, remains uncertain as well. In addition, a trophic factor or factors released by one or more of the cortical afferent systems may play an important role in receptor redistribution and this may be affected by the undercut procedure.

Cellular mechanisms underlying redistribution of neurotransmitter receptors

The striking alteration in the distribution of neurotransmitter receptors during development could have several causes. (a) Receptor redistribution occurs because of migration of neurons or glia with which the receptors are associated. (b) Receptor redistribution occurs because receptors are located on axons or dendrites and these processes are continuing to extend during postnatal development. The receptors are simply carried along by the extending processes and thus attain different distributions as a function of age. (c) Receptors are transiently expressed by certain populations of neurons or glial cells at certain stages of cortical development and are then eliminated and proliferate *de novo* in other neuronal populations at later stages. Receptor autoradiography is in general unable to distinguish between the possibilities outlined above. While the method is reliable and quantifiable, its spatial resolution is insufficient to visualize the individual somata or neuronal processes on which the receptors in question are located.

To obtain higher resolution information on this issue, we have used monoclonal antibodies against receptor binding sites to visualize neurotransmitter receptors and their alterations during postnatal development. Figure 4 illustrates binding of muscarinic receptors using the monoclonal antibody M35 (Van Huizen et al., 1988). The top panels of Figure 4 are low-power views comparing the binding of receptors in 28-day-old and adult cats. In the young kitten, whose binding pattern is illustrated at the top left of Figure 4, receptors are concentrated in Layer IV with some binding at the top of Layer II, (a pattern similar to, but not exactly the same as, that observed with [^{3}H]QNB binding). In the adult animal illustrated on the top right panel of Figure 4, the distribution favors superficial and deep layers, as is observed with [^{3}H]QNB autoradiography in normal adult animals. The lower panels of Figure 4 show that the binding in Layer IV of the young kitten cortex is concentrated around the somata of stellate cells. The binding clearly outlines the cell bodies with some dendritic involvement as well. The lower right panel of Figure 4 is a high-power view of Layer V binding in the adult animal. Here again the binding is concentrated on neuronal somata with some receptors clearly distributed along apical dendrites. In this case however it is *pyramidal* cells of Layer V that appear to be heavily labelled. Thus, in 28 day old kittens, stellate cells of Layer IV preferentially express muscarinic binding sites, whereas these neurons no longer express this binding site in adult animals. Instead pyramidal cells of Layer V and other pyramidal cells of the superficial layers (not shown here in high power) seem to preferentially express this receptor. It thus appears that receptors are transiently expressed by some neuronal populations during the critical period, and expressed by other neurons at later ages.

Suspicious coincidences and cortical plasticity mechanisms

It appears clear that parallelling the critical period for cortical plasticity is a remarkable alteration in the number and distribution of neurotransmitter receptors, especially those associated with neuromodulatory inputs to the visual cortex. This redistribution of receptors appears to involve transient expression of these receptor populations on some populations of cells early in life and then on different populations of cells later in life. It has also become clear that extracortical input can influence both the duration of the critical period and the timing and extent of receptor redistributions. How might the alteration of receptors that we have observed contribute to the mechanisms of cortical plasticity? The short answer is that we still do not know. However, one can imagine several distinct but non-exclusive mechanisms based on *coincidences* in the distribution of neurotransmitter receptors in animals of different ages. For instance, if one recalls the developmental distribution of muscarinic and nicotinic receptors (Figure 1) in the kitten visual cortex, it is clear that muscarinic receptors are vacating Layer IV at the same time as nicotinic binding sites are becoming more numerous in this layer. There is however a brief window of time, from about 20 to 35 days of postnatal age during which both these binding sites are concentrated within Layer IV. It is established that the nicotinic binding sites are located presynaptically, on LGN terminals, and that the muscarinic sites are postsynaptic (Prusky and Cynader 1988a; Prusky and Cynader 1988b; Shaw et al., 1987). The coincidence of pre- and post-synaptic sites in Layer IV during the height of the critical period provides a mechanism by

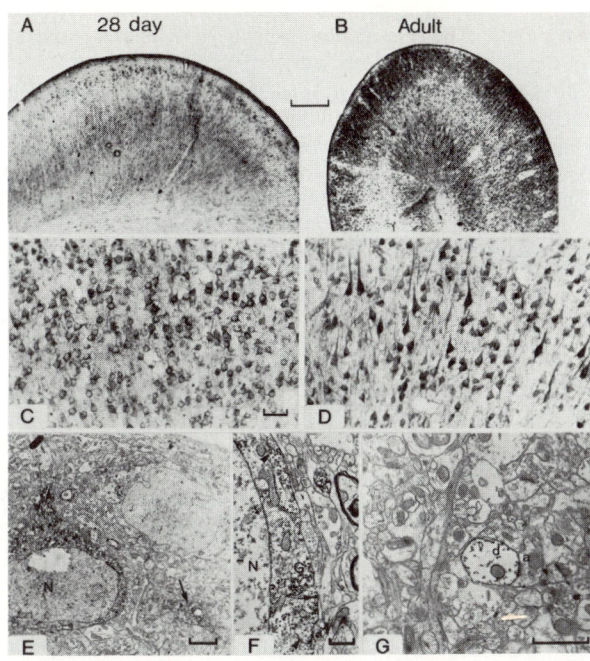

Figure 4. Immunocytochemical labelling of the muscarinic receptor in 28 day old (left panel) and adult (right panel) cat visual cortex using a monoclonal antibody. The upper panels represent low-magnification views of the pattern of binding in the crown of the visual cortex. Note that the immunostained cells are most dense in Layer IV in the young animal and are least dense in this layer of the adult animal. Scale bar = 0.5 millimeters. The lower two panels illustrate higher power light micrographs of tissue showing the densely labelled stellar cells in Layer IV of the young kitten (left panel) and of pyramidal cells in Layers V and VI of the adult cat cortex (right hand panel). Scale bar for lower panels = 50 microns.

which the same transmitter, acetylcholine, could simultaneously modulate both pre- and post-synaptic function with these two receptor populations in this layer. It could hence force correlations by coincident activation of pre- and post-synaptic targets. It may be that this forced correlation, when combined with an asymmetry in the inputs from the two eyes, is a necessary substrate of the mechanism by which ocular dominance modification occurs during the critical period. Similar effects may also occur with other classes of receptor such as $GABA_A$ and $GABA_B$ receptors.

A second, possibly interrelated mechanism refers to the striking concentration of several putative postsynaptic receptor binding sites within Layer IV during the critical period. Thus beta-adrenergic, dihydropyridine sensitive calcium binding sites, muscarinic, cholecystokinergic, opiatergic, and other binding sites are all found concentrated in Layer IV near the height of the critical period. In general when neurotransmitter receptors are stimulated they produce intracellular effects via activation of second messenger systems. There is evidence that simultaneous activation of the *same* second messenger system via two *different* receptors can cause much more intense activation than input via either receptor alone (Christ et al., 1988; Morrison et al., 1984). Strong activation of the second messenger system would then result in phosphorylation of intracellular substrates, leading to longer-term alterations of membrane proteins and/or ion channels. One can readily make the argument that plasticity should not occur in response to every ongoing stimulus. Rather only specific situations, which meet certain stringent requirements should enable long term neuronal modifications to take place. These requirements might include unequal input from the two eyes, and activity in two or more of the neuromodulatory receptor systems (Bear and Singer, 1986). The simultaneous presence of two or more of the receptor populations listed above on the same postsynaptic cells in cortical Layer IV would be crucial to the function of such a mechanism, and it may not function after the critical period because these receptor populations have vacated Layer IV.

CONCLUSION

The cortex normally develops according to rules that lead to an effective information processing system, appropriate to the visual world in which the organism was reared early in life. Unfortunately, with peripheral anomalies such as refractive errors, cataracts, and astigmatism, that visual world of the developing organism may not be the same as that experienced by a normally-sighted individual. Since cortical function can normally be altered only during the critical period, a major goal of current research has been to understand the mechanisms by which this occurs, and to discover ways in which the critical period can be reinstated later in life. We are still far from achieving this goal but our emerging understanding of the neurochemistry of the cortex and its transmitter systems during the critical period surely constrain any theory that can be constructed to account for the ways in which visual inputs shape visual processing mechanisms.

REFERENCES

Bear, M. and Singer, W. (1986) Modulation of visual cortex plasticity by acetylcholine and noradrenaline. *Nature*, 320: 172-176.

Christ, G.J., Goldfarb, J., Melman, A., Osman, R. and Maayani, S. (1988) Synergism between activated membrane receptors that trigger the same effector mechanism. *Soc. Neurosci. Abstr.*, 14: 111.

Cynader, M. (1983) Prolonged sensitivity to monocular deprivation in dark-reared cats: effects of age and visual exposure. *Dev. Brain Res.*, 8: 155-164.

Cynader, M. and Chernenko, G. (1976) Some factors influencing the development of ocular dominance in the cat striate cortex. *Association for Research in Vision and Ophthalmology*, Sarasota, FL. April.

Cynader, M. and Mitchell, D.E. (1977) Monocular astigmatism effects on kitten visual cortex development. *Nature*, 270: 177-178.

Cynader, M. and Mitchell, D.E. (1980) Prolonged sensitivity to monocular deprivation in dark-reared cats. *J. Neurophysiol.*, 43: 1026-1040.

Cynader, M. and Shaw, C. (1986) Mechanisms underlying binocular competition in cat visual cortex. In Kellar, E. and Zee, D. (eds.) *Adaptive Processes in Visual and Oculomotor Systems*. Pergamon Press, 53-61.

Cynader, M., Timney, B.N. and Mitchell, D.E. (1980) Period of susceptibility of kitten visual cortex to the effects of monocular deprivation extends beyond six months of age. *Brain Res.*, 191: 515-550.

Daw, N.W., Robertson, T.W., Rader, R.K., Vedeen, T.O., and Cosica, C.J. (1984) Substantial reduction of noradrenaline by lesions of adrenergic pathways does not prevent effects of monocular deprivation. *J. Neurosci.*, 4: 1354-1360.

Hubel, D.H. and Wiesel, T.N. (1962) Receptive fields, binocular interaction and functional architecture in the cat's visual cortex. *J. Physiol. London*, 160: 106-154.

Hubel, D.H. and Wiesel, T.N. (1970) The period of susceptibility to the physiological effects of unilateral eyelid closure in kittens. *J. Physiol. London*, 206: 419-436.

Hubel, D.H., Wiesel, T.N. and Levay, S. (1977) Plasticity of ocular dominance columns in monkey striate cortex. *Phil. Trans. Roy. Soc. London, Ser. B*, 278: 377-409.

Kasamatsu, T. and Pettigrew, J.D. (1979) Preservation of binocularity after monocular deprivation in the striate cortex of kittens treated with 6-hydroxydopamine. *J. Comp. Neurol.*, 185: 139-162.

Morrison, J.H., Magistretti, P.J., Benoit, R. and Bloom, F.E. (1984) The distribution and morphological characteristics of the intracortical VIP-positive cell: an immunohistochemical analysis. *Brain Res.*, 292: 269-282.

Prusky, G.T. and Cynader, M.S. (1988a) [^{3}H]Nicotine binding sites are associated with mammalian optic nerve terminals. *Visual Neurosci.*, 1: 245-248.

Prusky, G.T. and Cynader, M.S. (1988b) The distribution and ontogenesis of [3H]nicotine binding sites in cat visual cortex. *Dev. Brain Res.*, 39: 161-176

Prusky, G., Shaw, C. and Cynader, M.S. (1987) Nicotine receptors are located on lateral geniculate nucleus terminals in cat visual cortex. *Brain Res.*, 412: 131-138.

Reiter, R.O. and Stryker, M.P. (1988) Neural plasticity without postsynaptic action potentials: less active inputs become dominant when kitten visual cortex cells are pharmacologically inhibited. *Proc. Nat. Acad. Sci. U.S.A.*, 85: 3623-3627.

Shaw, C. and Cynader, M. (1984) Disruption of cortical activity prevents alterations of ocular dominance in monocularly deprived kittens. *Nature*, 308: 731-734.

Shaw, C., Prusky G. and Cynader, M. (1988) Surgical undercutting prevents receptor redistribution in developing kitten visual cortex. *Visual Neurosci.*, 1: 205-210.

Shaw, C., Needler, M.C. and Cynader, M. (1984) Ontogenesis of muscarinic acetylcholine binding sites in cat visual cortex: reversal of specific laminar distribution during the critical period. *Dev. Brain Res.*, 14: 295-300.

Shaw, C., Prusky, G., van Huizen, F. and Cynader, M. (1987) Cellular localization of receptor populations in cat visual cortex using quinolinic acid lesions. *Soc. Neurosci. Abstr.*, 13: 1046.

Shaw, C., Wilkinson, M.W. Cynader, M., Needler, M.C., Aoki, C. and Hall, S.E. (1986) The laminar distributions and postnatal development of neurotransmitter and neuromodulator receptors in cat visual cortex. *Brain Res. Bull.*, 16: 661-671.

Singer, W. and Rauschecker, J.P. (1982) Central core control of developmental plasticity in the kitten visual cortex. II. Electrical activation of mesencephalic and diencephalic projections. *Exp. Brain Res.*, 47: 223-233.

Van Huizen, F., Strosberg, A.D. and Cynader, M. (1988) Cellular and subcellular localization of muscarinic acetylcholine receptors during postnatal development of cat visual cortex using immunocytochemical procedures. *Dev. Brain Res.*, 44: 296-301.

Wiesel, T.N. and Hubel, D.H. (1963a) Single-cell responses in striate cortex of kittens deprived of vision in one eye. *J. Neurophysiol.*, 26: 1003-1017.

Wiesel, T.N. and Hubel, D.H. (1963b) Effects of visual deprivation on morphology and physiology of cells in the cat's lateral geniculate body. *J. Neurophysiol.*, 26: 978-993.

Wiesel, T.N. and Hubel, D.H. (1965) Comparison of the effects of unilateral and bilateral eye closure on cortical unit responses in kittens. *J. Neurophysiol.*, 28: 1029-1040.

DIRECTORS

B. Finlay (Ithaca, NY - USA)
J.M. Innocenti (Lausanne - Switzerland)
H. Scheich (Darmstadt - FRG)

CONTRIBUTORS

F. Bonhoeffer	Max-Planck-Institut für Entwicklungsbiologie, Spemannstr. 35, D-7400 Tübingen - FRG
V.S. Caviness, Jr.	Dept. of Neurology, Developmental Neurobiology, Massachusetts General Hospital, Harvard Medical School, Boston, MA 02114 - USA
M. Cynader	Dept. of Ophthalmology, University of British Columbia Vancouver, Vancouver, British Columbia - Canada
J.A. De Carlos	Instituto de Neurobiología, Santiago Ramón y Cajal (CSIC), Avenida del Doctor Arce 37, 28002 Madrid - Spain
J.A. Del Río	Unidad de Biología Celular, Facultad de Biología, Universidad de Barcelona, Diagonal 645, Barcelona 08028
M.E. Diamond	Neurobiology Section and Center for Neural Science, Brown University, Providence, RI 02912 - USA
J. Dörfl	Institute of Anatomy, University of Lausanne, 1005 Lausanne - Switzerland
F.F. Ebner	Neurobiology Section and Center for Neural Science, Brown University, Providence, RI 02912 - USA
I. Ferrer	Unidad de Neuropatología, Departamento de Anatomía Patológica, Hospital Príncipes de España, Hospitalet de LLobregat, Barcelona 08097 - Spain
B.L. Finlay	Dept. of Psychology, Uris Hall, Cornell University, Ithaca, NY 14853 - USA
B. Fritzsch	Scripps Institute of Oceanography, UCSD, Dep. of Neurosciences, A-001, La Jolla, CA 92093 - USA
D.O. Frost	Dept. of Neurology, Massachusetts General Hospital, Boston, MA 02114 - USA
J.-F. Gadisseux	Developmental Neurology Unit, University of Louvain Medical School, Brussels, B-1200 - Belgium
P.V. Hoogland	Department of Anatomy, Vrije Universiteit, 1007-MC Amsterdam - The Netherlands
G.M. Innocenti	Institute of Anatomy, Universite de Lausanne, 9 rue du Bugnon, 1005 Lausanne - Switzerland
H.J. Jerison	Dept. of Psychiatry and Biobehavioral Sciences, UCLA School of Medicine, 760 Westwood Plaza, Los Angeles, CA 90024 - USA
H.J. Karten	Dept. of Neurosciences, M-008, University of California, San Diego, La Jolla, CA 92093 - USA
A.H.M. Lohman	Dept. of Anatomy and Embryology, Vrije Universiteit, P.O. Box 7161, 1007 MC Amsterdam - The Netherlands
L. López-Mascaraque	Instituto de Neurobiología, Santiago Ramón y Cajal (CSIC), Avenida del Doctor Arce 37, 28002 Madrid - Spain
C. Métin	Laboratoire des Neurosciences de la Vision, Université de Paris, Paris - France
J.-P. Mission	Dept. of Developmental Neurobiology, University of Liege - Belgium
S. Pallas	Dept. of Brain and Cognitive Sciences, E25-618, Massachusetts Institute of Technology, Cambridge, MA 02139 - USA

J.D. Pettigrew	Vision, Touch and Hearing Research Center, University of Queensland, St. Lucia, 4067 - Australia
G. Prusky	Dept. of Biology, Yale University, New Haven, CT 06510 - USA
P. Rakic	Yale University School of Medicine, New Haven, CT 06510 - USA
H. Scheich	Zoological Institute, Technical University Darmstadt, Schnittspahnstr. 3, 6100 Darmstadt - FRG
C. Shaw	Dept. of Ophthalmology, University of British Columbia Vancouver, Vancouver, British Columbia - Canada
T. Shimizu	Dept. of Neurosciences, M-008, University of California, San Diego, La Jolla, CA 92093 - USA
W. J.A.J. Smeets	Dept. of Anatomy and Embryology, Vrije Universiteit, P.O. Box 7161, 1007 MC Amsterdam - The Netherlands
E. Soriano	Unidad de Biología Celular, Facultad de Biología, Universidad de Barcelona, Diagonal 645, Barcelona 08028
N.V. Swindale	Dept. of Ophthalmology, University of British Columbia, Vancouver, V5Z 3N9 - Canada
T. Takahashi	Dept. of Neurology, Developmental Neurobiology, Massachusetts General Hospital, Harvard Medical School, Boston, MA 02114 - USA
F. Valverde	Instituto de Neurobiologia, Santiago Ramón y Cajal, Doctor Arce 37, 28002 Madrid - Spain
F. Van Huizen	Dept. of Neuropharmacology, Organon International bv - The Netherlands
H. Van der Loos	Institute of Anatomy, University of Lausanne, 1005 Lausanne - Switzerland
J. Walter	Max-Planck-Institut für Entwicklungsbiologie, Spemannstr. 35, D-7400 Tübingen - FRG
E. Welker	Institute of Anatomy, University of Lausanne, 1005 Lausanne - Switzerland

INDEX

A1, *see* Auditory cortex
Alar plate
  evolutionary changes, 107
Allometry
  Cope's law, 13
  of cortex compared to brain, 33-34
  and cortical development, 28
  endocast measurement, 12
  gestational length and brain size, 34-35
  statistical methods, 11-14
Alz-50 immunoreactivity, 193-195
Alzheimer's disease, 193, 195
Auditory cortex
  cochleotopic mapping, 207
  compared to visual cortex, 206-207
  compared to Field L, 131-132
  electrophysiological properties, 207
  frequency band laminae, 130-131
  in gerbil, 128-131
  medial geniculate input, 129
  parcellation of, 131
  significance of multiple areas, 131-132
  tonotopic maps, 131, 207
  visual response properties induced in, 208-209
Auditory system
  best frequency analysis, 123
  binaural interactions in chickens, 125
  birds compared to mammals, 119
  relation to electroreception, 106
  isofrequency contours, 124
  in reptiles, 61
  tonotopy
    and dendritic structure, 125
    connections between maps, 132
  in chickens, 121-124
Avian telencephalon
  organization, 76
  visual areas in, 76
Axonal elongation
  avoidance reactions, 201
  constraints on duration, 50
  growth cones, 201
Axonal guidance, 199-202
Axonal transience
  role in developmental plasticity, 224

Barrel cortex (*see also*
    Somatosensory cortex)
  alterations with stimulation, 238-240
  contrast with turtle somatosensory cortex, 169
  contrast with visual cortex, 241
  description, 28, 168, 229
  development of, 230-231
  example of cortical module, 168
  GABA in, 232, 238
  immunohistochemical demonstration of, 232
  plasticity of efferents, 234
  spiny stellate cells in, 93
Basket cells (*See* Nonpyramidal cells)
Bats
  independent lineages, 155
  superior collicular organization, 155
Bergmann glia (*See* Radial glia)
Binocular vision (*See also* Ocular dominance), 137
  optic decussations in birds, 139
Blindsight, 82

Callosal connections
  constraints on early growth, 48
  early development, 43
  plasticity, 37
  transience
    across-species timing, 43
    axonal maturation, 45
    histochemistry, 45
    mechanisms of axon removal, 44
Carnivore brains
  compared with Creodonts, 14
Cat
  spiny stellate cells in, 93
  thalamic projections to cortex, 89
Causal explanations
  in development, 44, 46
  and complex networks, 47
Cell death, 47
  afferent removal and, 105
  Alz-50 as a marker for, 195
  callosal section and, 37
  in differentiation of cortex and hippocampus, 190
  in evolution of electroreception, 105

and hippocampal generation, 190
midbrain damage and, 37
patterns in cortex, 37-38
in subplate, 97
Chandelier cells (*See* Nonpyramidal cells)
Chiroptera (*See* Bats)
Cholinergic receptors in neocortex
development of, 246-248
laminar distribution, 246-248
muscarinic vs nicotinic, 246-248
redistribution after undercut, 249
Computational models
relevance to evolution, 47
relevance to cortical self-organization, 225
Cope's law, 13
Cortical areas (*See* Cytoarchitectonic areas)
Cortical columns
intrinsic circuitry, 96-97
periodicity
ocular dominance, 115
orientation, 115
coverage constraints, 115
relationship to ontogenetic columns, 25
Cortical cytoarchitecture (*See* Cytoarchitectonic areas; Neocortical cytoarchitecture)
Cortical neurogenesis
cell lineage
determined by retrovirus, 25
determined by transplants, 25
compared to hippocampus, 188
relationship to gestational length, 35-36
Cortical plate
diagram, 22
Cortical point image
definition, 111
Corticocortical projections
transience, 44
Coverage
algorithms for computation, 113-114
in avian auditory system, 127
definition, 113
and ocular dominance, 113-114
optimization, 116
in visual cortex, 114
Critical periods
for ocular dominance, 245-246
receptor binding in Layer IV, 251
shift from muscarinic to nicotinic receptors, 250
for specification of cortical areas, 232
Crocodiles (*See* Reptiles)
Cytoarchitectonic areas (*See also* Neocortical cytoarchitecture)
evolution of new areas, 205
historical overview, 87
induction by afferents, 206, 214-215
intrinsic circuitry, 206, 223
radial glia and, 180
Cytogenesis
of radial glia, 178-179

Diencephalon
divisions in reptiles, 59
Dorsal cortex in reptiles
homology with neocortex, 65
Dorsal lateral geniculate nucleus (*See* lateral geniculate nucleus)
Dorsal ventricular ridge
absence of descending pathways, 77
afferents, 60-64, 68, 77
behavioral effects of lesions, 79
description, in birds, 76
ectostriatum, 76-77
embryonic origin, 66
homology
Dorsal Ventricular ridge (continued)
with basal ganglia, 60 (continued)
with neocortex, 65-66, 68, 76
monoaminergic systems, 64
nomenclature, 60
proliferative zone for, 83
relationship to extrastriate cortex, 77-78
relationship to striatum, 66
Down syndrome, 193

Echolocation
and map parcellation, 132
Ectostriatum (*See also* Dorsal ventricular ridge)
behavioral effects of lesions, 79
efferent connections, 78
electrophysiology, 77
histochemistry, 77
subdivisions, 77
Electroreception
ampullary organs, 104
criteria for homology, 104
embryonic origin, 105
loss
in anurans, 104
role in diencephalic evolution, 107
and reinvention in teleosts, 105
and the auditory system, 106
neural representation of, 103
neuronal plasticity and evolution, 105-106
species variation, 103
tuberous receptors, 105-106
Encephalization (*See* Neocorticalization)
Endocasts
convolutions in, 5
early mammalian, 7
early placental mammals, 8
early primate, 28
of fish, 5
of multituberculates, 8
problems in quantification, 11-12

Ferrets, 206
Field L, 119-120
   amplitude modulation, 126-127
   best envelope frequency, 126
   binaural interaction, 125
   frequency band laminae, 123
   isofrequency contours, 124
   laminar organization
      comparison with auditory neocortex, 120
      electrophysiology, 120
      metabolic activity, 120
   multidimensional organization, 124, 127
   subdivisions, 120
   thalamic input, 119-120
   tonotopic organization, 120, 121-123
      demonstrated by metabolic activity, electrophysiology, 122
Frogs
   basilar papilla, 106
   evolution of auditory system, 106

GABA (Gamma amino-butyric acid)
   laminar differences in neocortex, 240
   mechanisms of alteration, 240
   neocortex compared to hippocampus, 187
   relationship to habituation, 241- 242
   role in cortical plasticity, 237
GABA-ergic neurons
   in mammalian neocortex compared to turtle, 160
GAD (glutamic acid decarboxylase) (*See* GABA)
Gekko
   forebrain organization, 59
Gestational length
   relationship to brain size, 34
Glycogen
   in radial glia, 176
Granule cells (*See* Nonpyramidal cells)
Growth cones
   causes of collapse, 201
   elongation of axons, 201
   lammelipodia in, 201-202
   of radial glia, 179

Hedgehog
   absence of Layer IV, 90
   allocortical organization, 90
   pattern of thalamic input, 89
   pyramidal and stellate cells, 96
Heterochrony
   and cortical evolution, 29
   and cortical neurogenesis, 22
   cytoarchitectonic areas, 29
   duration of axonal growth, 50
   and neurogenesis, 35
   and thickness of supragranular layers, 47
Hippocampus
   granule cells in, 185
   homologies of reptiles and mammals, 69
   intrinsic organization, 185
   neurogenesis of, 188-190
   pyramidal cells, 185

Insectivores
   ancestral classes, 7
Interhemispheric projections (*See* Callosal connections)
Invasion
   electroreceptive system by auditory fibers, 106-107
   altered sensory modalities in thalamus, 205-227
Lamination
   advantages, 79-80
   prevalence in vertebrates, 80
Lateral geniculate nucleus
   end-inhibition in cells, 80 (continued)
   homologous area in birds, 76, 140
   neurogenesis, 35
   receptive field properties, 141
   in reptile, 61
      homology with mammal, 66
      projections to dorsal cortex, 66
   visuotopy in owls, 141-142
   X and Y cells, 141
Lateral line
   innervating auditory nuclei, 107
Lateral pallium
   genesis in birds, 81-83
Lateral posterior nucleus of thalamus, 213
Layer I (of neocortex)
   thalamic afference
      in cat, 89
      in hedgehog, 89
      in primates, 89
      in rodent, 89
Layer IV (of neocortex)
   absence in hedgehog, 89
   afference
      in cat, 89
      in rodent, 89
   alteration by thalamic lesions, 37-38
   cell number, 37-38
   dendritic spines on cells, 91
   subdivisions in primates, 91
   thalamic terminals, 93
Limbic cortex
   neurogenesis, 47
Lizards (*See* Reptiles)
Long term potentiation, 241

Mammals
   first appearance, 8
   neocortical size in ancestral, 8
Marginal zone, 23
Maximal receptive field
   relationship to radial units, 166

Medial geniculate nucleus
  in ferrets, 211
  in gerbils, 131
  induction of visual projections, 208
Megabats, 153-154
Microbats, 153-154
Middle ear ossicles
  comparative anatomy, 104
  transformations of hyomandibular bone, 103
Middle temporal area (See visual system)
Minimal receptive field
  to demonstrate modules, 163
Mitotic cycle
  symmetric vs asymmetric division, 23-25, 34
  relationship to cortical area, 23-24
Modular units of neocortex
  definition, 161-162
  electrophysiological demonstration, 162-164
Modular units (continued)
  geometry, 164
  relationship to thalamic afferents, 161-162
Mongolian gerbil
  auditory system, 119

Neocortical cytoarchitecture
  area
    humans compared to monkeys, 21
    and neurogenesis, 47
    and proliferative units, 47
    and radial glial channels, 36, 47
  cell numbers
    unchanged by callosal damage, 37
    unchanged by midbrain damage, 37
  divisions
    in humans compared to monkeys, 21, 29
  thickness
    relationship to neurogenesis, 36-37
Neocortical plasticity
  role of neurotransmitter binding, 250
  shift from muscarinic to nicotinic, 250
  "suspicious coincidences", 250
Neocorticalization, 7, 11
  allometry of, 33-34
  archaic vs progressive species, 13-14
  in Carnivora, 14
  in Creodonts, 14
  functional advantages, 29
  heterochrony, 28, 47
  neurogenesis and, 47
  role of mitotic cycle, 28
  topographic mapping problems in, 205
Neural networks
  models of cortical plasticity, 225
Neurogenesis
  duration and relationship to brain size, 12
  relationship to gestational length, 34-35
  "sandwich gradients" in hippocampus, 188-189

  similarities in hippocampus and cortex, 185-189
  and visual system homologies, 81
Neuromodulatory systems, 251
Neuronal migration
  and alterations in radial glia, 177
  comparison of birds and mammals, 83
  radial glia and, 175
Nonpyramidal cells of hippocampus
  classification, 185-189
  neurogenesis, 188
  relationship to neocortical stellate cells, 187
Nucleus rotundus, 60-61, 77

Octavolateral system, 103
Ocular dominance
  critical periods, 245-246
  induced in auditory cortex, 209
  relationship to orientation columns, 114-116
Ocular dominance (continued)
  visual cortex plasticity, 245-246
    cholinergic mechanisms, 246
    monoaminergic mechanisms, 246
  in Wulst, 139-149
Olfactory bulbs, 5
  allometry, 13
  reduction in primates, 15
Ontogenetic columns, 22-26
  relationship to radial glia, 25
Optic decussation
  mammalian patterns, 153
  patterns in bats, 154-155
Orientation selectivity
  cortical coverage, 116
  genesis of, 225
  induced in somatosensory cortex, 220

Paleocortex, 13
Parallel evolution
  mammalian and avian visual systems, 151
Parallel processing
  in visual system, 82
Parcellation, 39, 106-107
  of mammalian auditory cortex, 131
Platypus
  electroreception in, 103
Point image size
  relationship to coverage uniformity in visual cortex, 114
Primates
  ancestral types, 7
  cytoarchitectonic areas in cortex, 21
  spiny stellate cells of, 93
  thalamic input to cortex, 91
Proliferative unit
  definition, 23
  diagram, 24

Protomap
    definition, 26
    relationship to ventricular zone, 26
Pulvinar, 213
Pyramidal cells
    continuity with stellate cells, 96
    functional identification, 92
    in hippocampus, 185
    muscarinic binding properties, 250
    thalamic input to, 91, 96
    transient receptor binding, 250

Radial glia
    cytogenesis, 176-177
    density changes in development, 178
    description, 175
    fasciculation of, 177
    growth cones, 179
    immunohistochemical markers, 176
    patterns in birds, 83
    and migrating neurons, 25, 175
    topology, 176
Radial unit
    absence in turtle cortex, 165
    anatomical evidence for, 165
    relationship to excitatory local cells, 167
    historical derivation, 87
    hypothesis, 21
    and radial glia, 180-181
    relationship to modular units, 165
Receptor redistribution
    during development, 246-250
    from glia to neurons, 250
Reeler mouse, 79
Regressive events
    role in telencephalic evolution, 108
Reptiles
    auditory system in, 61-62
    Chelonia, 59
    classification of, 59-60
    Crodilia, 59
    divisions of diencephalon, 59
    dorsal ventricular ridge, 59
    Rhynchocephalia, 59
    somatosensory system in, 62-63
    Squamata, 59
    telencephalon
        dorsal cortex efferents, 65
        intracortical connections, 65
    turtle telencephalon, 159
Retinotectal connections, 199-202
    in bats, 153-154
Rhinal fissure, 5, 9
Rodents
    Spiny stellate cells, 93
    thalamic projections, 89

"Sandwich gradients" in neurogenesis, 188
Scala naturae, 7, 47

Snakes (*See* Reptiles)
Somatosensory cortex (*See also* Barrel cortex)
    dependence on periphery, 232, 237
    induction of visual projections, 219
    mechanisms of adult plasticity, 237, 242
    in mouse, 229
    in reptiles, 62
    visual responses induced, 220
Spiny stellate cells (*See* Stellate cells)
Star pyramids (*See* Stellate cells)
Stellate cells, 91
    deviation from pyramidal cells, 96
    functional identification, 92-93
    heterogeneity, 91
    and hippocampal nonpyramidal cells, 187
    intrinsic connectivity, 93
    as a mammalian specialization, 161
    as organizers of radial units, 167
    in primates, 93
    proportions across species, 93
Stellate cells (continued)
    spine density, 91
    spiny stellate cells, 161
    thalamic input, 93
Stereopsis (*See* Binocular vision, Ocular dominance)
Striatum in reptiles, 70
Subplate, 22-23, 38, 97
    Alz-50 immunoreactivity, 193-197
    cell death in, 193
    cross-species difference, 97-98
    radial glial density in, 179-181
    role in cortical evolution, 98
Subventricular zone
    fate in mammals and nonmammals, 83

Tectofugal pathways in birds, 76
Tectum in reptiles, 70
Tegu lizard, 59
Telencephalon
    contrast of birds and mammals, 119
    divisions in reptiles, 64
    embryology across vertebrates, 81-82
Thalamic input to cortex
    in cat, 89
    functional units, 87
    in hedgehog, 89
    laminar distribution and evolution, 97
    partitioning, 87
    plasticity after lesions, 38-39, 206, 219
    in primates, 91
    in rodents, 89

in sensory areas, 87
  terminations on pyramidal cells, 91-96
  terminations on stellate cells, 96
Thalamic specification of cortex, 27, 38, 48, 214-215, 225-226
  in barrel fields, 232-234
Thalamo-telencephalic organization
  in agnathans, 107
  after loss of electroreception, 107
  homologies in reptiles and mammals, 161-169
  in teleosts, 107
Thalamic plasticity, 205, 219
Therapsids, 8
Topographic mapping
  in barrel cortex, 229-231
    role of GABA, 210
  comparison of visuotopy and somatotopy, 219-224
  comparison of visuotopy and tonotopy, 205-215
  critical periods for, 232
  development and plasticity in cortex, 199, 230
  molecular mechanisms, 199-200
Topographic mapping (continued)
  n-dimensional mapping, 213-214
  visuotopic map in
    auditory cortex, 210
Trophic dependence, 37, 105
Turtle dorsal cortex
  excitatory output cells, 159-160
  GABAergic cells in 160
  homologies with neocortex, 159-161
  inhibitory local cells, 160
Turtles (*See* Reptiles)

V1 (*See* Visual cortex)
V2 (*See* Visual cortex)
Ventricular zone, 22,175
  heterogeneity, 26
  proliferation, 25
  and radial units, 24
Ventrobasal nucleus, 161
Vernier acuity, 29
Vertebrate chronology, 8-11, 21
Visual cortex
  alteration by enucleation, 27
    changes in GABA, 237
  in blind mole rats, 205
  comparison with auditory cortex, 206-208
  comparison with somatosensory cortex, 220
  cortical magnification factor, 111
  coverage constraint, 111
  cytochrome oxidase in, 26
  description, 87
  difference in monkey and human, 21
  effect of thalamic lesions, 38
  hybrid cortex, 28
  multiple origins of, 80
  nicotinic labelling, 246-247
  ocular dominance columns, 111-112
  orientation columns, 111-112
  plasticity, 245-246
  point image size, 114
  topography, 114
  receptor redistribution
    in development, 247-248
    after undercutting, 248-249
  tectofugal and thalamofugal pathways
    functional segregation, 81-82
    neurogenesis, 81
  thalamic specification of, 27
  visuotopic mapping, 207
Visual system
  area MT, 82
  of birds
    tectofugal system damage, 82
    Wulst damage, 82
  compared to other sensory systems, 225-226
  parallel processing, 82

W cells, 210

Wulst
  binocular cells, 145-157
  cytoarchitecture in owls, 139
  homology with visual
    cortex, 76, 143, 149-152
  ocular dominance in 146
  receptive field classes, 143-145
  visuotopy, 141-142